ELEMENTARY MATHEMATICS FOR THE TECHNICIAN

ELEMENTARY MATHEMATICS FOR THE TECHNICIAN

EDWIN M. HEMMERLING

PROFESSOR AND DIVISION HEAD, EMERITUS
DIVISION OF SCIENCE, ENGINEERING, TECHNOLOGY, AND MATHEMATICS
BAKERSFIELD COLLEGE

McGRAW-HILL BOOK COMPANY

NEW YORK
ST. LOUIS
SAN FRANCISCO
DÜSSELDORF
KUALA LUMPUR
LONDON
MEXICO
MONTREAL
NEW DELHI
PANAMA
RIO DE JANEIRO
SINGAPORE
SYDNEY
TORONTO

Library of Congress Cataloging in Publication Data

Hemmerling, Edwin M
Elementary mathematics for the technician.

Published in 1965 under title: Elementary mathematics: arithmetic, algebra & geometry.
1. Mathematics—1961- I. Title.
QA39.2.H42 1974 512'.1 73-12825
ISBN 0-07-028074-6

ELEMENTARY MATHEMATICS FOR THE TECHNICIAN

9101112 KGPKGP 89

CONTENTS

PREFACE

This text has been written for students at two-year colleges and technical institutes who have inadequate mathematical backgrounds for the study of technical programs. It provides a foundation course in the essentials of arithmetic, the slide rule, elementary and advanced algebra, basic concepts of plane geometry, mensuration of plane and solid geometric figures, and logarithms, as well as a brief introduction to trigonometry.

The student who has mastered the contents of this book should be well prepared to handle the mathematical concepts in beginning technical courses. He should also be ready for further study in advanced technical mathematics courses.

This text is an outgrowth of the author's *Elementary Mathematics: Arithmetic, Algebra, and Geometry.* The earlier book has been enlarged and updated to accommodate the many fine suggestions given by its users. This book is designed to provide more than enough material for a two-semester course in technical mathematics.

The first four chapters cover elementary but fundamental concepts of arithmetic. These chapters may be omitted with no loss of continuity by students with adequate foundations in arithmetic. Those concepts in these chapters that are needed for an understanding of subsequent chapters are repeated in the chapters on algebra.

A study of the slide rule is presented early in the book in order that the student can use this instrument to solve and check the more complex problems involving multiplications, divisions, powers, and roots that appear throughout the book.

Concepts of significant figures and degrees of accuracy in measured quantities are used throughout the text. The student is taught to check the reasonableness of his answers, not only with respect to the magnitude of numerical answers but also with their indicated degrees of accuracy.

The emphasis throughout the book is on understanding rather than on memorizing. Careful explanations of the reasons behind mathematical operations are developed. The student is encouraged to determine the rationale for any solution to a problem.

Terms that may appear and need to be defined in the student's future studies in technology and the laws relating to them are intentionally left out of this book. The stated problems given here are drawn from experiences familiar to the student or within the sphere of his daily living and reading. Only general terms which should be in every student's vocabulary are used.

The intent is to give the student an ability to solve "type" problems, in anticipation that the methods he learns will be applicable to any specific areas of technology he encounters as he progresses in his studies. The student is taught to solve these problems by careful analysis and clear thinking, not by blind substitution in formulas.

Few logical proofs of geometric propositions are given. However, the student is encouraged to check propositions for reasonableness. He is then shown how the propositions can be used in determining additional properties of familiar geometric figures as they may be applied in daily observations and in his technical studies.

Chapter 24, "Trigonometry and Right Triangles," introduces the student to the application of practical trigonometry in technology. A four-place table of trigonometric functions may be used to teach the principles of interpolation.

Mathematical principles and operations are clearly illustrated by solutions to one or more carefully selected problems. Solution methods for difficult as well as elementary problems are provided.

An abundant supply of problems is given after each unit, enabling the instructor to vary his assignments from class to class. The easier problems are given at the beginning of each exercise. Answers to the odd-numbered problems appear in the back of the text. Answers to the even-numbered problems are available from the publisher.

EDWIN M. HEMMERLING

SECTION 1

BASIC ARITHMETIC COMPUTATIONS FOR TECHNICIANS

CHAPTER 1

OPERATIONS WITH NUMBERS

1-1 DEVELOPING A NUMBER SYSTEM Man's development from his primitive background to his present state of civilization is directly related to his progress in developing mathematics as a language. Primitive man took a great step forward when he was able to distinguish between a single object and several objects of the same kind. For thousands of years these people carried on simple conversations involving bartering and trading by means of gestures involving their fingers or hands. But their concepts of numbers did not go beyond two or three. Any quantity greater than two or three was treated as a group whose magnitude was roughly recognized by the size of the group.

The history of the evolution of number systems is a fascinating study which is beyond the scope of this text. The development of names and symbols for numbers led to many number systems. With these number systems man was able to keep written records of his countings.

1-2 SYMBOLS IN MATHEMATICS A *symbol* is a mark, a figure, or a word which represents an idea or an object. Generally there is little resemblance between the symbol and the idea or object it represents.

Many symbols are used in the study of mathematics. Some represent numbers, some represent operations with numbers, others compare numbers. Thus, 1, 5, 106, 5,392 and 1,000,000 represent numbers. Names or symbols for numbers are called *numerals.* The number of fingers on a person's hand can be named by the Arabic numeral 5 or the Roman numeral V. The English word "five," the Spanish word "cinco" (pronounced "sink-o"), and the German word "funf" (pronounced "fim-fth") name the number.

Other symbols used in arithmetic are those that indicate the four fundamental operations on numbers: addition, subtraction, multiplication, and division. The symbols are:

$+ \quad - \quad \times \quad \div$

FIG. 1-1 Operation symbols used in arithmetic.

Note that the expressions $3+2$, $10 \div 2$, and $11-6$ are numerals which designate the same number 5. The number named by the numeral is called the *value* of the numeral. Thus, we see that a number can be expressed with many different numerals.

To indicate that the numerical expressions $3+2$ and 5 have the same value, we use the symbol =, which is read "equals" or "is equal to." Thus, we may write the true statement

$3+2=5$

read, "Three plus two equals five." Such statement of equality is called an *equation.* It indicates that the expressions on the left and those on the right of the equality sign are different names (numerals) for the same number. Each of the following is a true statement of equality:

$3+2=10\div 2 \qquad 10\div 2=11-6$

1-3 WHOLE NUMBERS OR COUNTING NUMBERS The first numbers used by man were the whole or counting numbers. With these numbers he was able to associate the idea of numbers with groups or sets. For example, the groups of Fig. 1-2 have a common property. The common property is that for each object of one group there is an object in a second group, and for each object in the second group there is one object in the first group. Such pairing is termed *one-to-one correspondence.* We call this common property *number.* In Fig. 1-2 the particular number is 4.

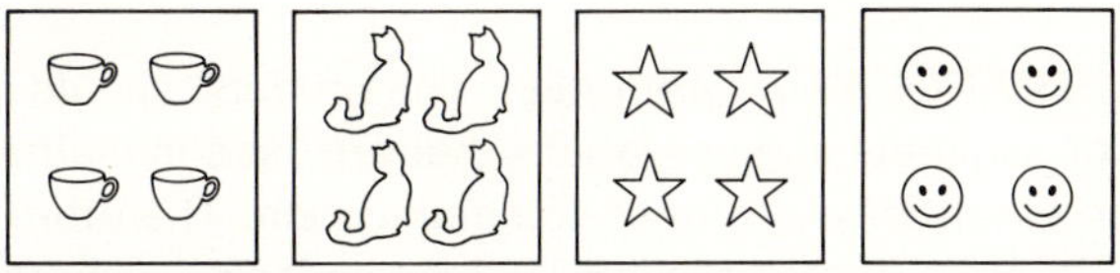

FIG. 1-2 Four.

Thus, for every set of things there is a number. For the fingers on one hand there is the number 5 (which can be expressed many ways, remember). For the desks in a room there is a number; for the cylinders in an automobile engine there is a number. If there is a single object in the group, the number is 1. If there are no objects in the group, the number is 0.

The numbers 0, 1, 2, 3, 4, 5, 6, . . . * are called *whole numbers.* They are also called *counting numbers* because they are used in counting. In Chap. 1 we will be discussing the properties of, and operations on, whole numbers. In the chapters that follow we will be dealing not only with whole numbers, but with numbers that are not counting numbers.

1-4 THE HINDU-ARABIC NUMERAL SYSTEM The present numeral system used by most civilized nations is known as the *Hindu-Arabic* system. It had its origin in India about 500 B.C. There are two unique features of this system which make it superior to all other systems. The introduction of the symbol for zero to represent the absence of quantity did not appear in the other notations that had been developed. The inclusion of the zero symbol in the numeral system made possible the feature which makes for the

* The symbol . . . (called ellipsis) following an arithmetic or algebraic expression is read "and so on" and indicates an indefinite continuation of the process used to obtain the expression to the left of the dots.

genius of the system, that of the *principle of place value.* Thus, for the number represented by 444, the right-hand 4 represents 4 ones, the next 4 represents ten times as much, and the left 4 represents one hundred times the right 4. Thus $444 = 4(100) + 4(10) + 4(1)$. Each digit in an Arabic numeral has not only its *face value* but also its *place* or *position value.*

The principle of digits having both face and position values makes it possible to represent any quantity, no matter how large, by using only a few symbols. This system also permits the development of easy rules for performing the four fundamental processes of addition, subtraction, multiplication, and division.

It should be noted that the system in common use today has 10 as its base; that is, ten symbols (0, 1, 2, 3, 4, 5, 6, 7, 8, 9) are used. However, equally effective systems could be and have been developed using other numbers as bases. Our modern high-speed computers have 2 as the base of the system used.

1-5 MATHEMATICS AS A LANGUAGE The student should realize that the symbols for concepts and operations that are used in mathematical computations are essentially a form of communication. Therefore, their use must be just as precise, understandable, and convincing as is the use of words. Every mathematical statement or operation expressed in symbolic form must be a complete statement. These statements must be understandable to any person who is familiar with the symbolisms and algorithms used. In solving a problem, the mere answer to the problem is not sufficient. The explanation, in symbolic form, of how it was determined is necessary. The student's written work should be a clear record of the method used to solve the problem.

1-6 ADDITION OF NUMBERS Addition is a procedure based upon the process of counting, where one seeks the total number of objects contained in several sets of objects. Thus, the child might add 5 and 5 by counting the number of fingers on both hands. A person might add the number of girls and boys in a given class by counting all the students in the class. It is important that the student recognize that only like quantities, that is, quantities that are expressed in terms of the same unit, can be added. Thus one could not add 2 feet (ft) and 3 quarts (qt). If the name of the unit is not attached to a number, it is called an *abstract number.* Abstract numbers represent the number of ones.

To find the sum of two or more numbers, one utilizes the principle of place value and adds only numbers which have the same place value.

Example 1-1 Find the sum of the abstract numbers 317 and 452.

SOLUTION We know that

$$317 = 3 \text{ hundreds} + 1 \text{ ten} + 7 \text{ ones}$$

and $452 = 4 \text{ hundreds} + 5 \text{ tens} + 2 \text{ ones}$

Place the numbers in a vertical arrangement so that the ones are in one column, the tens in the next column to the left, the hundreds next, and so on.

hundreds	tens	ones	
3	1	7	
4	5	2	
7	6	9	(sum)

Adding units of the same kind, we obtain 7 hundreds + 6 tens + 9 ones, which in symbolic form is written 769.

Example 1-2 Find the sum of 258 and 465.

SOLUTION Place the numbers in a vertical arrangement with like units in the same column.

hundreds	tens	ones
1	1	
2	5	8
4	6	5
7	2	3

Here 8 ones + 5 ones = 13 ones. We change the 13 ones to 1 ten and 3 ones and write 3 for the sum in the ones column and carry the 1 ten to the top of the tens column. 6 tens + 5 tens + 1 ten = 12 tens. Write 2 for the sum in the tens column and carry the 10 tens as 1 hundred to the top of the hundreds column. 4 hundreds + 2 hundreds + 1 hundred = 7 hundreds. Write 7 for the sum in the hundreds column.

Example 1-3 Combine into an equivalent single number $2{,}863 + 92 + 845 + 107$.

SOLUTION Write the numbers in a vertical arrangement with like units in the same column.

1	2	1	
2	8	6	3
		9	2
	8	4	5
	1	0	7
3	9	0	7

If you add the numbers in the ones column, you will get 17 ones or 1 ten and 7 ones. Write 7 for the sum in the ones column and carry the 1 ten. Adding the tens, we get 20 tens or 2 hundreds + no tens. Write 0 for the answer in the tens column and carry 2 hundreds. Next add the hundreds to get 19 hundreds or 1 thousand + 9 hundreds. Write 9 for the answer in the hundreds column and carry the 1 in the thousands column. Add 2 thousands + 1 thousand to get 3 thousands. Write 3 for the answer in the thousands column.

With practice, the student should be able to "carry" mentally (without writing the numbers down). A good way to check addition is to start at the bottom of a column and add upward once and then start at the top of the column and add downward. If the same answer is obtained using both methods, the answer is probably correct.

Example 1-4 Find the perimeter of *ABCD*.

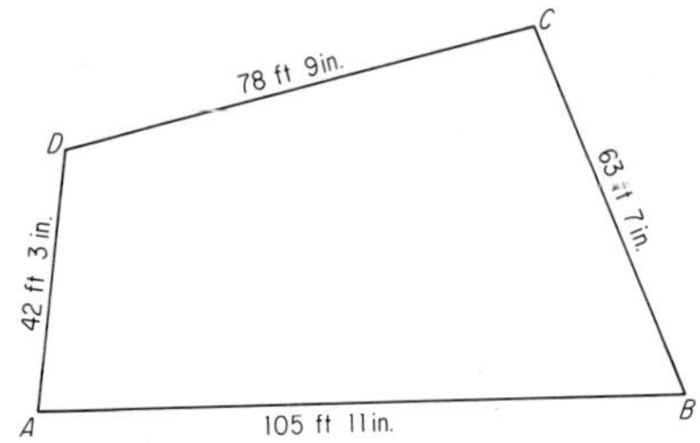

FIG. 1-3

SOLUTION The sum of the sides is the perimeter of the figure.

```
 42 ft  3 in.
 78 ft  9 in.
 63 ft  7 in.
105 ft 11 in.
-------------
288 ft 30 in.
```

Replacing 30 in. by 2 ft 6 in., we get

```
            288 ft
              2 ft 6 in.
            ----------
Perimeter = 290 ft 6 in.
```

1-7 PROPERTIES OF ADDITION WITH NUMBERS The student of technology has probably had considerable experience in adding numbers. The following are some important properties of addition he should have been led to discover in his work with numbers. First, when two counting numbers are added, the result is always a counting number. We say that counting numbers are closed under the operation of addition.

---

1 *If a and b are whole numbers, then* $a + b$ *is a whole number.*

A second property of addition concerns the number zero. Zero is a unique number in that when it is added to another number, the sum is always that second number. And when any number is added to zero the sum is that first number. Zero is called the *identity element under addition.*

2 *For all numbers a,* $a + 0 = a$ *and* $0 + a = a$.

The next property is quite obvious and results from the definition of addition. It states, for example, that the sum of 3 and 5 is identical to the sum of 5 and 3. This property is called the *commutative property of addition.* We say that addition of numbers is commutative.

3 *If a and b are numbers, then* $a + b = b + a$.

In considering the sum of three numbers, we discover another general property of addition. Suppose you are asked to add 9, 7, and 3. There are several ways to arrive at the correct sum. You could add 9 and 7 to get 16, and then add 16 and 3, which yields 19. A second method would be to add 9 to the sum of 7 and 3; that is, you would add 9 to 10 to get 19. In symbols, we have shown that $(9 + 7) + 3 = 9 + (7 + 3)$.* This example illustrates the *associative property of addition.*

4 *If a, b and c are numbers, then* $(a + b) + c = a + (b + c)$.

The method for checking the sum of a column of numbers by first starting at the bottom and adding upward and then starting at the top and adding downward is an application of the associative property of addition.

Exercise 1-1

Replace the indicated sums by single numbers:

1. $2 + 8 + 5$
2. $7 + 6 + 4$
3. $90 + 30 + 70 + 10$
4. $43 + 57 + 12 + 8$
5. $299 + 99 + 309 + 1{,}391$
6. $8 + 4 + 7 + 6 + 3$

Find the sum of each column, first adding upward, second adding downward to check:

7.	8.	9.
86	24	45
37	19	56
53	56	93

* The symbol ( ) is called a set of parentheses. It is used as a *grouping symbol;* that is, the numbers enclosed by the symbols are treated as a single entity.

10. 47
28
54

11. 72
34
68
59

12. 43
64
55
18

13. 51
27
46
19

14. 54
12
62
35

15. 385
603
170
444

16. 206
354
812
567

17. 862
286
313
505

18. 376
408
317
190

19. 6,735
2,076
6,439
2,856

20. 6,953
5,236
2,008
5,117

21. 2,135
716
8,032
544

22. 6,930
853
1,468
79

23. 4,671
50
679
8,523

24. 5,382
12,568
24
348
5,511

25. 95,057
1,983
3,624
807
99

26. 792,886
553
60,248
44
5,238,716

27. 8,560,243
1,799,207
4,153,648
7,134,089
5,624,107

28. Find the total length of fence required to enclose the parcel of land whose dimensions are shown in the figure.

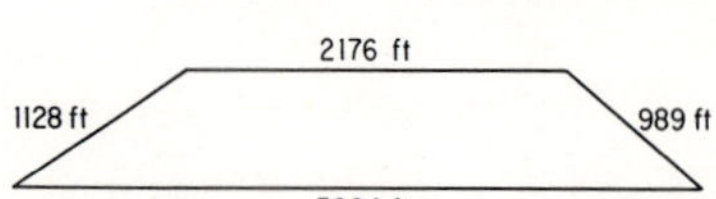

EXERCISE 28

29. The floor plan of a building is illustrated in the accompanying figure. What is the distance around the building?

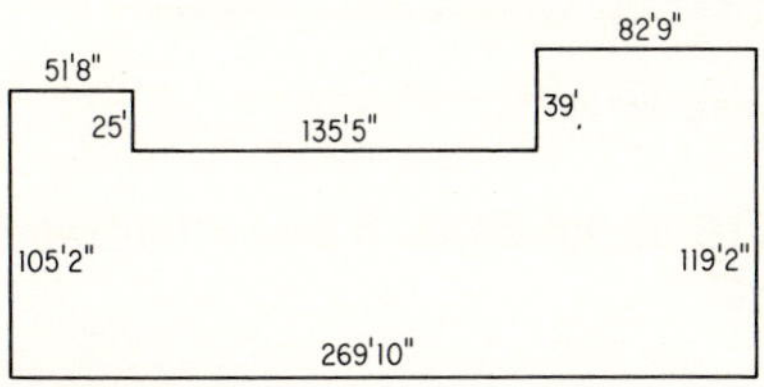

EXERCISE 29

30. Find the perimeter of the symmetric figure if $a = 78$ ft, $b = 48$ ft, $c = 123$ ft, $d = 39$ ft, $e = 107$ ft, $f = 64$ ft, $g = 65$ ft, $h = 54$ ft, $i = 98$ ft, $j = 40$ ft, $k = 46$ ft.

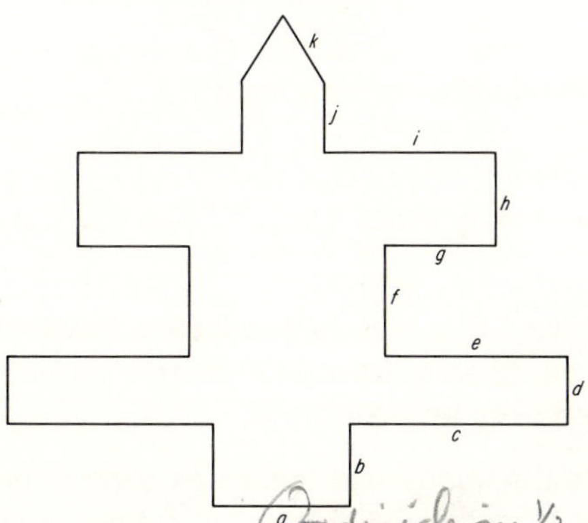

EXERCISE 30

31. Find the sum of 18 gallons (gal) and 3 qt, 7 gal and 1 qt, 2 qt and 5 gal. (*Hint*: 1 gal is equivalent to 4 qt.)
32. California has an area of 156,803 square miles (mi^2); Oregon has 96,350 mi^2; Washington has 66,977 mi^2. What is the total area of these three states that border the Pacific Ocean?
33. A cargo plane checked in loads of 2,867, 4,310, 5,862, 6,003, and 505 lb. What is its total pay load?
34. A salesman in the parts department of a television store had a sales record for a given week as follows: Monday, \$392.76; Tuesday, \$235.43; Wednesday, \$109.82; Thursday, \$85.07; Friday, \$293.87; Saturday, \$401.08. What were his total sales for the week?
35. Find the sum of the following time intervals: 3 hours (hr) 12 minutes (min), 22 hr 45 min, 15 hr 30 min, 9 hr 8 min. (Leave answer in a form that will have not more than 60 min in the total.)

1-8 SUBTRACTION OF WHOLE NUMBERS The process which is the inverse of addition is called *subtraction.* We can look at subtraction in the following three different ways.

1 Subtraction means to *take away.*

Example 1-5 John has $518 in the bank. If he withdraws $25, how much will he have left in the bank?

2 Subtraction means finding out *how much must be added* to a given number to obtain a second number.

Example 1-6 Bill is saving to buy a car which costs $2,500. He now has $1,650. How much more must he save?

3 Subtraction can be used in comparing two numbers to find out *how much more* (*or less*) one number is than the other.

Example 1-7 Double-belted, fiber glass tire A retails for $36. A four-ply, nylon tire B of the same size retails for $24. How much more does tire A cost than tire B?

The answer to Example 1-5 is that John will have $493 left. For Example 1-6, we find that Bill will need to save $850 more than he now has. In Example 1-7 tire A costs $12 more than tire B.

The symbol −, called the *minus sign,* indicates subtraction. For example, we subtract 5 from 9 to get 4. To subtract 5 from 9 means to find the number which must be added to 5 to get 9. Thus, $9-5=4$, because $4+5=9$. The number to be subtracted is called the *subtrahend.* The number from which it is subtracted is called the *minuend.* The result is called the *remainder* or *difference.*

To prove a subtraction, one adds the subtrahend and the remainder. If their sum is equal to the minuend, the subtraction is correct.

In subtraction, the subtrahend is written directly under the minuend, with the ones, tens, hundreds, etc., digits aligned in proper columns. In subtracting, we can subtract only ones digits from ones digits, only tens digits from tens digits, only hundreds digits from hundreds digits, etc. When the digits are subtracted from smaller digits, we use the borrowing method of subtraction. For example, when we write

Subtract:

$$\begin{array}{r} 683 \\ \underline{247} \end{array}$$

$$\begin{aligned}
\text{Since } 83 &= 8 \text{ tens} + 3 \text{ ones}\\
&= 7 \text{ tens} + 1 \text{ ten} + 3 \text{ ones}\\
&= 7 \text{ tens} + 10 \text{ ones} + 3 \text{ ones}\\
&= 7 \text{ tens} + 13 \text{ ones},\\
\text{we think } 683 &= 6 \text{ hundreds} + 7 \text{ tens} + 13 \text{ ones}\\
\underline{247} &= 2 \text{ hundreds} + 4 \text{ tens} + 7 \text{ ones}\\
\text{Difference } &= 436 \text{ or } 4 \text{ hundreds} + 3 \text{ tens} + 6 \text{ ones}
\end{aligned}$$

In the solution we "borrow" one of the tens digits in the minuend and replace it with 10 ones, which added to the 3 ones makes 13 ones. The answer will be 4 hundreds + 3 tens + 6 ones, or 436.

The problem can be written

$$\begin{array}{r} 6\overset{7}{\cancel{8}}{}^{1}3 \\ \underline{247} \\ 436 \end{array}$$

Example 1-8 Subtract 2,876 from 15,034.

SOLUTION

$$\begin{array}{rl} 15{,}034 & \text{(minuend)} \\ \underline{2{,}876} & \text{(subtrahend)} \end{array}$$

Since 6 (ones) cannot be subtracted from 4 (ones) we borrow 1 (ten) from 3 tens. But 1 ten equals 10 ones. Thus, the problem becomes

$$\begin{array}{r} 15{,}0\overset{2}{\cancel{3}}{}^{1}4 \\ \underline{2{,}876} \end{array}$$

We can now subtract 6 (ones) from 14 (ones). But we can't subtract 7 (tens) from 2 (tens). We would like to borrow 1 (hundred) from the 0 (hundred). This we can't do. So we borrow 1 (thousand) from 5 (thousands). This 1 thousand we convert into 9 hundreds and 10 tens. Now we can borrow 1 (ten) from the 10 (tens) and add it to the 2 (tens) and write

$$\begin{array}{rl} 1\overset{4}{\cancel{5}},\overset{9}{\cancel{0}}\overset{12}{\cancel{3}}{}^{1}4 & \\ \underline{2{,}876} & \\ 12{,}158 & \text{(difference)} \end{array}$$

It is not necessary to show the borrowing. With added experience the student should be able to do the borrowing mentally and write

$$\begin{array}{lrl} \text{Subtract:} & 15{,}034 & \\ & \underline{2{,}876} & \\ & 12{,}158 & \text{(the difference)} \end{array}$$

CHECK

$$\begin{array}{lr} \text{Add:} & 2{,}876 \\ & \underline{12{,}158} \\ & 15{,}034 \end{array}$$

Example 1-9 How much longer is 2,947 ft 3 in. than 1,563 ft 7 in.?

SOLUTION
Subtract:

2,947 ft	3 in.
1,563 ft	7 in.

Rewrite the problem:

Subtract:

2,946 ft	15 in.
1,563 ft	7 in.
1,383 ft	8 in.

CHECK
Add:

	1,563 ft	7 in.
	1,383 ft	8 in.
	2,946 ft	15 in.
or	2,947 ft	3 in. (sum)

Exercise 1-2

Subtract the bottom number from the top number and prove each answer by addition:

1. 98
   46

2. 764
   231

3. 876
   59

4. 6,842
   3,518

5. 3,567
   1,085

6. 8,439
   582

7. 63,854
   2,549

8. 49,009
   25,512

9. 80,808
   24,242

10. 30,028
    5,706

11. 853,086
    384,707

12. 7,005,300
    2,990,031

13. How much larger is 90,000 than 3,849?
14. From 8,349 subtract 5,761.
15. The electric meter at Mr. Brown's house read 82,763 kilowatt-hours (kwhr) on July 31. On Aug. 31 it read 84,108 kwhr. How much electricity was used during the month of August?
16. Mr. Smith bought a house valued at $25,350. Ten years later the house was valued at $21,780. How much did the property depreciate in value during the 10 years?
17. The stadium at Belleview City College has a seating capacity of 22,000. During a recent game all seats were occupied but a block containing 1,750 seats. How many were seated in the stadium?
18. The number of cars of a certain make sold during the first 3 months of this year were: January, 57,864; February, 42,357; March, 61,008. Last year's sales of the same car were: January, 49,038; February, 35,705; March, 53,444. Find the increase of sales for the first 3 months of this year over the comparable period last year.
19. Mr. Evans has $18,750 in the bank with which to build a home. His cost will be: lot, $3,475; house, $16,600; garage, $1,785; landscaping, $1,035; incidentals, $250. How much will he need to borrow before he can complete the home?
20. At the beginning of the day, the owner of an appliance store had $350 cash on hand. During the day he sold the following for cash: a television set, $212.50; one radio, $39.75; a second radio, $115.95; a toaster, $19.98; a percolator, $23.99. He paid out in cash: $237.50 to a bill collector, $112 for advertising, $50 to charity. How much cash should the owner have on hand at the end of the day?
21. A computing machine subtracted 48 from a certain number 16 times in succession leaving a final remainder of 19. Find the number from which the 48s were subtracted.
22. A 2 by 4 board 8 ft long is cut into two pieces. One piece has a length of 3 ft 7 in. How long is the other piece?
23. Tom Smith bought a car in August 1968. He sold the car in February 1974. How long did Tom own the car?

1-9 MULTIPLICATION Multiplication, like addition, is a mathematical operation that pairs two numbers to form a third number. Thus, we say that the numbers 4 and 6 under the operation of multiplication are associated with the number 24. The number 24 is called the *product* of the numbers 4 and 6. The numbers 4 and 6 are called *factors* of 24.

Multiplication can be interpreted as a short cut for repeated addition. To multiply 6 by 4, we add 6 to itself 4 times. In this case the 6 is called the *multiplicand,* the 4 is called the *multiplier* and the 24 is the *product.* The multiplier and the multiplicand are factors of the product.

There are several common notations to indicate multiplication. For example, each of the following expressions indicates multiplication of 6 by 4.

(a) $4 \times 6 = 24$

(b) $4 \cdot 6 = 24$

(c) $4\,(6) = 24$

(d) $(4)(6) = 24$

(e)
$$\begin{array}{rl} & 6 \text{ (multiplicand)} \\ \times & 4 \text{ (multiplier)} \\ \hline & 24 \text{ (product)} \end{array}$$

(f) Multiply
$$\begin{array}{rl} 6 & \text{(multiplicand)} \\ 4 & \text{(multiplier)} \\ \hline 24 & \text{(product)} \end{array}$$

Each of the above notations can be used to express the fact $6 + 6 + 6 + 6 = 24$. However, to find the product of two numbers by first changing each multiplication problem to an equivalent addition problem could well lead to ridiculously long and tedious manipulations. For example, 38×294 would mean adding thirty-eight 294s. We will need a better way to arrive at the product. But first we will list some properties of multiplication.

1-10 PROPERTIES OF MULTIPLICATION The student will note that the product of two whole numbers is always a whole number. This is called the *property of closure of multiplication.*

The set of counting numbers is closed under multiplication; that is, if a and b are counting numbers then so is $a \times b$.

In finding the product of two numbers, either number can be taken as the multiplicand and the other as the multiplier. Thus, $5 \times 7 = 7 \times 5$. This important property is called the *commutative property of multiplication.* It is often used in checking problems of multiplication. For example, to check that 19,432 is the product of 347 and 56, the student can solve for 347×56 and check the answer against that for 56×347.

For the numbers a and b: $a \times b = b \times a$. (Commutative property of multiplication)

The number 1 is unique in that when multiplied by any other number it gives that number. For example, $7 \times 1 = 7$, $13 \times 1 = 13$. Using the commutative property of multiplication we know that $1 \times 7 = 7 \times 1 = 7$ and $13 \times 1 = 1 \times 13 = 13$. The number 1 is called the *multiplicative identity element.*

The number 1 is a multiplicative identity element; that is, $a \times 1 = 1 \times a = a$ for all numbers a.

The product of three or more numbers does not depend upon the way the factors are grouped. Thus, $5 \times 6 \times 4 = (5 \times 6) \times 4 = 5 \times (6 \times 4)$. Thus, multiplication, like addition, is *associative.*

Multiplication of numbers is associative; that is, if a, b and c are numbers, then $(a \times b) \times c = a \times (b \times c)$.

Because of the associativity of multiplication we can, and usually will, write products such as $5 \times 6 \times 4$ without parentheses. No ambiguity should result. Combining the properties of commutivity and associativity enables us to show that $(5 \times 6) \times 4 = (4 \times 6) \times 5$.

The sum of two or more numbers may be multiplied by a given number by multiplying each of them by that number and then adding the products. For example, $4(5+2) = 4 \cdot 5 + 4 \cdot 2 = 20 + 8 = 28$. The same product can be obtained by multiplying 7 by 4. This property is called the *distributive property.*

The distributive property states that if a, b and c are numbers, then $a(b+c) = ab + ac$.

It should be noted that $4(5+2)$ does not equal (written $\neq$) $20 \times 8 = 160$. Thus,

$$4(5+2) = 20 + 8 = 28$$

$$4(5 \times 2) \neq 20 \times 8 = 160$$

$$4(5 \times 2) = 4 \times 5 \times 2 = 40$$

1-11 MULTIPLICATION OF WHOLE NUMBERS In multiplying two whole numbers, we usually place the ones of the multiplier under the ones of the multiplicand, the tens under the tens, etc. The procedure of multiplication is easier to illustrate than to describe. Consider the following:

Example 1-10 Multiply 347 by 6.

SOLUTION

$$\begin{array}{rl} 347 & \\ \times \quad 6 & \\ \hline 42 & (6 \times 7 \text{ ones} = 42 \text{ ones}) \\ 240 & (6 \times 4 \text{ tens} = 24 \text{ tens}) \\ 1\,800 & (6 \times 3 \text{ hundreds} = 18 \text{ hundreds}) \\ \hline 2{,}082 & \end{array}$$

or

$$\begin{array}{r} 347 \\ \times \quad 6 \\ \hline 2{,}082 \end{array}$$

The above solution uses a "crutch" to help in the carrying process. With practice, the student should become adept at carrying mentally.

Example 1-11 Multiply 347 by 56.

SOLUTION

```
       3 4 7 (multiplicand)
×        5 6 (multiplier)
    2  0 8 2 (6 × 347)
   1 7  3 5 0 (50 × 347)
   1 9 , 4 3 2 (product)
```

The 2,082 above, called a *partial product,* represents 2,082 ones while the partial product 17,350 represents 1,735 tens or 17,350 ones. The zero is usually deleted from the second partial product in the solution of the problem.

1-12 MULTIPLICATION OF DENOMINATE NUMBERS BY WHOLE NUMBERS In representing a physical quantity mathematically it is necessary to use a combination of a number to describe its magnitude and a name to indicate the unit in which the quantity is measured. Thus, 4 ft, 4 hr, and 4 lb describe the same number of different entities—units of length, time, and weight (or mass), respectively. In like manner, 4 in., 4 ft, and 4 mi denote the same number of units of the same entity—that of length. A number that names the quantity of a measurable unit is called a *denominate number.* Each of the numbers above are examples of *simple* denominate numbers. Denominate numbers expressed in terms of two or more units are called *compound* denominate numbers. Examples of compound denominate numbers are:

8 ft 5 in.

7 yd 2 ft 3 in.

5 hr 13 min 8 sec

4 gal 3 qt 1 pint (pt)

We have already had some experience with adding and subtracting compound denominate numbers (see Examples 1-4 and 1-9). The examples which follow illustrate how denominate numbers can be multiplied by whole numbers.

Example 1-12 Multiply 3 ft 4 in. by 5.

SOLUTION:

```
   3 ft  4 in.
×          5
  15 ft 20 in. = 16 ft 8 in.
```

Example 1-13 Multiply 7 hr 24 sec by 8.

SOLUTION:

```
   7 hr  24 sec
×             8
  56 hr 192 sec = 56 hr 3 min 12 sec
```

Exercise 1-3

Tell which property of numbers is being illustrated.

1. $1 \times 23 = 23$
2. $15 + 0 = 15$
3. $24 \times 11 = 11 \times 24$
4. $(2 + 5) + 8 = 2 + (5 + 8)$
5. $7 + 5 = 5 + 7$
6. $8(7 + 3) = 8(7) + 8(3)$
7. $(17 \times 5) \times 9 = 17 \times (5 \times 9)$
8. $4(9 + 3) = (9 + 3)4$
9. $8 + 3 = 3 + 8$
10. $5 \times (6 \times 7) = (5 \times 6) \times 7$
11. Find the product of 35 and 28.
12. Find the difference between 67 and 23.
13. Multiply 384 by 7.
14. Find the value of $8 \times 7 \times 5$.
15. Find the value of $30 \times 15 \times 20$.
16. Find the value of $12(17 + 5)$.
17. Find the value of $8(5 + 2)$.
18. Find the value of $8(6 + 3 + 4)$.
19. Show that $(8 \times 7) \times 3 = 8 \times (7 \times 3)$.
20. Show that $(36 \times 25) \times 24 = 36 \times (24 \times 25)$.
21. $\begin{array}{r} 384 \\ \times \quad 7 \\ \hline \end{array}$
22. $\begin{array}{r} 295 \\ \times \ 36 \\ \hline \end{array}$
23. $\begin{array}{r} 573 \\ \times 108 \\ \hline \end{array}$
24. $\begin{array}{r} 2{,}156 \\ \times \ 384 \\ \hline \end{array}$
25. $\begin{array}{r} 7{,}834 \\ \times 2{,}591 \\ \hline \end{array}$
26. $\begin{array}{r} 6{,}019 \\ \times 5{,}203 \\ \hline \end{array}$
27. 5×7 ft 6 in.
28. 8×9 lb 10 oz.
29. Multiply 8 yd 2 ft by 7.
30. Multiply 5 yd 2 ft 8 in by 20.
31. Multiply 14 hr 40 min by 30.
32. Multiply 5 gal 3 qt 1 pt by 4.
33. Find the product of 56 and 10 yd 1 ft 8 in.
34. Find the product of 73 and 3 hr 45 min 10 sec.
35. A wheel turns at a constant rate of 3,600 revolutions per minute (rpm). How many revolutions will it make in 24 hr?
36. Joe Turner is a traveling salesman who is given a travel allowance of 8 cents/mi. If he travels an average of 400 mi/week, what is his annual travel allowance?
37. In the figure, *A* and *B* represent plots of two parcels of land. Which has the larger area? How much larger?

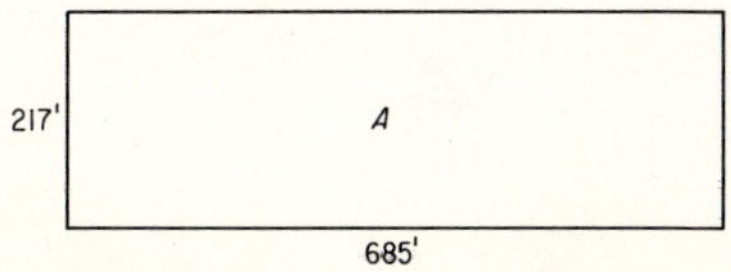

EXERCISE 37A

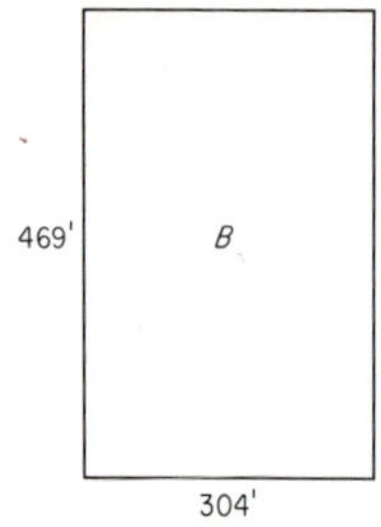

EXERCISE 37B

(*Hint:* The area of a rectangle is the product of its length and its width.)

38. Car X will travel 18 mi on a gallon of gasoline. Car Y will travel 15 mi on a gallon of gasoline. Car X holds 17 gal. Car Y holds 20 gal. Which car can travel farther on a full tank? How much farther?
39. An auditorium contains 24 rows of loges with 58 seats in each row, and 37 rows with 52 seats in each row for general admission. The balcony has 7 rows with 61 seats in each row. What is the seating capacity of the auditorium?
40. An oil tanker has a capacity of 105,370 barrels of crude oil. A barrel (of oil) is equivalent to 42 gal. How many gallons can the tanker carry?

1-13 DIVISION The fourth fundamental process with numbers is called division. Just as subtraction is the inverse of addition, so division is the inverse of multiplication. Thus to divide 12 by 3 is to determine the number (called the *quotient*) which multiplied by 3 (called the *divisor*) gives 12 (called the *dividend*).

There are several signs of division, among them $\div$, /, —, $\overline{)}$. Each symbol is read "divided by." Thus $12 \div 3 = 4$; $12/3 = 4$; $\frac{12}{3} = 4$; $\begin{array}{r} 4 \\ 3\overline{)12} \end{array}$. We can state: The quotient of two numbers is that number which when multiplied by the divisor gives the dividend. We know that $12 \div 3 = 4$ because $4 \times 3 = 12$.

Just as multiplication can be thought of as continued addition, division can be interpreted as continued subtraction. For example, $12 \div 3 = 4$ can be interpreted as follows:

$$\begin{array}{r} 12 \\ \underline{-\ 3} \\ 9 \\ \underline{-\ 3} \\ 6 \\ \underline{-\ 3} \\ 3 \\ \underline{-\ 3} \\ 0 \end{array}$$

This tells us that there are exactly four 3s in 12. This operation is sometimes called measuring. Our problem solution also tells us that if we divided 12 into three equal parts, the size of each part is 4. This operation is called *partition.* Thus division can be interpreted as:

1 The inverse of multiplication
2 Continued subtraction or measuring
3 Partition

1-14 LONG DIVISION Consider the problem 38,695 ÷ 42. This problem can be solved by the long-division method as follows.

Example 1-14

$$
\begin{array}{r@{}l}
 & 921 \\
42 & \overline{)38,695} \\
 & \underline{37\,800} = 900 \times 42 \\
 & 895 \\
 & \underline{840} = 20 \times 42 \\
 & 55 \\
 & \underline{42} = 1 \times 42 \\
 & 13 \text{ (remainder)}
\end{array}
\quad \left.\right\} = (900 + 20 + 1)42 = 921 \times 42
$$

It will be noted that we have found that there are exactly nine hundred 42s plus twenty 42s plus one 42 (total 921) in the dividend 38,695 with a *remainder* of 13 "left over." The process is usually shortened by deleting from the solution the zeros in 37,800 and 840. These zeros are "understood" but not written down. The complete solution (quotient) of the division is $921\frac{13}{42}$.

To check the result of a division, multiply the whole part of the quotient by the divisor and add the remainder to this product. If the result is equivalent to the dividend, the solution probably is correct. Thus, for the example above we have

CHECK

$$
\begin{array}{lr}
 & 921 \\
\times & 42 \\
\hline
 & 1\,842 \\
 & 36\,84 \\
\hline
 & 38,682
\end{array}
\qquad
\begin{array}{lr}
 & 38,682 \\
+ & 13 \\
\hline
 & 38,695
\end{array}
$$

1-15 DIVISION OF DENOMINATE NUMBERS BY WHOLE NUMBERS When dividing a denominate number by a whole number, the borrowing principle shown in Sec. 1-8 can often be used to simplify the solution.

Example 1-15 Divide 21 gal 3 qt by 4.

SOLUTION

$4\overline{)21 \text{ gal} + 3 \text{ qt}} =$

$4\overline{)(20 \text{ gal} + 1 \text{ gal}) + 3 \text{ qt}} =$ (since 21 gal = 20 gal + 1 gal)

$4\overline{)(20 \text{ gal} + 4 \text{ qt}) + 3 \text{ qt}} =$ (substituting 4 qt for 1 gal)

$4\overline{)20 \text{ gal} + (4 \text{ qt} + 3 \text{ qt})} =$ (by the property of associativity)

$4\overline{)20 \text{ gal} + 7 \text{ qt}} =$ (since 4 qt + 3 qt = 7 qt)

5 gal + $1\frac{3}{4}$ qt

Exercise 1-4

Divide and check by multiplication:

1. 9,800 ÷ 35
2. 1,898 ÷ 26
3. 32,724 ÷ 81
4. 12,285 ÷ 273
5. $\frac{66,468}{382}$
6. $\frac{81,621}{404}$
7. $\frac{522,252}{572}$
8. $\frac{229,329}{8,493}$
9. $\frac{19,272,708}{60,606}$
10. $\frac{27,484,392}{274}$
11. $222\overline{)1,001,001}$
12. $309\overline{)9,284,576}$
13. $3,704\overline{)19,013,501,714}$
14. $653\overline{)65,565,680}$
15. $4,309\overline{)1,771,503,153}$
16. $38,072\overline{)17,475,048}$
17. $4\overline{)35 \text{ ft } 7 \text{ in.}}$
18. $6\overline{)17 \text{ hr } 24 \text{ min}}$
19. Determine the rate of gas consumption (miles per gallon) for a car which travels 2,834 mi on 177 gal.
20. Bruce Marolf received the following monthly commissions for selling cars: January, $684.27; February, $513.24; March, $688.33; April, $591.34; May, $575.75; June, $521.17; July, $466.24; August, $450.85; September, $967.84; October, $1,358.84; November, $887.36; December, $581.26. What was his average monthly commission?
21. How many 80-lb sacks of fertilizer can be purchased for $2318.75 if each sack costs $43.75?
22. 64 nails of a certain size weigh 1 lb. How many pounds of nails must be purchased to surely obtain 2,750 nails? Express answer to the nearest whole number.
23. How many pieces of board 1 ft 3 in. in length can be cut from a 15-ft board?
24. A contractor needs 5 tons 382 lb of cement to do a certain job. How

many sacks of cement will he need if each sack contains 90 lb? (*Hint:* 1 ton = 2,000 lb.)

1-16 COMBINATIONS OF THE FOUR FUNDAMENTAL OPERATIONS Usually arithmetic and algebraic problems involve some combinations of the four fundamental operations. The order in which these operations are performed is critical. The following rules should be followed.

1 *A series of additions may be taken in any order.* For example, $2+5+3+7=7+2+5+3=3+7+5+2=17$.
2 *A series of subtractions must be performed in the order given.* Thus, $25-13-3=12-3=9$. It would be wrong first to subtract 3 from 13 to get 10 and then to subtract the 10 from 25.
3 *A series of multiplications may be taken in any order.* Thus, $8\times 2\times 5=8\times 5\times 2=2\times 5\times 8=80$.
4 *A series of divisions must be taken in the order given.* Thus, $480\div 24\div 4=20\div 4=5$. But $480\div 24\div 4\neq 480\div 6$.
5 When a problem involves the four fundamental operations, there is a definite order that must be followed in performing the operations. Common usage dictates the following rule: *Perform all the multiplications and divisions first in the order in which they are given. Then perform additions and subtractions in the remaining order.*

Example 1-16 Simplify $156-56\div 7\times 2+34$.

SOLUTION

$$156-56\div 7\times 2+34=$$
$$156-8\times 2+34=$$
$$156-16+34=$$
$$140+34=174$$

Example 1-17 Simplify $64+36\times 8\div 4\times 2-2$.

SOLUTION

$$64+36\times 8\div 4\times 2-2=$$
$$64+288\div 4\times 2-2=$$
$$64+72\times 2-2=$$
$$64+144-2=$$
$$208-2=206$$

1-17 SYMBOLS OF GROUPING It is often necessary or desirable to consider several terms as a single group upon which operations may be performed. In such cases, a symbol of grouping is used. Symbols for grouping include the parentheses ( ), the brackets [ ], the braces { }, and the bar, or vinculum, ———. Frequently, several symbols are used to indicate grouping of terms within a second group. If different symbols are used, it is easier to tell which terms are combined.

When no number is written before a symbol of grouping, it is understood to be $+1$ or -1 according to whether the sign preceding the symbol is plus or minus. Symbols are removed from or placed about a group according to the following rules:

1 Symbols of grouping preceded by a plus sign may be removed or inserted without changing the signs of the enclosed terms.
2 Symbols of grouping preceded by a minus sign may be removed or inserted if the sign of each of the enclosed terms is changed.
3 Symbols of grouping are generally removed by first removing the innermost pair according to the above rules, next the innermost pair of all that remain, and so on.

Example 1-18 Simplify $4 + \{9 - [-5 - (4 - 7) - 3]\}$.

SOLUTION

$$4 + \{9 - [-5 - (4 - 7) - 3]\} =$$
$$4 + \{+9 - [-5 - (+4 - 7) - 3]\} =$$
$$4 + \{9 - [-5 - 4 + 7 - 3]\} =$$
$$4 + \{9 + 5 + 4 - 7 + 3\} =$$
$$4 + 9 + 5 + 4 - 7 + 3 = 18$$

Example 1-19 Simplify $(6 - 2) - \{2 - [5 + 7 - \overline{1 - 2} + 8]\}$.

SOLUTION

$$(6 - 2) - \{2 - [5 + 7 - \overline{1 - 2} + 8]\} =$$
$$(6 - 2) - \{2 - [5 + 7 - 1 + 2 + 8]\} =$$
$$(6 - 2) - \{2 - 5 - 7 + 1 - 2 - 8\} =$$
$$6 - 2 - 2 + 5 + 7 - 1 + 2 + 8 = 23$$

Exercise 1-5

Simplify the following:

1. $112 + 48 \div 8 \times 3 - 67$
2. $120 \times 4 \div 4 - 4 + 4$
3. $263 - 63 \div 9 \times 7 - 8$
4. $18 \times 14 - 4 + 6 \div 5$
5. $288 \div 4 \div 6 \div 4$
6. $8 \times 4 \times 9 \div 3 \times 12$
7. $6 \div 6 \times 6 \times 6 \div 6$
8. $6 \div 6 \times 6 \div 6 \times 6$
9. $9 + (7 - 3)$
10. $4 - (7 - 5)$
11. $8 - (6 - 3)$
12. $(13 - 6) - 5$

13. $(8 - 4) - (3 - 7)$
14. $11 - [7 - (5 - 4)]$
15. $11 - [7 - (9 - 5) + 4]$
16. $25 - [9 - (7 + 2) + (5 - 2)]$
17. $28 - \{34 - [9 + (5 - 1)]\}$
18. $\{85 - [12 - (8 - 3)]\} - [10 - (2 - 4)]$
19. $42 - [17 - (5 - 4) - 11] + (6 - 2)$
20. $3 \times (2 \times 6) + (6 + 2) \div 4$
21. $3 \times (2 + 6) + (6 + 2) \div 4$
22. $3 + (2 \times 6) + (6 \times 2) - 4$
23. $13 - (3 \times 2) + (6 \times 2) \div 4$
24. $(13 - 3) \times (2 + 6) \div 4$
25. $13 - (2 \times 6) + (5 + 3) \div 2$

CHAPTER 2

FRACTIONS

2-1 COMMON FRACTIONS For many years the natural numbers, called *counting numbers* or *positive integers,* were sufficient for the needs of our ancestors. But as civilization advanced, there came a time when the idea of a part of a whole had to be considered. Lands and produce were partitioned, and concepts for these parts needed to be developed. This resulted first in the concepts of unit fractions, which we recognize as $\frac{1}{2}$, $\frac{1}{3}$, $\frac{1}{4}$, and so on. Symbols for these concepts have been found in the books used by Egyptians as early as 1650 B.C. Thus, if a board is divided into three equal parts, the fraction $\frac{1}{3}$ represents one of those equal parts (see Fig. 2-1). If a circle is

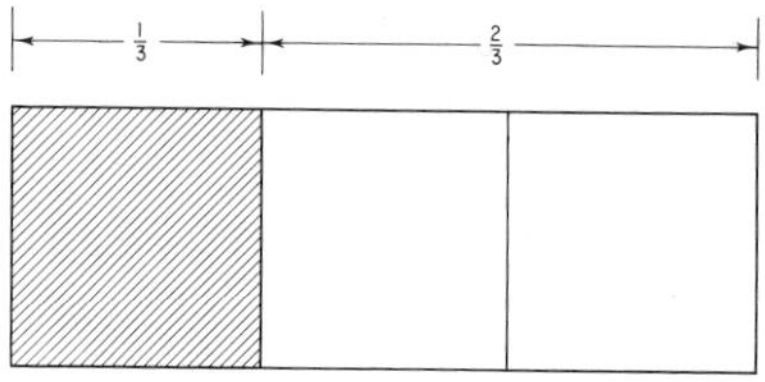

FIG. 2-1.

partitioned into four equal parts (see Fig. 2-2), one of the equal parts is represented by the unit fraction $\frac{1}{4}$.

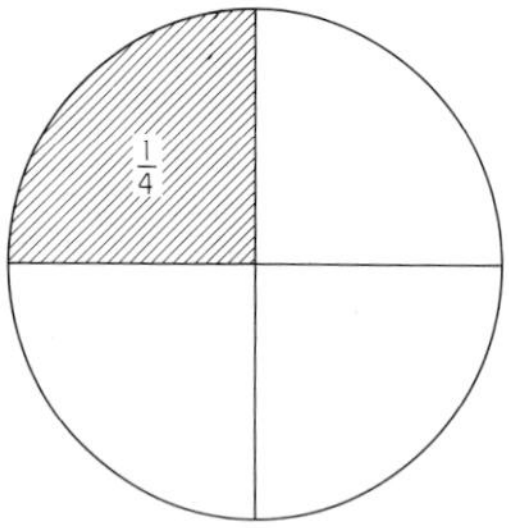

FIG. 2-2.

In Fig. 2-1 the unshaded portion of the figure, which is made up of two of the equal parts, is represented by $\frac{2}{3}$. The unshaded portion of Fig. 2-2 represents $\frac{3}{4}$ of the area of the circle. $\frac{5}{8}$ in. represents the length of five of eight equal parts into which an inch is divided (see Fig. 2-3).

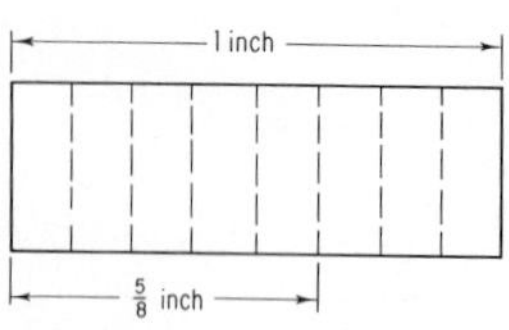

FIG. 2-3.

The number above the line of a fraction is called the *numerator* of the fraction. The number below the line is called the *denominator.* In the fraction $\frac{5}{8}$, the 5 is the numerator and the 8 is the denominator. Thus,

$$\text{A fraction} = \frac{\text{numerator}}{\text{denominator}}$$

$$= \frac{\text{number of parts we are using}}{\text{number of (equal) parts of whole}}$$

A *proper* fraction is a fraction whose numerator is less than its denominator. $\frac{1}{2}$, $\frac{3}{4}$, $\frac{7}{9}$ are proper fractions.

An *improper* fraction is a fraction whose numerator is equal to or greater than its denominator. $\frac{3}{3}$, $\frac{5}{5}$, $\frac{7}{6}$, and $\frac{11}{8}$ are improper fractions.

The numerator and denominator taken together are called the *terms* of the fraction.

Rational number is the name given to a fraction of whole numbers.

A number that can be written in the form a/b where a and b are counting numbers and $b \neq 0$ is called a rational number (*or fraction*).[1]

2-2 CHANGING THE FORM OF A FRACTION Equivalent fractions are fractions which represent the same part of a whole quantity. For example, $\frac{3}{4}$ of a square is equivalent to $\frac{6}{8}$ of the same square (see Fig. 2-4). $\frac{8}{12}$ of a dozen is

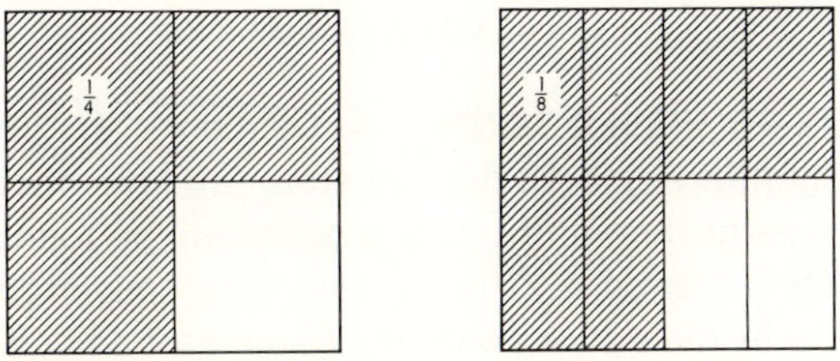

FIG. 2-4. $\frac{3}{4}$ is equivalent to $\frac{6}{8}$.

equivalent to $\frac{2}{3}$ of a dozen. Equivalent fractions can be obtained by applying the following fundamental principle of fractions: The value of a fraction is unchanged if both the numerator and the denominator are multiplied (or divided) by the same number, excepting zero.

If a, b, and c are numbers and b and c are different from zero, then

$$\frac{a}{b} = \frac{a \times c}{b \times c} \quad \textit{and} \quad \frac{a}{b} = \frac{a \div c}{b \div c}$$

---

[1] This definition will be adjusted in Chap. 7 to include negative integers.

Examples

2-1 $\frac{3}{5}=\frac{3\times 4}{5\times 4}=\frac{12}{20}$

2-2 $\frac{12}{20}=\frac{12\div 4}{20\div 4}=\frac{3}{5}$

A *mixed number* consists of a whole number plus a fraction. The mixed number $4\frac{2}{3}$ means $4+\frac{2}{3}$. In working with fractions it is often necessary to convert mixed numbers to improper fractions and improper fractions to mixed numbers. Examples of the processes follow.

Examples

2-3 $5\frac{2}{3}=5+\frac{2}{3}=\frac{5\times 3}{1\times 3}+\frac{2}{3}=\frac{15}{3}+\frac{2}{3}=\frac{17}{3}$

2-4 $\frac{18}{7}=18\div 7=2\frac{4}{7}$

2-3 REDUCTION TO LOWEST TERMS A fraction is said to be in its *lowest terms* if the numerator and denominator have no common divisor except 1. Thus, the fraction $\frac{12}{20}$, above, has been reduced to its lowest terms $\frac{3}{5}$ by dividing both the numerator and denominator by 4. Such a fraction is called a *prime fraction.* Reduction to lowest terms does not mean reducing to a mixed number. Thus, $\frac{18}{8}$ in lowest terms is $\frac{9}{4}$, not $2\frac{1}{4}$.

2-4 CONVERSION TO HIGHER TERMS A fraction can be converted to higher terms by multiplying both the numerator and the denominator by the same constant greater than unity. This process will be found useful in adding and subtracting fractions. The fraction $\frac{3}{7}$ is converted to the higher termed fraction $\frac{6}{14}$ by multiplying both the numerator and denominator by 2.

Exercise 2-1

Change each of the following fractions to equivalent fractions (fill in the blank spaces):

1. $\frac{3}{5}=\frac{}{25}$
2. $\frac{5}{8}=\frac{}{24}$
3. $\frac{3}{8}=\frac{12}{}$
4. $\frac{2}{3}=\frac{}{18}$
5. $\frac{16}{48}=\frac{1}{}$
6. $\frac{12}{64}=\frac{3}{}$
7. $\frac{13}{23}=\frac{}{92}$
8. $\frac{13}{18}=\frac{}{126}$

9. $\dfrac{705}{1{,}245} = \dfrac{\quad}{83}$

10. $\dfrac{7{,}896}{6{,}237} = \dfrac{376}{\quad}$

Change the improper fractions to equivalent mixed numbers:

11. $\dfrac{19}{3}$

12. $\dfrac{27}{4}$

13. $\dfrac{45}{4}$

14. $\dfrac{64}{15}$

15. $\dfrac{34}{24}$

16. $\dfrac{69}{48}$

17. $\dfrac{43}{12}$

18. $\dfrac{77}{42}$

19. $\dfrac{117}{36}$

20. $\dfrac{136}{56}$

Change the mixed numbers to equivalent improper fractions:

21. $1\frac{7}{8}$
22. $3\frac{4}{7}$
23. $4\frac{6}{11}$
24. $5\frac{8}{13}$
25. $2\frac{7}{9}$
26. $6\frac{9}{16}$
27. $7\frac{3}{4}$
28. $8\frac{1}{8}$
29. $9\frac{3}{5}$
30. $5\frac{11}{12}$

Reduce to lowest terms:

31. $\dfrac{18}{30}$

32. $\dfrac{48}{60}$

33. $\dfrac{160}{192}$

34. $\dfrac{252}{288}$

35. $\dfrac{56}{252}$

36. $\dfrac{159}{198}$

37. $\dfrac{140}{252}$

38. $\dfrac{288}{432}$

39. $\dfrac{6{,}216}{8{,}904}$

40. $\dfrac{20{,}160}{22{,}080}$

2-5 ADDITION AND SUBTRACTION OF FRACTIONS For the addition and subtraction of fractions we must remember the first law of addition and subtraction: *Only like quantities can be added or subtracted.* Two fractions are said to be *like* if they have the same denominator. Thus we can add $\frac{2}{9} + \frac{5}{9} = \frac{7}{9}$. In like manner, we can subtract $\frac{8}{11} - \frac{3}{11} = \frac{5}{11}$.

The sum (or difference) of two like fractions can be expressed as a single fraction whose numerator is the sum (or difference) of the numerators and whose denominator is the common denominator.

If a, b, and c are whole numbers and c is positive then

$$\frac{a}{c}+\frac{b}{c}=\frac{a+b}{c} \quad \text{and} \quad \frac{a}{c}-\frac{b}{c}=\frac{a-b}{c}$$

When the denominators are not alike, the fractions cannot be added directly. To attempt to add $\frac{3}{4}$ and $\frac{7}{10}$ would be similar to trying to add three quarters and seven dimes. In order to add unlike fractions, the fractions must first be written as equivalent fractions having the same denominator.

2-6 PRIME FACTORS You will recall that a *factor* of a number is an exact divisor of that number. Thus factors of 18 are 1, 2, 3, 6, 9, 18.

A *prime number* is a number which is exactly divisible only by itself and 1. The number 1 is not considered to be a prime number, because it has only itself as a divisor. Thus, the factors of 18 which are prime numbers are 2 and 3. These are called the *prime factors of 18.* The other factors (6, 9, 18) of 18 are not prime numbers because each is divisible, not only by 1 and itself, but also by 2 and/or 3.

To find the prime factors of a number one can start dividing the number by the smallest prime number 2 and continue dividing the consecutive quotients by 2 until he obtains a remainder not divisible by 2. He next will divide the last quotient (now a dividend) by 3. This process is repeated with increasing prime numbers as divisors until 1 is obtained as the final quotient.

Example 2-5 Find the prime factors of 840.

SOLUTION

$$\begin{array}{r|l} 2 & 840 \\ 2 & 420 \\ 2 & 210 \\ 3 & 105 \\ 5 & 35 \\ 7 & 7 \\ & 1 \end{array}$$

The number 840 can be expressed as the product of its prime factors: $840 = 2 \times 2 \times 2 \times 3 \times 5 \times 7$.

2-7 LEAST COMMON MULTIPLE The smallest possible number that can be divided exactly by two or more numbers is called the *lowest common multiple* of those numbers (abbreviated LCM). Often the lowest common multiple can be found by inspection. When inspection fails to reveal the LCM, one can always find it by the following method:

1 Arrange the numbers in a column and express each number as the product of prime factors.
2 Find the product of all the different prime factors, using each one the greatest number of times it appears as a factor of any one of the numbers. The product, thus found, is the least common multiple.

Example 2-6 Find the LCM of 12, 15, 21.

SOLUTION
$12 = 2 \times 2 \times 3$

$15 = 3 \times 5$

$21 = 3 \times 7$

The least common multiple is $2 \times 2 \times 3 \times 5 \times 7 = 420$.

Example 2-7 Find the LCM of 21, 24, and 42.

SOLUTION
$21 = 3 \times 7$

$24 = 2 \times 2 \times 2 \times 3$

$42 = 2 \times 3 \times 7$

The lowest common multiple is $2 \times 2 \times 2 \times 3 \times 7 = 168$.

2-8 LEAST COMMON DENOMINATOR The *least common denominator* (abbreviated LCD) of two or more fractions is the lowest common multiple of the denominators of those fractions. When unlike fractions are to be added or subtracted, the fractions must first be changed to equivalent like fractions. This involves finding the least common denominators of those fractions.

Example 2-8 Add $\frac{2}{3} + \frac{3}{4} + \frac{5}{6}$.

SOLUTION The LCD can be found, by inspection, to be 12.

$$\begin{aligned}\frac{2}{3} + \frac{3}{4} + \frac{5}{6} &= \frac{8}{12} + \frac{9}{12} + \frac{10}{12}\\ &= \frac{27}{12}\\ &= 2\frac{3}{12} \text{ or } 2\frac{1}{4}\end{aligned}$$

Example 2-9 Simplify $\frac{7}{12} + \frac{11}{30} - \frac{9}{20}$.

SOLUTION

$12 = 2 \times 2 \times 3$

$30 = 2 \times 3 \times 5$

$20 = 2 \times 2 \times 5$

The LCD is $2 \times 2 \times 3 \times 5$.

Then

$$\tfrac{7}{12} + \tfrac{11}{30} - \tfrac{9}{20} = \tfrac{35}{60} + \tfrac{22}{60} - \tfrac{27}{60}$$
$$= \tfrac{30}{60} \text{ or } \tfrac{1}{2}$$

Example 2-10 Simplify $5\tfrac{2}{3} + 3\tfrac{7}{12} - 2\tfrac{1}{4}$.

SOLUTION

$$5\tfrac{2}{3} + 3\tfrac{7}{12} - 2\tfrac{1}{4} = \tfrac{17}{3} + \tfrac{43}{12} - \tfrac{9}{4}$$
$$= \tfrac{68}{12} + \tfrac{43}{12} - \tfrac{27}{12}$$
$$= \tfrac{84}{12} \text{ or } 7$$

Example 2-11 Simplify $8\tfrac{1}{7} - 2\tfrac{3}{5} + 4\tfrac{1}{14} - 6\tfrac{7}{15}$.

SOLUTION

$7 = 7$

$5 = 5$

$14 = 2 \times 7$

$15 = 3 \times 5$

The LCD is $2 \times 3 \times 5 \times 7 = 210$.
Then

$$8\tfrac{1}{7} - 2\tfrac{3}{5} + 4\tfrac{1}{14} - 6\tfrac{7}{15} = \tfrac{57}{7} - \tfrac{13}{5} + \tfrac{57}{14} - \tfrac{97}{15}$$
$$= \frac{57 \cdot 2 \cdot 3 \cdot 5}{7 \cdot 2 \cdot 3 \cdot 5} - \frac{13 \cdot 2 \cdot 3 \cdot 7}{5 \cdot 2 \cdot 3 \cdot 7}$$
$$+ \frac{57 \cdot 3 \cdot 5}{14 \cdot 3 \cdot 5} - \frac{97 \cdot 2 \cdot 7}{15 \cdot 2 \cdot 7}$$
$$= \frac{1{,}710}{210} - \frac{546}{210} + \frac{855}{210} - \frac{1{,}358}{210}$$
$$= \tfrac{661}{210} \text{ or } 3\tfrac{31}{210}$$

Example 2-12 Add

$$\begin{array}{l} 9\tfrac{7}{8} \\ 3\tfrac{11}{16} \\ \underline{5\tfrac{3}{4}} \end{array}$$

SOLUTION This problem can be worked two ways.

$$\begin{array}{rl} 9\tfrac{7}{8} = & 9\tfrac{14}{16} \\ 3\tfrac{11}{16} = & 3\tfrac{11}{16} \\ 5\tfrac{3}{4} = & 5\tfrac{12}{16} \\ \hline & 17\tfrac{37}{16} = 19\tfrac{5}{16} \end{array}$$

ALTERNATE SOLUTION

$$\begin{aligned} 9\tfrac{7}{8} + 3\tfrac{11}{16} + 5\tfrac{3}{4} &= \tfrac{79}{8} + \tfrac{59}{16} + \tfrac{23}{4} \\ &= \tfrac{158}{16} + \tfrac{59}{16} + \tfrac{92}{16} \\ &= \tfrac{309}{16} \text{ or } 19\tfrac{5}{16} \end{aligned}$$

Example 2-13 Subtract

$$\begin{array}{l} 58\tfrac{5}{16} \\ \underline{17\tfrac{7}{12}} \end{array}$$

SOLUTION The least common denominator of the two fractions is 48. In this problem we shall have to borrow 1 from 58 and change it to 48 forty-eighths in order that the fractions can be subtracted.

$$\begin{array}{l} 58\tfrac{5}{16} = 58\tfrac{15}{48} = 57\tfrac{63}{48} \\ 17\tfrac{7}{12} = 17\tfrac{28}{48} = \underline{17\tfrac{28}{48}} \\ \qquad\qquad\qquad\quad 40\tfrac{35}{48} \end{array}$$

Exercise 2-2

Find the prime factors in Probs. 1 to 10:

1. 36
2. 42
3. 120
4. 252
5. 450
6. 1,400
7. 693
8. 1,470
9. 2,600
10. 3,960

Find the LCM in Probs. 11 to 20:

11. 4, 6, 15
12. 8, 9, 6
13. 16, 12, 15
14. 16, 18, 15
15. 16, 30, 75
16. 54, 45, 75
17. 14, 21, 35
18. 22, 21, 35
19. 56, 44, 42
20. 98, 50, 75

In Probs. 21 to 40 perform the indicated additions and subtractions:

21. $\frac{1}{3} + \frac{1}{3}$
22. $\frac{2}{5} + \frac{3}{10}$
23. $\frac{1}{3} + \frac{1}{4}$
24. $\frac{5}{6} - \frac{3}{6}$
25. $\frac{4}{5} - \frac{2}{10}$
26. $\frac{5}{6} - \frac{5}{8}$
27. $\frac{2}{7} + \frac{2}{3} + \frac{5}{7}$
28. $\frac{5}{8} + \frac{3}{4} + \frac{1}{2}$
29. $\frac{4}{4} + \frac{5}{8} - \frac{1}{6}$
30. $\frac{7}{10} - \frac{3}{20} + \frac{4}{5}$
31. $\frac{7}{15} + \frac{5}{6} - \frac{3}{4}$
32. $\frac{9}{16} + \frac{5}{18} - \frac{2}{15}$
33. $\frac{7}{54} + \frac{17}{45}$
34. $\frac{29}{45} - \frac{34}{75} + \frac{2}{15}$
35. $5\frac{5}{6} + 2\frac{1}{4} - 3\frac{7}{8}$
36. $4\frac{5}{16} + 6\frac{4}{21} + 8\frac{9}{35}$
37. $2\frac{7}{16} + 9\frac{5}{18} - 7\frac{4}{15}$
38. $87\frac{1}{2} + 33\frac{4}{15} - 26\frac{5}{6}$
39. $\frac{3}{98} + \frac{7}{50} + \frac{14}{75} - \frac{9}{30}$
40. $\frac{35}{56} - \frac{31}{44} - \frac{5}{42} + \frac{59}{77}$

41. Add: $\begin{array}{r} 9\frac{7}{9} \\ 16\frac{2}{3} \\ 13\frac{5}{6} \\ \hline \end{array}$

42. Add: $\begin{array}{r} 208\frac{15}{16} \\ 1{,}593\frac{11}{30} \\ 845\frac{4}{21} \\ \hline \end{array}$

43. Subtract: $\begin{array}{r} 28\frac{7}{8} \\ 13\frac{2}{3} \\ \hline \end{array}$

44. Subtract: $\begin{array}{r} 872\frac{11}{16} \\ 254\frac{11}{12} \\ \hline \end{array}$

45. Subtract: $\begin{array}{r} 39\frac{13}{54} \\ 24\frac{17}{45} \\ \hline \end{array}$

46. Subtract: $\begin{array}{r} 118\frac{9}{56} \\ 79\frac{17}{44} \\ \hline \end{array}$

47. $23\frac{5}{6}$ is how much greater than $15\frac{3}{4}$?
48. How much greater is $53\frac{5}{6}$ than $46\frac{7}{15}$?
49. What is the difference between $37\frac{7}{54}$ and $26\frac{19}{45}$?
50. A 15-ft board has cut from it lengths of $2\frac{3}{4}$, $1\frac{5}{6}$, $4\frac{1}{12}$, and $5\frac{1}{2}$ ft. How much of the board is left if the width of each saw cut is neglected?
51. A salesgirl on one day sold from one bolt of dry goods the following lengths of fabric: $4\frac{1}{2}$, $5\frac{2}{3}$, $12\frac{5}{6}$, 8, $18\frac{1}{2}$ yd. How many yards of the material did she sell?
52. How many feet of molding are required to go around a room $18\frac{3}{4}$ ft wide and $35\frac{2}{3}$ ft long?
53. Three months ago Mr. Panagos weighed $220\frac{1}{2}$ lb. Today he weighs $186\frac{3}{4}$ lb. How many pounds did he lose in the three months?
54. In preparing an 11- by 15-in. sheet in the drawing class, pupils were instructed to draw border lines 1 in. from the left edge and $\frac{3}{8}$ in. from the other three edges of the sheet. What will be the dimensions between the border lines?
55. Determine the length of the bolt shown.

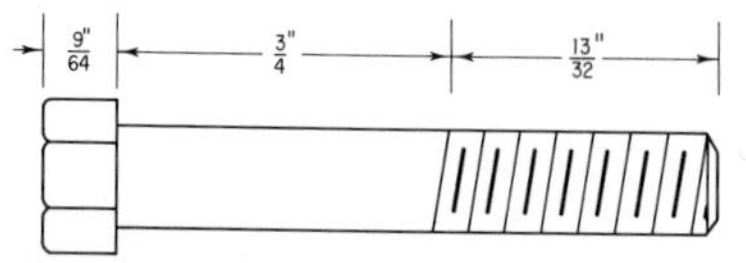

EXERCISE 55

56. Determine the length x.

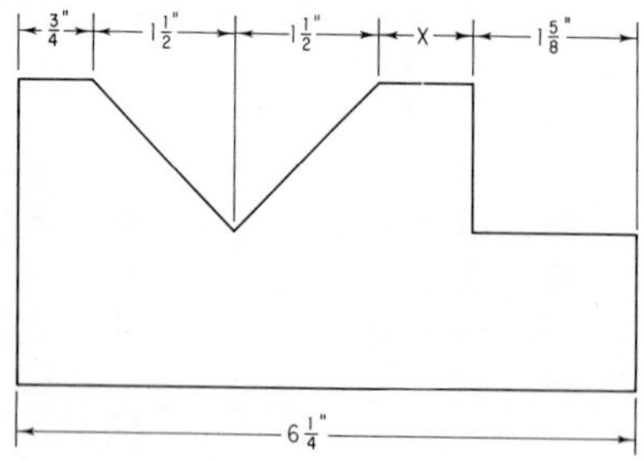

EXERCISE 56

57. Determine the length of the radius labeled x.

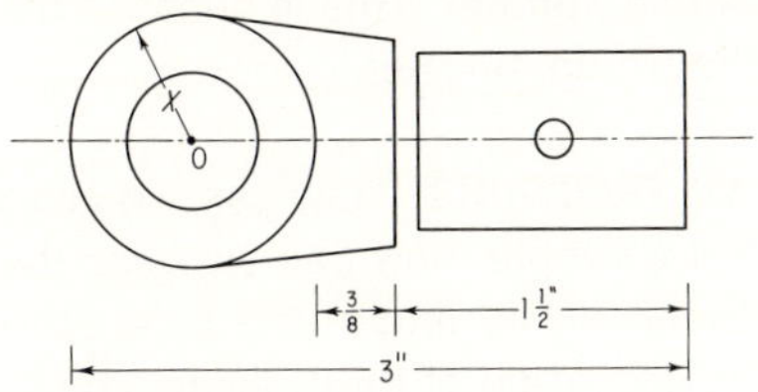

EXERCISE 57

58. Determine the dimension x.

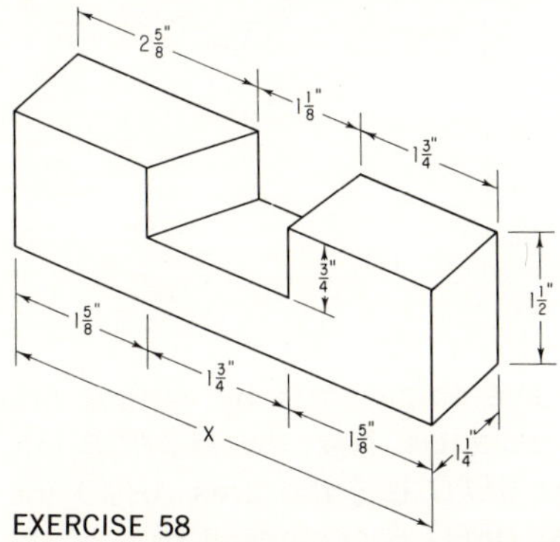

EXERCISE 58

2-9 THE PRODUCT OF A WHOLE NUMBER AND A FRACTION In Sec. 1-9 we noted that the multiplication of two numbers can be interpreted as a short cut for repeated addition of equal quantities. We can use this interpretation in finding the product of a whole number and a fraction. To multiply a fraction by a whole number, for example, $\frac{3}{8}$ by 4, means to add the fraction a whole number of times. Thus,

$$4 \times \tfrac{3}{8} = \tfrac{3}{8} + \tfrac{3}{8} + \tfrac{3}{8} + \tfrac{3}{8} = \tfrac{12}{8} = \tfrac{3}{2}$$

But how do we interpret the operation of multiplying 4 by $\frac{3}{8}$? To say $\frac{3}{8} \times 4$ means to "add the number 4 three-eighths of a time" doesn't make sense.

To multiply a whole number by a fraction, we apply the commutative property of multiplication (see Sec. 1-10) to show that $\frac{3}{8} \times 4$ is equivalent to $4 \times \frac{3}{8}$. Thus, $\frac{3}{8} \times 4 = 4 \times \frac{3}{8} = \frac{12}{8} = \frac{3}{2}$. $\frac{3}{8} \times 4$ can be read "$\frac{3}{8}$ of 4." Thus, to find the product of a whole number and a fraction, we multiply the numerator of the fraction by the whole number and divide the product by the denominator of the fraction.

If a, b, and c are whole numbers, then

$$a \times \frac{b}{c} = \frac{b}{c} \times a = \frac{a \times b}{c}$$

It will be noted that the product of a whole number and a proper fraction is always smaller than the whole number. This is because the product represents a fractional part of the whole number.

2-10 THE PRODUCT OF TWO FRACTIONS How can we give meaning to the product of two fractions? For example, how can we describe $\frac{2}{3} \times \frac{4}{5}$?

The product of two fractions can be illustrated by a simple drawing (see Fig. 2-5). Let us consider the operation of multiplying $\frac{4}{5}$ by $\frac{2}{3}$. Let a rectangle

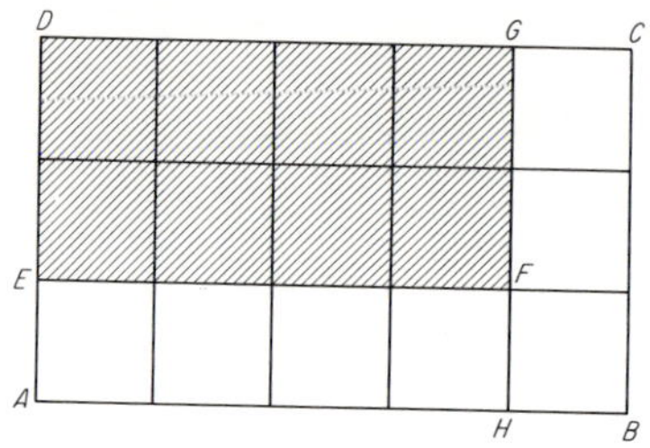

FIG. 2-5.

ABCD be partitioned into five equal parts by vertical lines. The area of *AHGD* is $\frac{4}{5}$ the area of the rectangle. Next divide *AHGD* into thirds by horizontal lines. Then the area *DEFG* is $\frac{2}{3}$ the area *AHGD* (or $\frac{2}{3}$ of $\frac{4}{5}$ of area *ABCD*). It can be seen that *DEFG* is composed of 8 of the 15 equal rectangles that make up *ABCD*. Then area $DEFG = \frac{8}{15}$ area *ABCD* or $\frac{2}{3} \times \frac{4}{5} = \frac{8}{15}$.

The product of two fractions is a fraction whose numerator is the product of the numerators and whose denominator is the product of the denominators.

If a, b, c and d are whole numbers and if neither b nor d is zero, then

$$\frac{a}{b} \times \frac{c}{d} = \frac{a \times c}{b \times d}$$

Examples

2-14 $\frac{7}{8} \times \frac{5}{12} = \frac{35}{96}$

2-15 $\frac{5}{3} \times \frac{7}{9} = \frac{35}{27} = 1\frac{8}{27}$

2-11 CANCELING The product of two fractions should be reduced to lowest terms whenever possible. This process can often be shortened by applying the fundamental principle of fractions before multiplying the numerators and the denominators. For example, in finding the product of $\frac{8}{15}$ and $\frac{7}{12}$, we can divide both the numerator and the denominator by 4 before multiplying the numerators and the denominators. Thus, we could write

$$\frac{8}{15}\times\frac{7}{12}=\frac{\overset{2}{\cancel{8}}\times 7}{15\times\underset{3}{\cancel{12}}}=\frac{14}{45}$$

instead of writing

$$\frac{8}{15}\times\frac{7}{12}=\frac{8\times 7}{15\times 12}=\frac{56}{180}=\frac{14}{45}$$

As another form, we could solve the problem as

$$\frac{8}{15}\times\frac{7}{12}=\frac{\overset{2}{\cancel{8}}}{15}\times\frac{7}{\underset{3}{\cancel{12}}}=\frac{14}{45}$$

This procedure is called *canceling.* It is especially helpful in determining the product of several fractions.

Examples

2-16 $\frac{5}{13}\times\frac{7}{15}\times\frac{11}{14}=\frac{\overset{1}{\cancel{5}}}{13}\times\frac{\overset{1}{\cancel{7}}}{\underset{3}{\cancel{15}}}\times\frac{11}{\underset{2}{\cancel{14}}}$

$$=\frac{11}{78}$$

2-17 $3\tfrac{1}{5}\times 7\tfrac{3}{4}\times 2\tfrac{1}{12}=\frac{\overset{1}{\cancel{16}}}{\underset{1}{\cancel{5}}}\times\frac{31}{\underset{1}{\cancel{4}}}\times\frac{\overset{5}{\cancel{25}}}{\underset{3}{\cancel{12}}}$

$$=\frac{155}{3}\text{ or }51\tfrac{2}{3}$$

The student is cautioned not to make the following common type of mistake.

$$\frac{\overset{1}{\cancel{2}}+3}{5+\underset{1}{\cancel{2}}}=\frac{4}{6}\qquad\text{NOT VALID}\qquad\frac{\overset{2}{\cancel{4}}+\overset{1}{\cancel{5}}}{\underset{2}{\cancel{10}}+\underset{3}{\cancel{6}}}=\frac{3}{5}$$

A quick check will reveal that the first fraction should be reduced to $\frac{5}{7}$ (not $\frac{4}{6}$) while the second fraction should be reduced to $\frac{9}{16}$ (not $\frac{3}{5}$). In the first instance 1 was subtracted from both the numerator and the denominator. In the second, $2+4$ or 6 was subtracted from the numerator and $8+3$ from the denominator. This is not permissible.

Example 2-18 Multiply 3 lb $5\tfrac{1}{2}$ oz by 7.

SOLUTION

$$\begin{array}{l} \quad 3 \text{ lb} \quad 5\tfrac{1}{2} \text{ oz} \\ \underline{\times \qquad\quad 7} \\ 21 \text{ lb } 38\tfrac{1}{2} \text{ oz} = 21 \text{ lb} + (32 \text{ oz} + 6\tfrac{1}{2} \text{ oz}) \\ \qquad\qquad\quad = 21 \text{ lb} + (2 \text{ lb} + 6\tfrac{1}{2} \text{ oz}) \\ \qquad\qquad\quad = (21 \text{ lb} + 2 \text{ lb}) + 6\tfrac{1}{2} \text{ oz} \\ \qquad\qquad\quad = 23 \text{ lb } 6\tfrac{1}{2} \text{ oz} \end{array}$$

Exercise 2-3

Find the following products and reduce to lowest terms:

1. $\frac{5}{6} \times \frac{3}{5}$
2. $\frac{2}{3} \times \frac{5}{6}$
3. $\frac{7}{8} \times \frac{5}{12}$
4. $\frac{2}{3} \times \frac{5}{7}$
5. $\frac{7}{9} \times \frac{3}{4}$
6. $\frac{3}{4} \times 6$
7. $12 \times \frac{2}{3}$
8. $21 \times \frac{3}{7}$
9. $\frac{2}{3} \times \frac{8}{9}$
10. $\frac{4}{5} \times \frac{2}{3} \times \frac{3}{8}$
11. $\frac{18}{25} \times \frac{5}{9} \times \frac{3}{4}$
12. $\frac{9}{8} \times \frac{2}{3} \times \frac{12}{7}$
13. $\frac{12}{5} \times \frac{5}{16} \times \frac{4}{3}$
14. $\frac{7}{9} \times \frac{2}{5} \times \frac{3}{8}$
15. $\frac{16}{45} \times \frac{47}{66} \times \frac{90}{164} \times \frac{92}{141}$
16. $\frac{4}{5} \times \frac{11}{14} \times \frac{25}{48} \times \frac{16}{55}$
17. $12 \times 1\frac{5}{6}$
18. $30 \times 2\frac{9}{16}$
19. $1\frac{1}{5} \times 2\frac{1}{2}$
20. $4\frac{4}{5} \times 2\frac{1}{2}$
21. $7\frac{1}{3} \times 2\frac{4}{7}$
22. $5\frac{3}{4} \times 2\frac{6}{7}$
23. $2\frac{3}{4} \times 3\frac{3}{7} \times 5\frac{3}{5}$
24. $6\frac{1}{4} \times 12\frac{2}{5} \times 8\frac{2}{3}$
25. $12 \times 3\frac{5}{8}$ in.
26. 9×5 lb $6\frac{3}{4}$ oz
27. 10×5 ft 7 in.
28. 50×2 hr 45 min 10 sec
29. If 20 pieces of work can be machined in 8 hr, how many hours will it take to machine 12 pieces of work?
30. Pump *A* can fill a reservoir in 12 hr and pump *B* can fill it in 8 hr. What fractional part of the full reservoir will result if together pump *A* operates 2 hr and pump *B* 3 hr?

2-12 DIVISION OF FRACTIONS The process of division of fractions follows easily from multiplication and the fundamental principle of fractions.

A fraction or mixed number divided by a whole number can be interpreted as *partition*. Consider the problem $\frac{4}{5} \div 2$. We learned in Sec. 1-13 that dividing by 2 is to find the size of each of two equal parts into which a quantity can be partitioned. Since $\frac{2}{5} + \frac{2}{5} = \frac{4}{5}$, we know that $\frac{4}{5} \div 2 = \frac{2}{5}$. To divide a number by 2 is to take one-half of the number, to divide a number by 3 is to take one-third of the number, and so on. Thus, $\frac{4}{5} \div 2 = \frac{1}{2} \times \frac{4}{5} = \frac{4}{5} \times \frac{1}{2} = \frac{2}{5}$ and $\frac{4}{5} \div 3 = \frac{1}{3} \times \frac{4}{5} = \frac{4}{5} \times \frac{1}{3} = \frac{4}{15}$. It will be noted that, when a given fraction is divided by a number larger than 1, a number smaller than the given fraction will result.

An integer divided by a fraction or mixed number can be interpreted as *measurement*. For example, $4 \div \frac{1}{3}$ can mean how many times $\frac{1}{3}$ is contained in 4. We know that each unit has 3 thirds. Then 4 units must have 4×3 or 12 thirds. This can be written $4 \div \frac{1}{3} = 4 \times \frac{3}{1} = 12$. Here we notice that, when a given integer is divided by a proper fraction, the quotient will always be larger than the given fraction.

Consider the problem $\frac{2}{3} \div \frac{3}{4}$. A possible instance in which such a problem might develop could be: A contractor determines that he will need $\frac{3}{4}$ ton of cement to build a retaining wall. He only has $\frac{2}{3}$ ton of cement. What fractional part of the wall can he complete with the available cement? We know that $\frac{2}{3} \div \frac{3}{4}$ can be written $\frac{\frac{2}{3}}{\frac{3}{4}}$. Such a fraction, in which either or both terms are themselves fractions, is called a *complex fraction*. If we apply the fundamental principle of fractions to multiply both terms of the complex fraction by $\frac{4}{3}$, our complex fraction will reduce to an equivalent simple fraction. Thus, we have

$$\frac{2}{3} \div \frac{3}{4} = \frac{\frac{2}{3}}{\frac{3}{4}} = \frac{\frac{2}{3} \times \frac{4}{3}}{\frac{3}{4} \times \frac{4}{3}} = \frac{\frac{8}{9}}{1} = \frac{8}{9}$$

The contractor can complete $\frac{8}{9}$ of the wall.

2-13 RECIPROCALS AND DIVISION OF FRACTIONS Two numbers are said to be *reciprocals* (or multiplicative inverses) if their product is 1. For example, the reciprocal of 2 is $\frac{1}{2}$, $\frac{1}{3}$ and 3 are reciprocals, $\frac{3}{4}$ and $\frac{4}{3}$ are reciprocals. It will be noted that the reciprocal of a whole number is a unit fraction and the reciprocal of a unit fraction is an integer. If N is a number, then N and $1/N$ are reciprocals. The reciprocal of a fraction is often called the *inverted* fraction. It should be clear that any integer N can be expressed as the improper fraction $N/1$.

From the three illustrations (of Sec. 2-12) of divisions of fractions we can deduce the following generalization. To divide by a fraction, invert the fraction and multiply.

If a, b, c, and d are whole numbers, $b \neq 0$, $c \neq 0$ and $d \neq 0$, then

$$\frac{a}{b} \div \frac{c}{d} = \frac{a}{b} \times \frac{d}{c}$$

Since the division of a fraction by a fraction is a purely mechanical operation, the student sometimes makes the mistake of inverting the wrong fraction. The pupil should always check his answer to see if it is sensible. For example, we know that $\frac{3}{5} \div \frac{8}{9} = \frac{27}{40}$ is reasonable since if $\frac{8}{9}$ is larger than $\frac{3}{5}$, the quotient must be less than 1. If, on the other hand, we inverted the $\frac{3}{5}$ and multiplied to get the answer $\frac{40}{27}$, we should realize that the answer is not reasonable.

Examples

2-19 $\frac{18}{25} \div 4 = \frac{\overset{9}{\cancel{18}}}{25} \times \frac{1}{\underset{2}{\cancel{4}}} = \frac{9}{50}$

2-20 $27 \div \frac{3}{7} = \overset{9}{\cancel{27}} \times \frac{7}{\underset{1}{\cancel{3}}} = 63$

2-21 $\frac{8}{17} \div \frac{16}{51} = \frac{\overset{1}{\cancel{8}}}{\underset{1}{\cancel{17}}} \times \frac{\overset{3}{\cancel{51}}}{\underset{2}{\cancel{16}}} = \frac{3}{2} = 1\frac{1}{2}$

2-22 $\frac{8}{17} \div 1\frac{5}{51} = \frac{8}{17} \div \frac{56}{51} = \frac{\overset{1}{\cancel{8}}}{\underset{1}{\cancel{17}}} \times \frac{\overset{3}{\cancel{51}}}{\underset{7}{\cancel{56}}} = \frac{3}{7}$

2-23 $2\frac{14}{15} \div 8\frac{2}{9} = \frac{44}{15} \div \frac{74}{9} = \frac{\overset{22}{\cancel{44}}}{\underset{5}{\cancel{15}}} \times \frac{\overset{3}{\cancel{9}}}{\underset{37}{\cancel{74}}} = \frac{66}{185}$

Example 2-24 Divide 7 lb $11\frac{1}{2}$ oz by 4.

SOLUTION

$4\overline{)7\text{ lb } 11\frac{1}{2}\text{ oz}} =$

$4\overline{)(4\text{ lb} + 3\text{ lb}) + 11\frac{1}{2}\text{ oz}} =$

$4\overline{)4\text{ lb} + (48\text{ oz} + 11\frac{1}{2}\text{ oz})} =$

$4\overline{)4\text{ lb } 59\frac{1}{2}\text{ oz}} =$

1 lb $14\frac{7}{8}$ oz

Exercise 2-4

Divide and leave the answer in lowest terms:

1. $\frac{6}{7} \div 3$
2. $\frac{15}{17} \div 5$
3. $4 \div \frac{3}{8}$
4. $24 \div \frac{12}{35}$
5. $4 \div 3\frac{1}{2}$
6. $17 \div 2\frac{1}{3}$
7. $2\frac{7}{9} \div 3$
8. $13\frac{2}{3} \div 7$
9. $\frac{3}{7} \div \frac{4}{5}$
10. $\frac{9}{11} \div \frac{15}{33}$
11. $\frac{12}{25} \div \frac{8}{15}$
12. $\frac{49}{240} \div \frac{14}{30}$

13. $7\frac{1}{2} \div 5\frac{1}{4}$

14. $5\frac{2}{3} \div 4\frac{1}{7}$

15. $26\frac{4}{7} \div 3\frac{5}{6}$

16. $17\frac{2}{3} \div 34\frac{3}{4}$

17. $\dfrac{8}{\frac{2}{3}}$

18. $\dfrac{17}{\frac{3}{4}}$

19. $\dfrac{\frac{11}{12}}{2}$

20. $\dfrac{\frac{16}{17}}{8}$

21. $\dfrac{\frac{7}{8}}{\frac{4}{3}}$

22. $\dfrac{\frac{15}{64}}{\frac{45}{8}}$

23. $\dfrac{\frac{27}{32}}{2\frac{1}{4}}$

24. $\dfrac{\frac{25}{2}}{2\frac{2}{3}}$

25. $\dfrac{12\frac{1}{4}}{\frac{7}{30}}$

26. $\dfrac{13\frac{4}{5}}{\frac{9}{20}}$

27. $\dfrac{7\frac{5}{12}}{3\frac{5}{8}}$

28. $\dfrac{1\frac{5}{23}}{1\frac{5}{64}}$

29. $\frac{8}{3} \times \frac{15}{14} \div \frac{4}{7}$

30. $\frac{8}{3} \div \frac{15}{14} \times \frac{4}{7}$

31. $\frac{8}{3} \div \frac{4}{7} \times \frac{15}{4}$

32. $\frac{8}{3} \times \frac{4}{7} \div \frac{15}{4}$

33. $\dfrac{\frac{2}{3} \div \frac{3}{4}}{\frac{4}{5} \div \frac{5}{6}}$

34. $\dfrac{\frac{16}{21} \div \frac{2}{7}}{\frac{3}{5} \div \frac{12}{35}}$

35. $\dfrac{\frac{2}{3} + \frac{3}{4}}{\frac{4}{5} + \frac{5}{6}}$

36. $\dfrac{\frac{8}{15} + \frac{1}{3}}{\frac{11}{12} - \frac{3}{4}}$

37. Divide 7 ft 9 in. by 5.
38. Divide 23 min 40 sec by 2.
39. Divide 15 lb 8 oz by 6.
40. Divide 9 hr 18 min 27 sec by 4.
41. A barrel holds $31\frac{1}{2}$ gal. How many barrels of oil are contained in an oil tanker that holds 34,000 gal?
42. In building a brick wall it was found that the average height of the brick and mortar was $2\frac{7}{8}$ in. How many layers of brick and mortar will be required to build a brick wall $7\frac{2}{3}$ ft high?
43. It is desired to cut from a 24- by 45-in. sheet of aluminum smaller rectangular pieces whose dimensions are $2\frac{2}{3}$ by $3\frac{3}{4}$ in. How many pieces can be cut from the sheet?
44. How many pieces of aluminum whose dimensions are $2\frac{3}{4} \times 3\frac{2}{3}$ in. can be cut from the sheet of Prob. 43? How much waste material will be left over?
45. A boy buys pencils at 3 for 10 cents. How many must he sell at 5 for 20 cents in order to make a profit of $1?

CHAPTER 3
DECIMAL FRACTIONS

3-1 EXTENSION OF THE NUMBER SYSTEM: DECIMALS We have discussed the structure of our number system which uses only 10 different symbols. The real significance of the system lies in the place-value concept. Each digit has not only its own value but also a place value, depending upon its position in the sequence of symbols which represent the number. For example, the number 3,184 means $3(1{,}000) + 1(100) + 8(10) + 4(1)$. Thus the 3 represents three thousands, the 1 represents one hundred, the 8 represents eight tens, and the 4 represents four ones. The set of such numbers is called the *decimal system* because 10 symbols are used to represent all numbers.

It is interesting to note that until 400 years ago the system allowed only for an unlimited number of digits to the left of the units digit but stopped with the ones digit on the right. Not until the sixteenth century is there any record of attempts to extend the system with digits to the right of the unit digit. Credit for such an extension of the system is usually given to the Belgian mathematician Simon Stevin. Stevin published his ideas in 1585. He reasoned that, if each place in the number system has a value ten times the value of the next place to the right, then, conversely, each place must have a value one-tenth the next place to the left. When the place-value concept was extended to the right of the ones place, the usefulness of our system was greatly increased.

With the extension of the number system it became necessary to find some means of determining the value of each digit in a given number. This was done by using a symbol as a separatrix between the unit and a part of the unit. Various symbols have been used. We are most familiar with the dot (.) placed immediately to the right of the units digit.

Many European countries, e.g., Germany, France, Italy, Belgium, and the Scandinavian countries, use a comma as the separatrix. In England a dot is used, but it is placed in the middle of the line of writing. Thus π is approximated in England as $3 \cdot 1416$.

In England our decimal point is used to stand for multiplication. Thus, in England $3\,.\,1416 = 3 \times 1416$.

In the Scandinavian countries it is customary to print the fractional part in smaller print than the integral part. In America we often write $\$86^{\underline{25}}$ for \$86.25 to lessen the chance of forgery by altering the position of the decimal.

3-2 READING AND WRITING DECIMAL FRACTIONS While the decimal separates the fractional part from the integral part of a number, the value of a digit in a given number is determined by the number of spaces it is to the left or right of the unit place. Figure 3-1 illustrates the relationship between the corresponding values of digits on opposite sides of the units digit. The tens place is one place to the left of the ones place, the tenths place is one place to the right of the ones place, and so on.

Ten thousands	Thousands	Hundreds	Tens	Ones		Tenths	Hundredths	Thousandths	Ten-thousandths	Hundred-thousandths
2	6	8	3	4	.	1	7	9	3	8

FIG. 3-1.

3-3 DECIMAL FRACTIONS AND RATIONAL NUMBERS In Chap. 1 it was pointed out that the same number may be represented by many different symbols. Thus, $3 + 1$, $\frac{12}{3}$ and 2×2 each represent the same number. We can also say that $\frac{1}{4}$ and 0.25 are symbols for the same number. (In the next chapter we will find that the same number can also be expressed as 25%.) Other symbols for the number are: $3 - 2\frac{3}{4}$, $\frac{3}{12}$ and $2 \times \frac{1}{8}$. Numbers that can be expressed as a ratio of two integers are called *rational numbers.*

The numbers which can be written in the form a/b where a is an integer and b is a positive integer are called rational numbers.

3-4 CHANGING A COMMON FRACTION TO A DECIMAL FRACTION To convert the symbol for a common fraction to the symbol for an equivalent decimal fraction we need merely to divide the numerator by the denominator. All common fractional quantities have decimal equivalents. Consider the following:

(a) $\frac{7}{16} = 0.4375$ (b) $\frac{5}{6} = 0.8333...$ (c) $\frac{4}{7} = 0.571428571428...$

A decimal that "comes out evenly" with a finite number of places, like 0.4375 in (a) is called a *terminating decimal.* The decimal numbers in (b) and (c) are nonterminating; however, it will be noted that if the division is carried out far enough a digit-sequence group repeats over and over in the quotient. In example (b) the 3 is repeated; in example (c) the digit-sequence combination 571428 will repeat forever. These numbers can be written $\frac{5}{6} = 0.8333... = 0.8\overline{3}$; $\frac{4}{7} = 0.571428571428... = 0.\overline{571428}$. Actually a terminating decimal also has a repeating block. Thus, $\frac{7}{16}$ can be written 0.43750000... or $0.4375\overline{0}$. We consider the zeros as the repeating part.

For all practical purposes, divisions that do not produce a terminating decimal are "stopped" one digit beyond the desired accuracy and "rounded off" to the desired place (see Sec. 3-11).

All terminating decimals and all repeating decimals are rational numbers and can be expressed in the form a/b, where a is an integer and b is a positive integer.

3-5 CHANGING A DECIMAL FRACTION TO A COMMON FRACTION Occasionally it becomes necessary to change a decimal fraction to an equivalent common fraction. If we know how to read decimal fractions, we should have little trouble writing the equivalent common fractions.

Examples

3-1 0.0384 is read "three hundred eighty-four ten-thousandths."
Hence $0.0384 = \frac{384}{10{,}000} = \frac{24}{625}$.

3-2 13.275 is read "13 and 275 thousandths."
Hence $13.275 = 13 + \frac{275}{1{,}000} = 13 + \frac{11}{40} = \frac{531}{40}$.

Exercise 3-1

1. Write in decimal form:
   (a) Six hundred forty-two thousandths.
   (b) Fourteen and one hundred eight thousandths.
   (c) Eighteen and four thousandths.
   (d) Fifty-six thousandths.
   (e) Eighty-seven and six hundred five ten-thousandths.
   (f) Two thousand two and two hundred thousandths.
   (g) $8\frac{3}{10}$
   (h) $47\frac{13}{1{,}000}$
   (i) $\frac{7}{10} + \frac{3}{100}$
   (j) $\frac{9}{10} + \frac{4}{100} + \frac{5}{1{,}000} + \frac{1}{10{,}000}$
   (k) $4 + \frac{8}{10} + \frac{3}{1{,}000} + \frac{9}{10{,}000}$
2. Write in words:
   (a) 0.385
   (b) 0.207
   (c) 0.00436
   (d) 0.68
   (e) 0.1738
   (f) 42.027
   (g) 4,500.0017
   (h) 298,060.0304
3. Arrange the following set of numbers according to size, from the largest to the smallest: 0.01083, 0.1083, 1.083, 1.803, 0.01803, 0.10380, 0.0108, 1.08.

Write the following rational numbers as terminating or repeating decimal fractions.

4. $\frac{3}{4}$ 7. $\frac{3}{125}$ 10. $\frac{57}{75}$ 13. $\frac{163}{110}$
5. $\frac{5}{8}$ 8. $\frac{43}{20}$ 11. $\frac{5}{9}$ 14. $\frac{1}{44}$
6. $\frac{13}{20}$ 9. $\frac{13}{8}$ 12. $\frac{263}{999}$ 15. $\frac{5}{99}$

Convert the following decimal fractions to common fractions (assume the fractions are terminating decimals).

16. 0.375 19. 0.0625 22. 4.125 25. 5.08
17. 0.14 20. 0.6875 23. 8.375 26. 8.4375
18. 0.128 21. 0.71875 24. 3.24 27. 3.5625

3-6 ADDITION AND SUBTRACTION OF DECIMAL FRACTIONS We have observed that only items of the same kind can be added or subtracted. So it is in adding decimal fractions we can add only tenths to tenths, hundredths to hundredths, and so on. This should pose no problem in adding decimal fractions if we arrange the numbers in columns with corresponding decimal places in the same column.

Example 3-3 Find the sum of 11.384, 5.06, 0.8, 4.107.

SOLUTION

```
11.384          11.384
 5.06            5.060
 0.8      or     0.800
 4.107           4.107
------          ------
21.351          21.351
```

Some people prefer to add the zeros, as in the second example, as an aid in keeping their columns straight. Subtraction of decimals is performed in like manner.

It should be apparent that the addition of decimal fractions is often considerably easier than the addition of equivalent common fractions. The same problem above expressed in common fractions would be:

Add $11\frac{48}{125} + 5\frac{3}{50} + \frac{4}{5} + 4\frac{107}{1,000}$

Example 3-4 Subtract 35.462 from 269.7.

SOLUTION

```
 269.7   =  269.700    Attaching zeros does not change the value of the number.
− 35.462 = − 35.462
            -------
            234.238    (difference)
```

3-7 MULTIPLICATION OF DECIMAL FRACTIONS We can arrive at the rule for multiplying decimal fractions inductively by considering some sample problems. Consider the problem of finding the product 3.4 × 2.76. We might write $3.4 \times 2.76 = \frac{34}{10} \times \frac{276}{100} = \frac{9{,}384}{1{,}000} = 9.384$. In like manner, $0.28 \times 0.319 = \frac{28}{100} \times \frac{319}{1{,}000} = \frac{8{,}932}{100{,}000} = 0.08932$. Note that the number of zeros in the denominator of the product of the common fractions is equal to the sum of the zeros in the denominators of the factors. Hence, to find the product of two decimal fractions, multiply the numbers together as though they were whole numbers and then put the decimal point as many places from the right in the answer as the sum of the number of decimal places in the factors.

Our problems can, then, be solved as

$$\begin{array}{lr} & 2.76 \\ \times & 3.4 \\ \hline & 1\ 104 \\ & 8\ 28 \\ \hline & 9.384 \end{array} \qquad\qquad \begin{array}{lr} & 0.319 \\ \times & 0.28 \\ \hline & 2552 \\ & 638 \\ \hline & 0.08932 \end{array}$$

Exercise 3-2

Add each column of numbers:

1. 37.4
   45.03
   28.005

3. 6.045
   1.008
   0.57

5. 18.07
   2.534
   7.006
   3.1

7. 13.
   283.45
   50.505
   110.9

2. 5.004
   23.068
   0.51

4. 47.
   3.8
   12.504

6. 0.00394
   2.51
   0.509
   2.5704

8. 0.63
   8.471
   27.8
   0.059
   34.038

Find the indicated sums:

9. 8.236 + 0.243 + 0.81 + 0.056
10. 5.03 + 0.516 + 39.4 + 17.23
11. 0.039 + 2.1 + 3.09 + 0.057 + 1.0006
12. 6.8 + 0.354 + 2.76 + 8.08 + 1.001
13. 3.76 + 0.461 + 2.46 + 40.03 + 12.5
14. 8.003 + 250.01 + 24.61 + 8.703 + 181.3
15. 6.01 + 8.333 + 2.0729 + 0.0424 + 3.8
16. 12.02 + 2.112 + 0.3224 + 0.0433 + 0.0054
17. 102.102 + 20.230 + 4.007 + 0.506 + 0.089

Find the indicated differences:

18. 0.92
    −0.65

19. 4.506
    −0.982

20. 7.000
    −5.308

21. 18.000
    −11.918

22. $\begin{array}{r} 8.4 \\ -6.37 \\ \hline \end{array}$
23. $\begin{array}{r} 5.36 \\ -2.7 \\ \hline \end{array}$
24. $\begin{array}{r} 5.8 \\ -0.0023 \\ \hline \end{array}$
25. $\begin{array}{r} 8 \\ -3.0004 \\ \hline \end{array}$

26. $9.2 - 5.8$
27. $18.35 - 7.9$
28. $13.04 - 7.435$
29. $12.019 - 8.37$
30. $318.295 - 283.0417$
31. $824 - 0.307$

Find the difference between the numbers:

32. 0.124 and $\frac{1}{8}$
33. 0.3 and $0.\overline{3}$
34. 0.14 and $0.\overline{14}$
35. 0.101 and $\frac{1}{9}$

Find the indicated products:

36. 0.375×10
37. 12.354×100
38. $4.962 \times 1{,}000$
39. $1.00504 \times 10{,}000$
40. 37.8×25.4
41. 6.007×2.12
42. 803.2×1.014
43. 49.15×0.03426
44. 246.1358×0.00555
45. 3.0303×0.30303
46. $(0.305)(0.21)(8.37)$
47. $(2.76)(0.345)(0.0108)$
48. A cubic foot of water weighs 62.425 lb. What will 55 cu ft of water weigh?
49. Gasoline is 0.67 times as heavy as water. What will 55 cu ft of gasoline weigh (see Prob. 48)?
50. How much will 18.9 gal of gasoline cost if each gallon costs 42.3 cents?

3-8 DIVISION OF DECIMAL FRACTIONS To divide a decimal fraction by a decimal fraction we can first apply the fundamental principle of fractions to reduce the problem to the quotient of two numbers with the denominator a whole number. Let us illustrate with an example.

Example 3-5 Divide 81.357 by 0.26.

SOLUTION

$$81.357 \div 0.26 = \frac{81.357}{0.26} = \frac{8{,}135.7}{26} = 312.9^{+}$$

The + symbol is sometimes used to indicate that the answer is a little more than the indicated number. It will be noted that the dividend and divisor (numerator and denominator) were each multiplied by 100. This resulted in moving the decimal point the same number of places to the right in both the dividend and the divisor. The symbol ∧, called a caret, is often used to indicate the new locations of the decimals in the division. The solution form appears below.

```
          312.9
0.26∧)81.35∧7
      78
       3 3
       2 6
         75
         52
         237
         234
           3
```

3-9 EXACT AND APPROXIMATE NUMBERS In computing, one must distinguish between an exact and an approximate number. *Counting numbers are exact.* If the number of students in a class were counted to be 32, the number would be an exact number. When we say that a book contains 386 pages, we mean exactly 386—no more and no less. We know we are dealing with exact numbers when we say that 250 dimes are equivalent in value to 25 dollars.

In measuring, one uses numbers in a different way from in counting. Suppose we are asked to measure the sides of a room. If one uses a foot rule graduated into hundredths of a foot, he might obtain, for example, the following measurements: 40.37, 25.13, 40.36, and 25.12 ft. These numbers would not be exact but would be *approximate.* Each dimension is found to *the nearest hundredth of a foot.* In this instance we would know that the actual first dimension would not differ from the measured value by more than $\frac{1}{2}$ of 0.01 ft. This number is often written 40.37 ± 0.005 ft to show that 0.005 ft is the greatest *possible* error in measurement.

All numbers which are measurements are approximate numbers. Numbers other than measurements are also approximations. For example, we often use 3.1416 as the approximate value for π, the ratio of the circumference of a circle to its diameter. Again, we might use 1.414 as the approximate value for $\sqrt{2}$.

In using a number such as 57,000 it is not clear if this number is accurate to the thousands, hundreds, tens, or ones. Thus, when we state that the distance from the earth to the sun is 93,000,000 miles, the degree of accuracy in this measurement is not clear. It might be off by $\frac{1}{2}$(1,000,000) mi, $\frac{1}{2}$(10,000) mi, $\frac{1}{2}$ mi, and so on.

In this book we shall assume that a number such as 93,000,000 will imply accuracy to the nearest 1,000,000, 37,400 will imply accuracy to the nearest 100, and so on. It should be clear that the size of a number has nothing to do with the accuracy of that number. Thus, while 8,634,000 is a million million times as large as 0.000008634, they both have the same degree of accuracy. Each is accurate to four significant figures, or stated another way, each is accurate to one part in 8,634.

3-10 ZERO AND SIGNIFICANT FIGURES It is important in solving problems to consider the accuracy of data and the consequent accuracy of solutions derived from such data. The accuracy of a computational solution can be no greater than the data from which it is derived. One method of indicating degrees of accuracy is the use of significant figures or significant digits as shown in the previous examples.

It is important to understand what is the meaning of significant figures. Let us give examples from which we can deduce certain rules concerning significant figures.

Example 3-6

63,000 is (usually) correct to *two* significant digits.
0.0063 is correct to *two* significant digits.
6.3×10^4 is correct to *two* significant digits.
5.8715 is correct to *five* significant digits.
587.15 is correct to *five* significant digits.
6,040 is correct to *three* significant digits.
0.6040 is correct to *four* significant digits.
60.40 is correct to *four* significant digits.
0.00006040 is correct to *four* significant digits.

We can summarize the notion of significant figures by the following:

1 Nonzero digits are always significant.
2 Zeros at the end of a whole number may or may not be significant. Usually they serve only as place holders and are not significant.
3 Zeros at the beginning of a pure decimal are not significant but serve only as place holders.
4 Zeros between nonzero digits are significant.
5 Zeros occurring at the end of a decimal fraction are significant.

3-11 ROUNDING NUMBERS Solutions to problems frequently are carried to more significant figures than the data from which they are derived. This practice not only is a waste of both time and energy but also is not justifiable in terms of its implied accuracy. Thus, for example, if we were asked to find the area of a circle whose radius was measured to be 8.4 ft, we would be inaccurate in solving the problem thus: $A = \pi R^2 = 3.1416(8.4 \text{ ft})^2 =$

221.671296 ft^2. If the original data are given to two significant figures, the answer can be expected to be accurate to only two significant figures. Our computed area would be more realistic if it were determined to be 220 ft^2.

Much time can be saved, without loss of accuracy, by *rounding off* numbers. A number is rounded off by dropping one or more digits at its right end. The digits are replaced by zeros when it is necessary in order to keep the decimal point in its proper place.

Consider the following examples of rounded numbers:

Example 3-7

3.14159 rounded to five digits is 3.1416.
3.14159 rounded to three digits is 3.14.
6.15 rounded to two digits is 6.2.
6.25 rounded to two digits is 6.2.
6.35 rounded to two digits is 6.4.
6.37482 rounded to three digits is 6.37.
6.37582 rounded to three digits is 6.38.

The method for rounding numbers can be given by the following rules.

1 If the digit to be dropped is less than 5, no change is made in the part retained.

2 If the digit to be dropped is more than 5, the digit on its immediate left is increased by 1.

3 If the digit to be dropped is 5 and is also the last significant digit of the number, round the number so its rightmost digit will be an even digit. If the digit to the left of the one to be dropped is already even, make no change to that digit; if it is odd, raise the digit to its left to the next even digit.

4 If several digits are dropped from a number and the left-hand digit of these digits is less than 5, the digit on the left of the dropped digits should not be changed. If the leftmost digit of those to be dropped is 5 or larger, the digit to the left of the dropped digits should be increased by 1.

It should be apparent that, in rounding off any set of decimals of type 3 above, in about half of the numbers the digits to the left of the dropped digit will be *decreased* to an even number while the remaining half will be *increased* to an even number. This is done to avoid a cumulative increase (or decrease) in a series of rounding off of numbers in a given problem solution.

Frequently the quotient of two decimal fractions cannot be expressed as a simple decimal fraction without a remainder. In practice it is customary to carry the division out to the desired number of significant digits. Then, if the remainder is less than half the divisor, it is dropped, but if the remainder is greater than half the divisor, the quotient is increased by one digit in the last place of the quotient.

Example 3-8 Divide 0.083 by 2.46, and round off the quotient to the nearest thousandth.

SOLUTION

```
              0.0337
2.46‸)0.08‸3000
          738
           920
           738
           1820
           1722
```

Rounding off to the nearest thousandth, the quotient is 0.034.

Exercise 3-3

1. In the following, tell which numbers are exact and which are approximate:
   (a) Water freezes at 32°F.
   (b) Tom is 6 ft 3 in. tall.
   (c) Mrs. Jones purchased a dozen eggs and 3 lb of beans.
   (d) The moon is 240,000 mi from the earth.
   (e) Atmospheric pressure at sea level is 14.7 lb/in.^{2}
   (f) There are 1,728 in.^{3} in a cubic foot.
   (g) Iron will melt at 1535°C.
   (h) Dry air by volume is 12% oxygen.
   (i) Bill ran the 100-yd dash in 9.4 sec.
   (j) $\sqrt{5} = 2.236$.
   (k) Three cans of coffee cost \$2.49.
   (l) In 1963 the 25 acres of land cost \$75,000.
2. In each of the following pairs of numbers indicate which, if any, has the greater degree of accuracy:
   (a) 27,300; 20,370
   (b) 3,620; 0.362
   (c) 0.001053; 5,027
   (d) 92,350; 0.0004118
   (e) 20,063; 4.15×10^3
   (f) 8,007; 23.59
3. Between what two numbers do each of the following measurements lie? (*Note:* 3.57 lies between 3.565 and 3.575.)
   (a) 87.24 ft
   (b) 0.309 in.
   (c) 12.4 yd
   (d) 85,000 gal
   (e) 0.0001037 centimeter (cm)

4. Round off the following numbers to the nearest hundredths, the nearest tenths, and the nearest thousandths:
   (a) 0.28634
   (b) 3.1416
   (c) 1.0732
   (d) 8.55555
   (e) 27.8350
   (f) 0.9005
5. Round off the following to the indicated accuracy:
   (a) 87,176 to the nearest hundred
   (b) 4,645 ft to the nearest 10 ft
   (c) 20,390 bottles to the nearest 1,000 bottles
   (d) 3,015 trees to the nearest 10 trees
   (e) 99,291 gal to the nearest 100 gal
   (f) 183,279,000 people to the nearest million people

Divide the following:

6. $0.07\overline{)0.83}$ to the nearest hundredth
7. $0.35\overline{)0.0264}$ to the nearest thousandth
8. $0.26\overline{)87}$ to the nearest hundred
9. $9.03\overline{)4.53921}$ to the nearest hundredth
10. $2.234\overline{)891.273941}$ to the nearest thousandth
11. $70.05\overline{)6235.2909}$ to the nearest thousandth
12. $0.0136\overline{)0.000059}$ to the nearest thousandth
13. $0.00237\overline{)5.218499}$ to the nearest thousand
14. A shaft should have a diameter of $1\frac{5}{8}$ in. It was measured to be 1.704 in. How much too large was it?
15. What is the area of the figure?

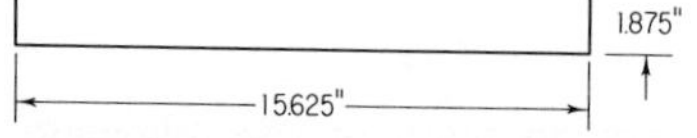

EXERCISE 15.

16. A dealer buys 640 gal of paint for \$4.86/gal and sells it for \$1.75/qt. How much is his profit?
17. A prime rib roast weighing 12.35 lb costs \$20.43. What is the cost of the roast to the nearest cent per pound?

18. What was the average mileage (miles per gallon) achieved by a car if it traveled 625 mi on 42.7 gal of gasoline?

3-12 COMPUTATION WITH APPROXIMATE DATA The results of any computation which uses one or more approximate numbers must likewise be an approximate number. The accuracy of the computed number will depend upon the accuracy of each number involved in the calculation. The result will be no more accurate than the least accurate number used in the calculations.

In computing with approximate numbers one must also consider the purpose of the answer to the computation. Thus, if one were interested in calculating the volume of a cylindrical water storage tank, he would be foolish to measure the diameter and height of the silo accurate to 0.001 in. On the other hand, many precision pieces of equipment must be measured to tolerances of 0.00001 in. However, most technical calculations involve measured data with degrees of accuracy that can be handled with a good 10-in. slide rule. In calculating with approximate numbers it is well to eliminate all calculations which do not affect the final result.

3-13 ADDITION AND SUBTRACTION OF APPROXIMATE NUMBERS Before we give a rule for adding and subtracting approximate numbers, let us consider the problem of adding the approximate numbers 2.354, 1.08, and 5.7. Let us tabulate the least possible values, the greatest possible values of the numbers and their sums and then the computed sum of the numbers, as follows:

Least possible value	Greatest possible value	Measured value
2.3535	2.3545	2.354
1.075	1.085	1.08
5.65	5.75	5.7
9.0785	9.1895	9.134

It will be noted that our sum could range from 9.0785 to 9.1895. Each of the three answers has the same value only when rounded off to two places (or fewer). Thus, we can be certain that our answer is accurate to not more than two places. Hence, our answer should be 9.1.

Let us next carry through the addition of the same numbers taken to one more place accuracy than the least accurate of the original numbers.

$$\begin{array}{l} 2.35 \\ 1.08 \\ \underline{5.7} \\ 9.13 \end{array}$$

Again our total will be 9.1, accurate to two places. Our working rule then will be:

PROCEDURE *In adding or subtracting approximate numbers:*

1 *Round off each number to one more place than the least accurate of the given numbers.*

2 *Round off the sum or difference of these numbers to the same number of places as the term having the least accuracy.*

Example 3-9 Add 75.3592, 254.08, 30.291, 8.6.

SOLUTION

$$\begin{array}{r} 75.36 \\ 254.08 \\ 30.29 \\ 8.6 \\ \hline 368.33 \end{array}$$

The answer should be 368.3.

Example 3-10 Subtract 28.56 − 13.4072.

SOLUTION

$$\begin{array}{r} 28.56 \\ 13.407 \\ \hline 15.153 \end{array}$$

The answer should be 15.15.

3-14 MULTIPLICATION AND DIVISION OF APPROXIMATE NUMBERS Suppose the floor area of a room is to be computed from its measured dimensions of 24.7 and 15.8 ft. Considering the range of values possible for the dimensions of the room, we compute

1 The smallest possible area = 24.65 ft × 15.75 ft = 388.2375 ft^2.
2 The largest possible area = 24.75 ft × 15.85 ft = 392.2875 ft^2.
3 The computed area = 24.7 ft × 15.8 ft = 390.26 ft^2.

It should be clear that we can be certain that the area must lie between 388.2375 and 392.2875 ft^2. The three areas do not have the same whole-number part, much less the same decimal part. However, the value 3.90×10^2 is about in the middle of the range of possible values. It also has the same number of significant figures as does each measured dimension.

PROCEDURE *To find the product or quotient of two approximate numbers:*

1 *Round off each number to not more than one more significant place than the least accurate of the numbers.*

2 *Round off the product or quotient to the same number of significant digits contained in the least accurate of the numbers used.*

3-15 SCIENTIFIC NOTATION Very large and very small numbers are frequently encountered in technical and scientific work. Illustrations of such numbers follow.

The speed of light in a vacuum is about 984,000,000 ft/sec. The number of molecules in a quart of any gas under ordinary conditions of temperature and pressure is approximately 25,000,000,000,000,000,000,000. This number is an extremely large number, so large that if a million of these molecules could flow out of a quart container each second, it would take almost 800 million years to empty the flask! The diameter of a hydrogen atom is about 0.0000000001058 meters (m). The wavelength of some x-rays is about 0.00000000582 cm.

Writing such large and such small numbers in ordinary notation often proves cumbersome or inconvenient, particularly when they must be handled frequently or with operations involving multiplication and division. Fortunately there is a more convenient notation, known as *scientific notation,* to represent such numbers. To write a number in scientific notation it is written as a product of a number between 1 and 10 and a power of 10. We know that $5{,}600 = 5.6 \times 1{,}000 = 5.6 \times (10 \times 10 \times 10)$. Mathematicians often write $10 \times 10 \times 10$ in the shorter form 10^3. The number 1,000 is said to be the *third power of 10.*

Example 3-11 Write 984,000,000 in scientific notation.

SOLUTION

$$\begin{aligned} 984{,}000{,}000 &= 9.84 \times 100{,}000{,}000 \\ &= 9.84 \times 10^8 \end{aligned}$$

(This number has an accuracy of three significant figures.)

Example 3-12 Write 28,050 in scientific notation.

SOLUTION

$$\begin{aligned} 28{,}050 &= 2.8050 \times 10{,}000 \\ &= 2.8050 \times 10^4 \end{aligned}$$

(Written this way, the number has five-significant-figure accuracy.)

Negative powers of 10 are needed in expressing in scientific notation numbers less than 0.1. Mathematicians will often write

$$0.001 = \frac{1}{10} \times \frac{1}{10} \times \frac{1}{10} = \frac{1}{10^3} \quad \text{as } 10^{-3}$$

Thus, 0.001 is said to be the *negative three* power of 10. We will discuss negative numbers in more detail in Chap. 5. Negative exponents will be explored more completely in Chap. 19.

Example 3-13 Express 0.0000000001058 km in scientific notation.

SOLUTION

$$\begin{aligned} 0.0000000001058 \text{ km} &= 1.058 \times 0.0000000001 \text{ km} \\ &= 1.058 \times 10^{-10} \text{ km} \end{aligned}$$

We noted in Sec. 3-9 that the accuracy of denominate numbers (such as 93,000,000 mi or 0.015080 gal/sec) is often not clear when written in the normal form. The degree of accuracy can be indicated by writing the numbers in scientific form. Thus, 9.3×10^7 mi would mean that the number is accurate to two *significant* figures, i.e., to the nearest 0.1×10^7 or 1,000,000 mi; 9.300×10^7 mi would mean that the number has four-significant-figure accuracy and is accurate to the nearest 0.001×10^7 or 10,000 mi. The number 1.508×10^{-2} has four-significant-figure accuracy and 1.5080×10^{-2} has five-significant-figure accuracy.

Exercise 3-4

In Probs. 1 to 8 determine the least possible and the greatest possible values for the answer. Then give the most reasonable answer.

1. Add 4.15, 8.762, 0.03, 1.2.
2. Add 0.0043, 0.105, 0.82, 0.059.
3. Subtract $87.05 - 23.147$.
4. Subtract $0.09253 - 0.0416$.
5. Multiply 2.7×5.4.
6. Multiply 1.53×2.64.
7. Divide 8.9 by 5.1.
8. Divide 27.6 by 3.04.

In Probs. 9 to 21 all numbers are approximate (unless stated otherwise). Using the working rules for computations with approximate numbers, perform the indicated operations.

9. Add $215.3 + 3.086 + 7.21 + 89$.
10. Add $10.503 + 2.43 + 8.0769 + 33.76$.
11. Add $649.11 + 22.212 + 98 + 76.3$.
12. Add $2.34 \times 10^3 + 1.659 \times 10^2 + 5.69 \times 10^2$.
13. Subtract $8.7 - 0.24921$.
14. Subtract $63.452 - 0.089092$.
15. Subtract $(2.47 \times 10^6) - (8.36 \times 10^5)$.

16. Multiply 314.6 × 2.7.
17. Multiply 0.039254 × 5.61.
18. Multiply 0.351762 × 82 (consider 82 as an exact number).
19. Divide 25.3 by 6.47.
20. Divide 3749.62 by 268.53.
21. Divide 803.4 by 56 (consider 56 as an exact number).

In Probs. 22 to 29 write the numbers in scientific notation.

22. 3,750
23. 287,600
24. 12.3
25. 437
26. 0.065
27. 0.803
28. 0.0000512
29. 0.0107

In Probs. 30 to 37 write the numbers from scientific notation to ordinary notation.

30. 5.1×10^{3}
31. 8.3×10^{7}
32. 2.13×10^{8}
33. 7.01×10^{5}
34. 1.54×10^{-3}
35. 4.27×10^{-6}
36. 1.08×10^{-9}
37. 5.130×10^{-7}

CHAPTER 4
PERCENT

4-1 THE MEANING OF PERCENT In our study thus far we have learned that fractions such as $\frac{67}{100}$ and 0.67 are two ways of expressing the same thing. They mean, in this instance, 67 parts out of 100 equal parts or 67 out of a group of 100. We shall now study a third way, called percent, of writing the same thing.

The term *percent* comes from the Latin *per centum,* meaning "by the hundred." Thus, 67 percent means 67 out of each hundred. The symbol % stands for the word "percent." Any fraction having 100 as a denominator is a percent, whether it is written as a common fraction, as a decimal fraction, or with the percent sign. It is interesting to note that the symbol % contains two zeros. One might consider the zeros to stand for the zeros in 100.

Today percent has a much broader meaning than the original notion of "by the hundred." Now we not only speak of 5 percent and 67 percent, for example; but we deal with such improper fractions as 325 percent and 1,200 percent. These latter percents would have no meaning according to the original definition, but considering them as fractions having 100 as a denominator gives them meaning. Thus, 325 percent is equivalent to $\frac{325}{100}$ and 1,200 percent is equivalent to $\frac{1,200}{100}$.

The arithmetic dealing with percent is sometimes called *percentage.* However, the terms *percentage* and *percent* are often used interchangeably. Percentages are used extensively in many walks of life. Local, county, state, and Federal taxes are collected on a percentage basis. Percentage increase or decrease is used in expressing death rates, birth rates, population trends, school enrollments, assessed valuations, costs of living, sales records, employment and unemployment conditions, and so on.

Perhaps one of the greatest advantages of using percents in preference to common and decimal fractions is the ease of comparing sizes of things expressed as percents. Common fractions cannot be compared for size

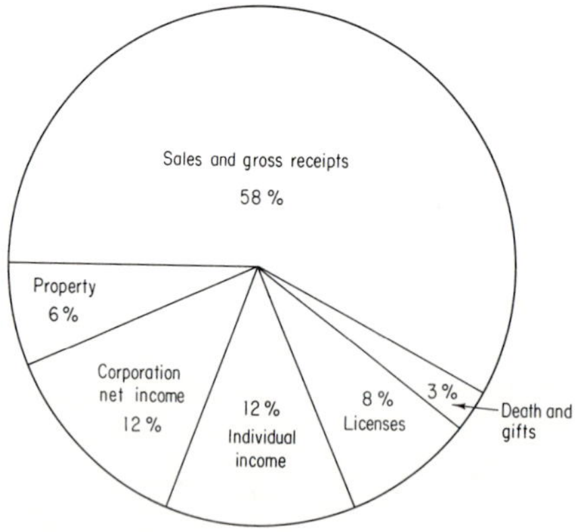

FIG. 4-1.

unless they have been converted to equivalent fractions with a common denominator. Percents always have the same denominator. For example, if you were given a choice between investing in Company *A* which will return a profit of $\frac{16}{25}$ of your investment each year and Company *B* which will return $\frac{13}{20}$ of your investment each year, you probably would have to think twice before deciding which will give you the greater profit. On the other hand, if you were told that Company *A* will return 64 percent and Company *B* will return 65 percent of your investment each year, your decision would be easy.

The classroom teacher often converts test scores to percents because of the ease of comparing the scores.

A glance at the circular graph of Fig. 4-1 and the percentages given provides a ready means of comparison of the major sources of recent tax collections in the state of California.

4-2 CHANGING PERCENTS TO DECIMAL FRACTIONS A great advantage of the use of percent lies in the simple way it can be used to describe situations and solve problems which occur often enough to be common. However, any computations with percent involve first changing the percent to a common fraction or to a decimal fraction. Therefore, we will need to learn how to change from percents to equivalent common or decimal fractions.

Since percent is simply a kind of fraction, we need only apply what we have already learned about fractions to make our conversions. In Chap. 3 we considered converting common fractions to decimal fractions and vice versa.

From the definition of percent, we know that

$$7 \text{ percent} = \frac{7}{100} = 0.07$$

or that $62\frac{1}{2}$ percent $= \dfrac{62\frac{1}{2}}{100} = \dfrac{125}{200} = 0.62\frac{1}{2}$ or 0.625.

PROCEDURE *To reduce a percent to a decimal fraction:*

1 *Rewrite the given number, omitting the % sign.*

2 *Move the decimal point two places to the left. (Divide by 100.)*

Example 4-1

$6\% = 0.06$

$75\% = 0.75$

$140\% = 1.40$

$5.6\% = 0.056$

$13\frac{3}{4}\% = 0.13\frac{3}{4} = 0.1375$

$\frac{3}{4}\% = 0.00\frac{3}{4} = 0.0075$

4-3 CHANGING A DECIMAL FRACTION TO A PERCENT To change a decimal fraction to an equivalent percent we need only to reverse the procedure of the last paragraph.

PROCEDURE *To reduce a decimal fraction to a percent:*

1 *Copy the given number.*

2 *Move the decimal point two places to the right. (Multiply by 100.)*

3 *Annex the percent sign, %.*

Example 4-2

$$0.09 = 9\%$$

$$0.375 = 37.5\%$$

$$5.70 = 570\%$$

$$234.27 = 23{,}427\%$$

4-4 CHANGING A PERCENT TO A COMMON FRACTION If the student will remember that the word "percent" means hundredths, he should have no difficulty in accepting the following rule.

PROCEDURE *To change a percent to a common fraction:*

1 *Make a fraction by writing the given number as the numerator and the number 100 as the denominator.*

2 *Reduce to lowest terms.*

Example 4-3

$$3\% = \tfrac{3}{100}$$

$$42\% = \tfrac{42}{100} = \tfrac{21}{50}$$

$$5\tfrac{1}{2}\% = \frac{5\frac{1}{2}}{100} = \frac{11}{200}$$

$$62.5\% = \tfrac{62.5}{100} = \tfrac{625}{1{,}000} = \tfrac{5}{8}$$

$$750\% = \tfrac{750}{100} = \tfrac{15}{2}$$

4-5 CHANGING A COMMON FRACTION TO A PERCENT Since we know how to change a common fraction to a decimal fraction and a decimal fraction to a percent, we can combine the two processes to get the following rule.

PROCEDURE *To change a common fraction to a percent:*

1 *Change the common fraction to a decimal by dividing the numerator by the denominator, finding the quotient to two decimal places.*

2 *Rewrite the quotient, omitting the decimal point.*

3 *Annex the percent sign, %.*

Example 4-4

$\frac{2}{5} = 0.40 = 40\%$

$\frac{7}{8} = 0.87\frac{1}{2} = 87\frac{1}{2}\%$

$\frac{18}{25} = 0.72 = 72\%$

$\frac{37}{4} = 9.25 = 925\%$

$5\frac{1}{3} = 5.33\frac{1}{3} = 533\frac{1}{3}\%$

Exercise 4-1

Change each from a percent to a decimal fraction:

1. 11%
2. 60%
3. 8.4%
4. 135%
5. $137\frac{1}{2}\%$
6. $87\frac{1}{2}\%$
7. $\frac{1}{4}\%$
8. 52.13%
9. 0.4%
10. 0.04%
11. 0.007%
12. $\frac{9}{16}\%$
13. 374.5%
14. $0.5\frac{3}{4}\%$
15. $0.06\frac{1}{4}\%$

Change each to a percent:

16. 0.08
17. 0.009
18. $0.83\frac{1}{3}$
19. $0.2\frac{1}{2}$
20. 3
21. 0.0004
22. 0.235
23. 7.6
24. $0.6\frac{1}{4}$
25. 12
26. 0.3079
27. $0.08004\frac{1}{2}$
28. $1\frac{3}{8}$
29. 283
30. 72.9
31. $\frac{7}{12}$
32. $\frac{4}{5}$
33. $\frac{5}{16}$
34. $\frac{7}{32}$
35. $\frac{9}{64}$
36. $\frac{6}{7}$
37. $\frac{13}{20}$
38. $5\frac{3}{10}$
39. $7\frac{11}{12}$
40. $\frac{13}{1,000}$
41. $2\frac{5}{16}$
42. $\frac{74}{25}$
43. $\frac{89}{50}$
44. $\frac{33}{16}$
45. $\frac{185}{8}$

Change to common fractions or mixed numbers:

46. 36%
47. 125%
48. 44%
49. 548%
50. 75%
51. 1,500%
52. $12\frac{1}{2}\%$
53. $87\frac{1}{2}\%$
54. $83\frac{1}{3}\%$
55. $41\frac{1}{4}\%$
56. $3\frac{2}{7}\%$
57. $47\frac{3}{4}\%$
58. $\frac{1}{5}\%$
59. $\frac{2}{3}\%$
60. $\frac{4}{9}\%$

4-6 THE THREE CASES OF PERCENTAGE Consider the problem, "What is 25 percent of 60?" To solve this problem, the percent is first changed to a decimal or common fraction. A fractional part of 60 is to be determined. The fraction has a denominator. The answer to the question can be found as follows: 25% of 60 = $\frac{25}{100} \times 60 = 15$. In this example, the $\frac{25}{100}$ is called the *rate;* the 60 is called the *base;* and 15, the product of the rate and the base, is called the *percentage.* The three quantities are related by the equation

$$\text{Rate} \times \text{base} = \text{percentage} \tag{4-1}$$

If any two of the three quantities of Eq. (4-1) are known, the third quantity can be found. By selecting different pairs of known values the following three types of percentage problems are obtained: (1) given the rate and the base, to find the percentage; (2) given the base and the percentage, to find the rate; (3) given the rate and the percentage, to find the base. We will now consider in more detail each of the three types.

4-7 DETERMINING A PERCENT OF A NUMBER We solved the first type of percentage problem in the last section when we found 25 percent of 60.

PROCEDURE *To determine a specified percent of a number:*

1 *Change the percent to a decimal or common fraction.*

2 *Multiply the given number by this fraction.*

Example 4-5 Find $37\frac{1}{2}$% of \$1,200.

SOLUTION

$37\frac{1}{2}\%$ of \$1,200 =

$0.375 \times \$1,200 = \$450$

Example 4-6 Find $6\frac{1}{3}$% of 240.

SOLUTION

$6\frac{1}{3}\%$ of 240 =

$$\frac{6\frac{1}{3}}{100} \times 240 =$$

$$\frac{19}{300} \times 240 = 15\frac{1}{5}$$

4-8 DETERMINING THE PERCENT ONE NUMBER IS OF ANOTHER Frequently a question of the type "What percent of 75 is 15?" is asked. In this case the base and the percentage are known and the rate is to be determined. The relationship of these quantities can be written as a "multiplier statement": Rate × 75 = 15. Division is the converse of multiplication. Consequently the "multiplication statement" can be converted to an equivalent "division statement." Thus,

$$\text{Rate} = \frac{15}{75} = \frac{1}{5} = 0.20 = 20 \text{ percent}$$

As a general statement, if rate × base = percentage, then

$$\text{Rate} = \frac{\text{percentage}}{\text{base}}$$

PROCEDURE *To determine the percentage one number is of another:*

1. *Form a fraction with the percentage the numerator and the base the denominator.*
2. *Change the fraction to a percent.*

Example 4-7 33 is what percent of 55?

SOLUTION

Percentage = 33

Base = 55

$\text{Rate} = \frac{33}{55} = \frac{3}{5} = 0.60 = 60\%$

Example 4-8 11 is what percent of 40?

SOLUTION

$\text{Rate} = \frac{11}{40} = 0.275 = 27.5\%$

4-9 DETERMINING A NUMBER WHEN A SPECIFIED PERCENT IS KNOWN Consider the question, "Eight percent of what number is 56?" In this case the percentage (56) and the rate (8 percent) are known and the base is to be calculated. Using the relationship among the three numbers, we get

8 percent × base = 56

If division is treated as the converse of multiplication, the base can be computed as follows: $\text{Base} = \frac{56}{8\%} = \frac{56}{0.08} = 700.$

$$\text{Base} = \frac{\text{percentage}}{\text{rate}}$$

PROCEDURE *To determine a number when a percent of it is known:*

1. *Form a fraction with the percentage as numerator and rate as denominator.*
2. *Convert the rate to a decimal fraction or a common fraction.*
3. *Determine the quotient.*

Example 4-9 50% of what number is 80?

SOLUTION

Percentage = 80

Rate = 50%

$$\text{Base} = \frac{80}{50\%} = \frac{80}{0.50} = 160$$

Example 4-10 $\frac{3}{16}$% of what number is 12?

SOLUTION

$$\text{Base} = \frac{12}{\frac{\frac{3}{16}}{100}} = \frac{12}{\frac{3}{16} \times \frac{1}{100}} = \frac{12}{\frac{3}{1,600}}$$

$$= \frac{(12)(1,600)}{3} = 6,400$$

Exercise 4-2

1. Determine 8% of 400.
2. Determine 72% of 25.
3. 20% of 48 is what number?
4. What number is 12% of 360?
5. 250% of $1,800 is how much?
6. 60 is what percent of 150?
7. If 40% of 284 is N, what is N?
8. 36 is what percent of 48?
9. 15% of what number is 120?
10. What percent of 70 is 210?
11. 32 is what percent of 960?
12. Determine a number such that 40% of it is 14.
13. Determine a number such that 74% of it is 145.
14. Eighty-six and thirty-five hundredths is 45% of what number?
15. Two hundred sixty-four and three hundred eighty-seven thousandths is 25% of what number?
16. 18.4 is what percent of 368?
17. 13.2 is what percent of 6.6?
18. If $12\frac{1}{2}$% of a number is 108, what is the number?
19. If $37\frac{1}{2}$% of a number is 441, what is the number?
20. $137\frac{1}{2}$% of $18.30 is how much?
21. 0.8% of 423,450 is how much?
22. 222 is $16\frac{1}{4}$% of what number?
23. $16\frac{2}{3}$% of what number is 72?
24. 54% of what number is 0.0345?
25. If $5\frac{3}{4}$% of a given number is 0.184, what is the number?
26. If $13\frac{3}{8}$% of a given number is 0.0928, what is the number?
27. $168 is $33\frac{1}{3}$% of what cost?
28. $83\frac{1}{3}$% of 426 is how much?
29. If $37\frac{1}{2}$% of a number is 46, what is the number?
30. 960 is 125% of what number?
31. Determine 23.16% of 0.0392.
32. Determine a number such that 284% of it is 356.

33. $87\frac{1}{2}\%$ of what weight equals 13 lb?
34. What percent of a foot is an inch?
35. What percent of an inch is a foot?
36. Which is larger: 6% of 308 or 7% of 289? How much larger?
37. 4,500% of 15 is how much?
38. 262% of what number is 0.308?
39. What percent of 833 is 175?

4-10 APPLICATIONS OF PERCENT Percents arise in numerous situations encountered in business, engineering, and technology. The use of percentage (or percent) is very common in newspapers and magazines. The radio and television media assume of listeners an understanding of percentage and similar commonly used mathematical concepts.

The greatest difficulty in solving problems in percentage seems to come from the inability to determine the "percent of what" of the situation. Several examples of the more common uses of percentages will now be given.

4-11 SALES TAX Many cities and states levy a sales tax, which is based on a percent of the sales price.

Example 4-11 The retail sales tax in a state is 4%. Mr. Smith purchased a metal lathe, priced \$535, and a three-jaw chuck, priced \$52.50. What was the amount of the sales tax he had to pay to the state?

SOLUTION

$$\begin{aligned}\text{Total sales} &= \$535 + \$52.50\\ &= \$587.50\end{aligned}$$

Then

$$\begin{aligned}\text{Sales tax} &= 4\% \text{ of } \$587.50\\ &= 0.04 \times \$587.50\\ &= \$23.50\end{aligned}$$

4-12 DISCOUNT The regular or full price of an article is called the *list price.* Sometimes articles are sold below the list price. For example, an article may be marked or priced at \$75 but sold for \$60. The difference between the list price and the *sale price* or *net price* is called the *discount.* In this example the sale price is \$60 and the discount is \$15. The discount is often quoted as a percent of the list price and is called the *discount rate.* Thus, the discount rate in this situation would be $\frac{15}{75} = \frac{1}{5} = 20$ percent.

Example 4-12 If a set of four tires originally priced at \$132.80 is discounted 15% in a state where the sales tax rate is $3\frac{1}{2}\%$, what is the total cost of the four tires?

SOLUTION A The discount rate is based on the list price.

$$\begin{aligned}\text{Discount} &= 15\% \text{ of } \$132.80\\ &= 0.15 \times \$132.80\\ &= \$19.92\end{aligned}$$

$$\begin{aligned}\text{Sale price} &= \$132.80 - \$19.92\\ &= \$112.88\end{aligned}$$

$$\begin{aligned}\text{Sales tax} &= 3\tfrac{1}{2}\% \text{ of } \$112.88\\ &= 0.035 \times \$112.88\\ &= \$3.95\end{aligned}$$

$$\begin{aligned}\text{Purchase price} &= \$112.88 + \$3.95\\ &= \$116.83\end{aligned}$$

SOLUTION B Since the list price is reduced by 15%, the sale price will be 85% of the list price.

$$\begin{aligned}\text{Sale price} &= 85\% \text{ of } \$132.80\\ &= 0.85 \times \$132.80\\ &= \$112.88\end{aligned}$$

$$\begin{aligned}\text{Purchase price} &= \$112.88 + 0.035 \times \$112.88\\ &= \$116.83\end{aligned}$$

4-13 COMMISSION Salesmen are often paid a "flat salary" augmented by a sum of money, called a *commission,* based on the total sale value of the goods they sell. The commission rate is frequently given as a percent of the sale price.

Example 4-13 A salesman received $645.25 in commissions for selling $12,905 worth of cars. Under what commission rate was he working?

SOLUTION

$$\text{Commission rate} \times \text{total sales} = \text{commission}$$

or

$$\begin{aligned}\text{Commission rate} &= \frac{\text{commission}}{\text{total sales}}\\ &= \frac{\$645.25}{\$12{,}905.00}\\ &= \tfrac{1}{20}\\ &= 5\%\end{aligned}$$

4-14 PROFIT AND LOSS If an article cost a merchant $10 and he sold it for $12, his profit would be $12 − $10 or $2. That is,

$$\text{Profit} = \text{selling price} - \text{cost}$$

or

$$\text{Selling price} = \text{cost} + \text{profit}$$

and

$$\text{Cost} = \text{selling price} - \text{profit}$$

The profit can also be expressed as (1) a percent of the cost or (2) a percent of the selling price. In the preceding example, the profit on the cost would be $\frac{2}{10} = 0.20 = 20$ percent. The profit based on the selling price would be $\frac{2}{12} = \frac{1}{6} = 0.16\frac{2}{3} = 16\frac{2}{3}$ percent. While both methods are used, the more common practice is to base the profit on the selling price.

In the problems that follow, it will be assumed that the percent profit is based upon the selling price.

Example 4-14 What must be the selling price on a tractor which costs the merchant \$8,400 for him to realize a profit of 25% on the sale price?

SOLUTION

$$25\% \text{ of the selling price} = \text{profit}$$

then $$75\% \text{ of the selling price} = \text{cost}$$

As a multiplication statement,

$$0.75 \times \text{selling price} = \text{cost}$$

As an equivalent division statement,

$$\text{Selling price} = \frac{\text{cost}}{0.75} = \frac{\$8{,}400}{0.75} = \$11{,}200$$

4-15 EFFICIENCY The efficiency of a machine is defined as its output divided by its input. Usefully the efficiency is expressed as a percent. It should be clear that both output and input must be expressed in the same units. Thus,

$$\text{Efficiency, in } \% = \frac{\text{output}}{\text{input}} \times 100$$

Since it is impossible to get out of any machine as much as is put into it, the efficiency must always be less than 100 percent.

Example 4-15 An electric motor receives an input of 7,500 watts and delivers 7.5 horsepower (hp) output. What is the efficiency of the motor? (*Note:* 1 hp is equivalent to 746 watts.)

SOLUTION In order to use the defining equation for efficiency, the output and input must be expressed in the same units.

$$\text{Output} = 7.5 \text{ hp} = 7.5 \times 746 \text{ watts}$$

Then

$$\text{Efficiency} = \frac{7.5 \times 746 \text{ watts}}{7{,}500 \text{ watts}} \times 100 = 74.6\%$$

4-16 SIMPLE INTEREST When a person borrows money from a bank, he must pay an *interest* for the use of that money. The money borrowed is called the *principal.* The sum of the principal and the interest is called the *amount.* These quantities are related by the formula

Interest = principal × rate per year × time in years

Example 4-16 Mr. Ryan borrowed $12,000 at 6% simple interest for 3 years. How much interest did he owe at the end of 3 years? What amount was due at the end of 3 years?

SOLUTION

$$\begin{aligned}\text{Interest} &= \text{principal} \times \text{rate} \times \text{time}\\ &= \$12{,}000 \times \tfrac{6}{100} \times 3\\ &= \$2{,}160\end{aligned}$$

$$\begin{aligned}\text{Amount} &= \text{principal} + \text{interest}\\ &= \$12{,}000 + \$2{,}160\\ &= \$14{,}160\end{aligned}$$

Exercise 4-3

1. A gun listed at $129.50 was sold for $99.95. What was the percent of discount (to the nearest whole number)?
2. If the sales tax is 3%, what tax must be paid on a purchase of $135.95?
3. Two years ago Mr. Owens had a salary of $7,200. Last year he was given a 10% cut in salary. This year his salary was increased 10%. What is his present salary?
4. Paul Turner is paid a salary of $2,500 a year plus a commission of 5% on his sales. What must be the amount of his sales to guarantee him an annual salary of $7,500?
5. Goe's variety store buys a box of 24 candy bars for 75 cents and sells them for 5 cents each. What is the percent of profit when based upon the selling price?
6. A certain motor is rated 80% efficient. The output of the motor is 30 hp. What is the input of the motor in watts? (*Note:* 1 hp = 746 watts.)
7. A salesman earned $500 in commissions during the month of October. If his commission rate was 8%, what was the value of his sales?
8. Mr. Bishop purchased goods listed at $276.50. He was given a 20% discount for paying cash. The state taxed him 2% of the purchase price. What cash did he have to give for the goods?
9. If the sales tax on $579.95 is $20.30, what is the tax rate?
10. The sales tax on a jig saw is $0.62. The retail sales tax rate is $2\frac{1}{2}$%. What is the price of the saw before tax? After tax?

11. An interior decorator receives 15% commission on all sales to his clients. One of his clients purchased the following merchandise at 5% off list price: broadloom carpet, list \$1,500; draperies, list \$1,200; chesterfield, list \$450; lamps, list \$125; fixtures, list \$75. How much did the client pay for the merchandise? How much was the decorator's commission?
12. If lumber lists at \$65 per 1,000 board feet and is offered at \$57.20 per 1,000 board feet, what is the rate of discount?
13. A hi-fi set is listed at \$459.95. The set is discounted 20% on a storewide fire sale. What is the selling price?
14. What selling price must a merchant put on an article which costs him \$27.50 if he is to realize a profit of 40% on the list price?
15. If a merchant wishes to realize a profit of \$12 at a profit of 15% on the list price of an article, what must he list as the sale price of the article?
16. A man borrows \$1,800 at 6% simple interest per year for 30 months. How much interest will he owe at the end of the 30 months? What is the total amount he will owe at that time?
17. A man bought a house listed at \$24,000. He paid 25% down and took out a mortgage bearing 6% simple interest for the remainder. What was his annual interest on the mortgage?
18. An aluminum solder is composed of the following metals by weight: 78% tin, 9% aluminum, 8% zinc, 5% cadmium. How many pounds each of aluminum, zinc, and cadmium must be added to 234 lb of tin in making the solder? How many pounds of solder can be made from this amount of metal?
19. A manufacturer produces an article for \$5.60. He sells it to a wholesaler at a profit of 20% on his sale price. The wholesaler sells the article to a retailer at a profit of 25% on his sale price. The retailer sells the article to a consumer at a profit of 30% on his sale price. What does the consumer pay for the article?
20. A salesman is paid a salary of \$2,500 a year plus a commission of 5% on his sales. What must be the amount of his sales if he is to receive a total salary of \$8,500 per year?
21. The alloy Wood's metal used in overhead safety sprinkler systems has the following composition: bismuth, 50%; lead, 25%; tin, $12\frac{1}{2}$%; cadmium, $12\frac{1}{2}$%. How many pounds of the alloy can be made if only 50 lb of cadmium is available?
22. Mr. Brown owns a house worth \$10,000. He sells it to Mr. Black at a 10% profit based on the worth of the house. Mr. Black sells the house back to Mr. Brown at a 10% loss. How much does Mr. Brown make on the deal?

23. A rise of 600 ft is required to get a railroad line over a mountain. The grade can be kept down by curving the track around the mountain peak. Find the length of track required to produce a grade such that the rise will be 2% of the length of the track.
24. A man has $10,000 to invest. He invests $4,000 at 5% and $3,500 at 4%. At what rate must he invest the remainder to have a yearly income of $500 from his three investments?

CHAPTER 5

SIGNED NUMBERS

5-1 POSITIVE AND NEGATIVE NUMBERS We have discussed mathematical operations involving natural numbers and fractions. For centuries these were the only kinds of numbers needed. In our discussions thus far we have not encountered a situation in which a greater positive number is subtracted from a lesser number. Nor have we needed numbers which indicate opposites.

Many concepts in our daily lives have their opposites. For example, we use the opposites glad and sad, bitter and sweet, hard and soft, pretty and ugly, fast and slow, etc. This dual notion also exists in mathematical concepts. Consider the opposites \$8 earned and \$8 spent, 5° above zero versus 5° below zero, 25 miles east and 25 miles west, 10 lb gain versus 10 lb loss, etc.

Mathematicians often represent these ideas symbolically by the use of a plus (+) or a minus (−) sign. They may represent

Eight dollars earned as +\$8

Eight dollars spent as −\$8

Five degrees above zero as +5°

Five degrees below zero as −5°

Twenty-five miles east as +25 mi

Twenty-five miles west as −25 mi

and so on.

Numbers preceded by a plus sign are called *positive* numbers. They are greater than zero. Numbers preceded by a minus sign are called *negative* numbers. They are less than zero. Positive and negative numbers taken together are called *signed* numbers. Zero is neither positive nor negative. If no sign is written before a number, the number is understood to be positive.

5-2 THE NUMBER SCALE A graduated horizontal (or vertical) line is useful in visualizing the relative sizes of signed numbers (Fig. 5-1). We can select any arbitrary point on the line, called the *origin,* and label it 0. Then using any convenient unit of length, we can mark off equally spaced points to the right (or above) and left (or below) of the origin. The points to the right (or up) we designate as positive numbers, and those to the left (or down) as negative. Note, again, that the origin 0 is neither positive nor negative.

It can be shown that every real number can be represented by a single point on the number scale of Fig. 5-1 and every point of the line represents only one real number. Such matching is termed a *one-to-one corre-*

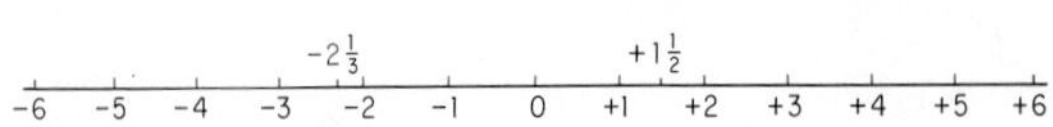

FIG. 5-1.

spondence between the numbers and points on the line. Any number which is either a positive number, a negative number, or zero is known as a *real number.* Such numbers can be plotted on a number line. The point of the line paired with a number is called the *graph* of that number. The number paired with a point is called the *coordinate* of that point. Thus, in Fig. 5-2,

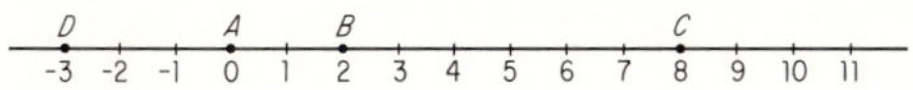

FIG. 5-2.

point D is the graph of -3; point B is the graph of 2; 0 is the coordinate of point A, and 8 is the coordinate of point C. The number $+4$ is twice as far from the origin as is $+2$; -3 is three times as far from the origin as is $+1$. The numbers -3 and $+1$ are on opposite sides of the origin along the number scale.

5-3 ORDER AND SYMBOLS OF INEQUALITY Real numbers have a natural order. Figure 5-3 indicates the natural ordering. For example, we know that

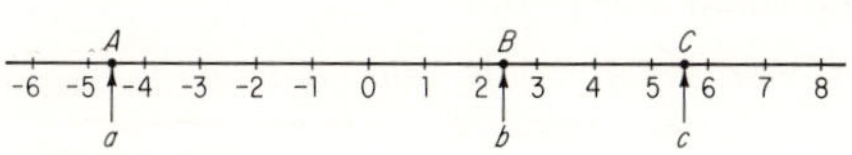

FIG. 5-3.

$+5$ is greater than $+3$ and that $+6$ is greater than $+5$. Note on the horizontal scale that $+5$ is on the right of $+3$ and that $+6$ is to the right of $+5$. It should be apparent that, if we are dealing with positive number points, the greater of two numbers will lie on the right of the lesser number. In the same way, we can say that, given two points representing negative numbers on a horizontal scale, the greater number point is to the right of the lesser number point. For example, the point representing -2 is to the right of the point representing -3. Thus -2 is greater than -3. Stated as a rule:

Given p and q are coordinates of points P and Q on a horizontal number scale. If P is on the right of Q, then p is greater than q; if P is on the left of Q, then p is less than q.

Thus, in Fig. 5-3, b is greater than a and c is greater than b. The symbols $>$ and $<$ are commonly used for the words "greater than" and "less than," respectively. Thus, $5 > 3$, $3 < 5$, $2 > 0$, $-3 < 0$, $-4 < 1$, $-2 < 0 < 1$. In Fig. 5-3, $b > a$ and $c > b$ or $a < b < c$.

5-4 ABSOLUTE VALUE The denominate numbers $+4$ ft, -4 ft, $+4°$, $-4°$, $+\$4$ and $-\$4$ have a common quality. Real numbers have both magnitude and direction. Each of the numbers above have the same magnitude, namely 4. The magnitude of a real number is called its *absolute value.* The symbol for absolute value is | |. Thus, $|+5| = 5$, $|-5| = 5$, $|-3| = 3$, etc.

Intuitively, we may think of the absolute value of a number as the distance of the graph of that number from the origin where the distance is expressed as a positive number (see Fig. 5-4). The absolute value of -5

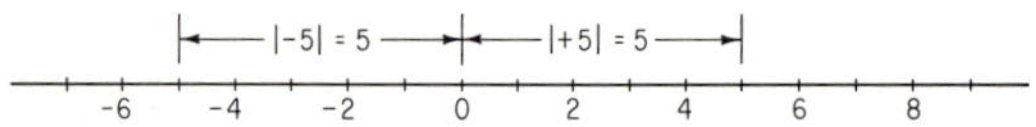

FIG. 5-4. $|-5| = |+5|$.

equals the absolute value of $+5$, since both points representing the numbers are equally distant from the origin.

The absolute value of a number is always positive or 0. It is 0 only if the number itself is 0. The absolute value of a number is never negative.

Given N is a real number.

If N is a positive number, then the absolute value of N is N.

If N is 0, then the absolute value of N is 0.

If N is a negative number, then the absolute value of N is the negative of N.

The following examples will be cleared up as we progress through the rest of this chapter.

Example 5-1

$$|+9| = +9$$
$$|-7| = +7$$
$$|-(+3)| = |-3| = +3$$
$$-|+6| = -(+6) = -6$$
$$-|-6| = -(+6) = -6$$

Exercise 5-1

1. What is the coordinate of *B*? of *D*?
2. What is the coordinate of the point halfway between *B* and *D*?
3. What is the coordinate of the point 7 units to the left of *D*?

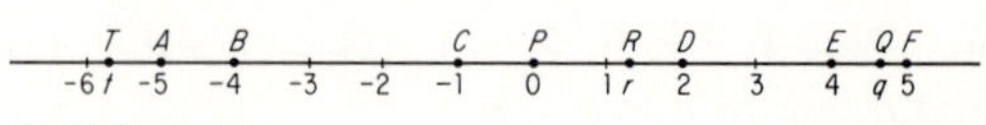

PROBS. 1–20.

4. What is the coordinate of the point 3 units to the right of *C*?
5. What is the coordinate of the point midway between *C* and *F*?
6. What is the coordinate of the point midway between *D* and *F*?

7. What is the coordinate of the point midway between C and E?
8. What is the coordinate of the point midway between A and C?

In Probs. 9–20, which number is the lesser?

9. r, t
10. q, t
11. q, r
12. $-3, -6$
13. $2, 0$
14. $-1, 0$
15. $-15, -25$
16. $0, -7$
17. $-5, -3$
18. $-6, 0$
19. $0, -3$
20. $0, 9$

Use the symbols $>$, $=$, or $<$ to compare the pairs of numbers:

21. $-24 \;.\; ? \;.\; 60$
22. $1 \;.\; ? \;.\; -1$
23. $0 \;.\; ? \;.\; -4$
24. $1+1 \;.\; ? \;.\; 1-1$
25. $1 \times 0 \;.\; ? \;.\; 1 \times 1$
26. $0 \times 4 \;.\; ? \;.\; 0+4$
27. $|-5| \;.\; ? \;.\; |-4|$
28. $|-2| \;.\; ? \;.\; |-3|$
29. $24+1 \;.\; ? \;.\; 5 \times 5$
30. $3 \times 7 \;.\; ? \;.\; 7 \times 3$
31. $\frac{4}{3} \;.\; ? \;.\; \frac{3}{4}$
32. $-\frac{2}{3} \;.\; ? \;.\; -\frac{3}{2}$
33. $|+5| + |+3| \;.\; ? \;.\; |5+3|$
34. $|+5-3| \;.\; ? \;.\; |5| - |3|$
35. $|+5|\,|-3| \;.\; ? \;.\; 16-1$
36. $0 \times 5 \;.\; ? \;.\; 0$
37. $\frac{1}{5} \times 6 \;.\; ? \;.\; \frac{5}{6}$
38. $4\frac{2}{3} \;.\; ? \;.\; 4+\frac{2}{3}$

5-5 ADDITION OF SIGNED NUMBERS Let us consider the addition of signed numbers. One way of illustrating addition of signed numbers is to use the idea of gain and loss. Consider the four additions

	+\$5 (gain)	−\$5 (loss)	+\$5 (gain)	−\$5 (loss)
	+\$2 (gain)	−\$2 (loss)	−\$2 (loss)	+\$2 (gain)
Sum:	+\$7 (gain)	−\$7 (loss)	+\$3 (gain)	−\$3 (loss)

Another interpretation of the addition of signed numbers can be made using a number line. We will let positive numbers represent displacements to the right and negative numbers represent displacements to the left. Our task will be to determine the final displacements. The four sums above are illustrated in this manner in Fig. 5-5.

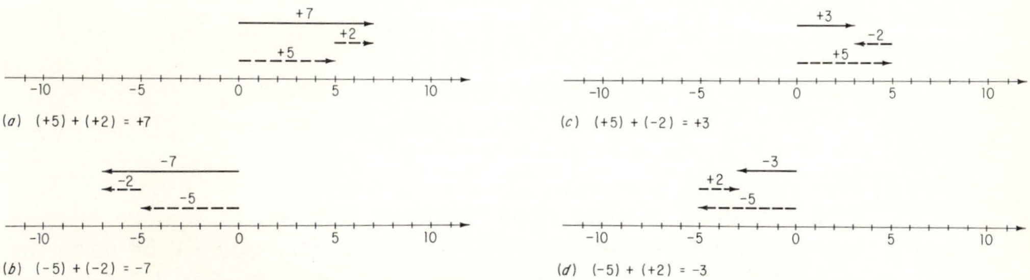

FIG. 5-5. Addition of signed numbers.

These examples suggest the following procedures for addition of signed numbers:

PROCEDURE *To add two numbers with like signs, find the sum of their absolute values and prefix the common sign to the sum.*

PROCEDURE *To add two numbers with unlike signs, find the difference between their absolute values and prefix the sign of the number having the greater absolute value to the difference.*

5-6 DUAL USE OF PLUS AND MINUS SIGNS In previous chapters in this book we have used the + and − signs to indicate addition and subtraction. That is, they were used as *signs of operation,* telling what mathematical operation is to be performed.

In this chapter we have shown that these signs are also used to indicate opposites, positive and negative numbers. In these cases they are used as *signs of quality* and indicate the *kind* of number, such as +5, −3, −2, +1.

This dual meaning often causes confusion for the beginning student. With practice, however, the confusion should disappear. Examples 5-2 illustrate various ways of writing the operation of addition. Note the dual use of the + sign.

Examples The following are equivalent forms.

5-2a Add:
$$\begin{array}{r} +5 \\ \underline{+3} \\ +8 \end{array} \text{ (sum)} \qquad \begin{aligned} 5+3 &= 8 \\ (+5)+(+3) &= +8 \\ 5+(+3) &= 8 \end{aligned}$$

5-2b Add:
$$\begin{array}{r} -5 \\ \underline{-3} \\ -8 \end{array} \text{ (sum)} \qquad (-5)+(-3) = -8$$

5-2c Add:
$$\begin{array}{r} -5 \\ \underline{+3} \\ -2 \end{array} \text{ (sum)} \qquad \begin{aligned} -5+3 &= -2 \\ (-5)+(+3) &= -2 \\ (-5)+3 &= -2 \end{aligned}$$

5-2d Add:
$$\begin{array}{r} +5 \\ \underline{-3} \\ +2 \end{array} \text{ (sum)} \qquad \begin{aligned} (+5)+(-3) &= 2 \\ 5-3 &= 2 \\ 5+(-3) &= 2 \end{aligned}$$

To add three or more signed numbers, add all the positive integers together, add all of the negative integers together (you will be using the commutative and associative properties of numbers, see Sec. 1-7, when you do this), and then add the two resulting sums.

Example 5-3

$$\begin{aligned}(+7)+(-6)+(-3)+(+5)+(-8) &= [(+7)+(+5)]+[(-6)+(-3)+(-8)]\\ &= (+12)+(-17)\\ &= -5\end{aligned}$$

Example 5-4 Add

$$\begin{array}{r} +23 \\ -16 \\ -38 \\ +27 \\ \underline{+31} \end{array}$$

SOLUTION:

$$\begin{array}{rrrl} +23 & -16 & +81 & \\ +27 & \underline{-38} & \underline{-54} & \\ \underline{+31} & -54 & +27 & \text{(answer)} \\ +81 & & & \end{array}$$

Exercise 5-2

Find the sum in each of the following:

1. $\begin{array}{r} +13 \\ \underline{+35} \end{array}$

2. $\begin{array}{r} -24 \\ \underline{+65} \end{array}$

3. $\begin{array}{r} -15 \\ \underline{-12} \end{array}$

4. $\begin{array}{r} +36 \\ \underline{-14} \end{array}$

5. $\begin{array}{r} 0 \\ \underline{-13} \end{array}$

6. $\begin{array}{r} -15 \\ \underline{0} \end{array}$

7. $\begin{array}{r} 47 \\ \underline{-47} \end{array}$

8. $\begin{array}{r} -86 \\ \underline{10} \end{array}$

9. $\begin{array}{r} 15 \\ -\ 5 \\ \underline{-10} \end{array}$

10. $\begin{array}{r} -2 \\ -7 \\ \underline{5} \end{array}$

11. $\begin{array}{r} +47 \\ -35 \\ \underline{-23} \end{array}$

12. $\begin{array}{r} 11 \\ 34 \\ \underline{-79} \end{array}$

13. $\begin{array}{r} 2 \\ -3 \\ -8 \\ \underline{4} \end{array}$

14. $\begin{array}{r} 49 \\ -16 \\ -32 \\ \underline{-11} \end{array}$

15. $\begin{array}{r} -23 \\ -38 \\ -17 \\ \underline{+88} \end{array}$

16. $\begin{array}{r} 15 \\ -26 \\ +34 \\ \underline{-\ 8} \end{array}$

17. $(+18)+(+12)$

18. $(-23)+(-15)$

19. $(-71)+(+26)$
20. $(+59)+(-85)$
21. $(8)+(-43)$
22. $(+2)+(+7)+(-3)$
23. $(-16)+(+4)+(-18)+(-3)$
24. $87+(-25)+6+(-35)$
25. $(+75)+(-23)+(-34)+(-11)$
26. $(+8.5)+(-1.3)+(-2.7)+(-3.0)$
27. $(+5.8)+(-8.6)+(-7.3)+(9.2)$
28. $(+\frac{2}{3})+(-\frac{1}{4})+(-\frac{1}{6})+(-\frac{1}{2})$
29. $(\frac{3}{4})+(-\frac{2}{3})+(-\frac{7}{12})+(\frac{1}{2})$
30. $(-6\frac{3}{5})+(-2\frac{2}{3})+(5\frac{1}{2})+(8\frac{5}{6})$

For Probs. 31 to 40, replace each question mark by the appropriate numeral.

31. $(+3)+(?)=+12$
32. $(+9)+(?)=(+5)$
33. $(-12)+(?)=-20$
34. $(-13)+(?)=-8$
35. $(+7)+(?)=-11$
36. $(?)+(-7)=-13$
37. $(?)+(-7)=13$
38. $(?)+(+11)=-25$
39. $(?)+(-16)=+38$
40. $(?)+(-17)=-20$

5-7 SUBTRACTION OF SIGNED NUMBERS Subtraction is a process inverse to that of addition. To subtract a number b from a number a we must find the number c which when added to b gives a. Thus $a-b=c$, if, and only if, $b+c=a$.

Suppose in buying a tube of glue costing 67 cents you gave the clerk a one dollar bill. In all probability, in giving you your change, the clerk will give you three pennies, a nickel, and a quarter while saying "67 cents, 70 cents, 75 cents, a dollar." The clerk by so doing will have subtracted 67 from 100 by counting what must be added to 67 to get 100.

Consider the following examples:

Example 5-5 Subtract +5 from +3.

SOLUTION Our problem can be written several ways:

Subtract: +3 (minuend)
+5 (subtrahend) or $3-5=[?]$ or $(+3)-(+5)=[?]$
[ ] (difference)

To find the answer we try to determine the number which when placed in the [ ] will make the following addition statements true. Or

Add: [?] (addend)
+5 (addend) or $(+5)+[?]=(+3)$
+3 (sum)

Applying the procedures of adding signed numbers, we can determine that -2 is the missing number.

Example 5-6 Subtract +5 from -3.

SOLUTION

Subtract:	-3 (minuend)		Add:	[?] (addend)
	$+5$ (subtrahend)	is equivalent to		$+5$ (addend)
	[?] (difference)			-3 (sum)

Alternately,

$(-3) - (+5) = [?]$ is equivalent to $(+5) + [?] = (-3)$

Replacing the ? by -8 will make the above true, hence -8 is the answer.

Example 5-7 Subtract -5 from -3.

SOLUTION

Subtract:	-3		Add:	[?] (addend)
	-5	is equivalent to		-5 (addend)
	[?] (difference)			-3 (sum)

Alternately,

$(-3) - (-5) = [?]$ is equivalent to $(-5) + [?] = (-3)$

Using our procedures for adding signed numbers will give $+2$ as the answer.

Example 5-8 Subtract -5 from $+3$.

SOLUTION:

Subtract:	$+3$		Add:	[?] (addend)
	-5	is equivalent to		-5 (addend)
	[?]			$+3$ (sum)

or

$(+3) - (-5) = [?]$ is equivalent to $(-5) + (?) = (+3)$

The answer is $+8$.

The last four examples can be summarized by the following set of statements:

If $(+3) - (+5) = ?$, then $(+3) + (-5) = ?$

If $(-3) - (+5) = ?$, then $(-3) + (-5) = ?$

If $(-3) - (-5) = ?$, then $(-3) + (+5) = ?$

If $(+3) - (-5) = ?$, then $(+3) + (+5) = ?$

The foregoing illustrative examples lead to the following:

PROCEDURE *To subtract two signed numbers, change the sign of the subtrahend (to get its opposite or additive inverse). Then add the changed subtrahend (the inverse) to the minuend.*

5-8 THE ADDITION-SUBTRACTION ALTERNATIVES We noted in Sec. 5-6 that the + and − signs have dual meanings. In the four examples just discussed the dual meanings should be apparent. It is important when first learning the process of addition and subtraction to keep clearly in mind the distinctive meanings of the signs. Another factor which may add to the confusion to the beginning student is the common practice of omitting the algebraic sign before a positive number. Thus, for example, we often write 3 instead of +3.

For the expression $19 + 4 + 6$, it should be clear that three positive quantities are being added. The sum equals 29. But for the expression $19 - 4 - 6$, one might ask, "Do we subtract 4 from 19 and then subtract 6 from the remainder, or do we add −4 and −6 to 19?" Careful examination for the answer to the question will reveal that either interpretation will yield the same answer, 9. Hence:

$$19 - 4 - 6 = +19 - (+4) - (+6) = +19 + (-4) + (-6) = 9$$

Example 5-9

$$-3 - 6 = (-3) - (+6) = (-3) + (-6) = -9$$
$$-8 - 5 - 2 = (-8) - (+5) - (+2) = (-8) + (-5) + (-2) = -15$$
$$\begin{aligned} 7 - 9 - 4 + 3 - 5 &= 7 - (9) - (4) + 3 - (5) \\ &= 7 + (-9) + (-4) + 3 + (-5) \\ &= 10 + (-18) \\ &= -8 \end{aligned}$$
$$\begin{aligned} 8 - (-3) - 3 - (-6) &= 8 + (+3) + (-3) + (+6) \\ &= 17 + (-3) \\ &= 14 \end{aligned}$$
$$\begin{aligned} -8 - 3 + 7 - (-9) &= -8 + (-3) + 7 + 9 \\ &= 16 - 11 \\ &= 5 \end{aligned}$$

Example 5-10 Combine to a single number $(+34) - (-16) - (+25) + (-43) - (-61)$.

SOLUTION Change the problem to one involving only the operation of addition of signed numbers.

$$\begin{aligned} &(+34) - (-16) - (+25) + (-43) - (-61) = \\ &(+34) + (+16) + (-25) + (-43) + (+61) = \\ &(+111) + (-68) \qquad = 43 \end{aligned}$$

Exercise 5-3

Subtract the bottom number from the top number:

1.	3.	5.	7.
$+8$	$+12$	0	83
-5	$+18$	-18	417

2.	4.	6.	8.
-7	-64	0	-39
-3	$+37$	-25	-107

Subtract the top number from the bottom number:

9.	11.	13.	15.
13	-83	78	6.32
-46	-107	87	-5.79

10.	12.	14.	16.
27	0	103	-1.07
23	-2	-230	-4.35

Find each of the following differences:

17. $(-8)-(+5)$
18. $(2)-(7)$
19. $(+11)-(-11)$
20. $(-14)-(-16)$
21. $(0)-(-39)$
22. $(-37)-(+26)$
23. $(-1)-(+18)$
24. $(23)-(-64)$
25. $(-87)-(0)$

Find the value of the letter which makes each equation a true statement:

26. $x-7=5$
27. $y-(-3)=-6$
28. $z-(-4)=0$
29. $(11)-x=-12$

Combine each of the following into a single number:

30. $(+3)+(-8)-(+4)$
31. $(-4)-(+7)+(-9)$
32. $(-8)-(-6)+(-7)-(+4)$
33. $(-5)-(+6)-(-7)-(+2)$
34. $(-12)-(-10)-(+8)+(+24)$
35. $(-36)+(15)-10-28$
36. $-36+15-10-28+7$
37. $(+24)-(-12)-(+12)+(-12)$
38. $(-2)-(+3)+(-4)-(-5)+(-6)$
39. $(-1.72)-(-3.04)+(+2.97)+(-5.40)-(+0.09)$
40. $(0.05)+(-0.03)-(-8.09)+(+9.06)-(+5.35)$

5-9 MULTIPLICATION OF SIGNED NUMBERS You will recall that one of the definitions we gave for multiplication was that of repeated addition. With this definition and the properties of multiplication given in Sec. 1-10 we can find the product of any two numbers. Thus, $3(-4)=-4-4-4=-12$ and $2(-5)=-5-5=-10$.

Consider the product $-4(3)$. Applying the commutative law for multiplication (see Sec. 1-10), we know that $-4(3) = 3(-4) = -4 - 4 - 4 = -12$.

The product of two numbers with unlike signs is a negative number and is found by taking the negative of the product of the absolute values of the two numbers. That is, if a and b are real numbers with unlike signs, then $ab = -(|a|\ |b|)$.

Now consider the product of two numbers with like signs. We already realize that the product of two positive numbers is a positive number; e.g., $(+3)(+4) = +12$. Recalling the distributive property of multiplication, we can determine a rule for multiplying two negative numbers. Let us find the product of -4 and -5. First look at

$$(-4)[(5) + (-5)] = -4[0] = 0$$

By the distributive law,

$$(-4)[(5) + (-5)] = (-4)(5) + (-4)(-5)$$

or $$-4(0) = -20 + (-4)(-5)$$

and $$0 = -20 + (-4)(-5)$$

This tells us that $(-4)(-5)$ is a number which added to -20 gives 0. There is only one such number. That number is $+20$. Therefore it must be true that

$$(-4)(-5) = +20$$

The product of two numbers with like signs is a positive number and is found by multiplying the absolute values of the two numbers. That is, if a and b are real numbers with like signs, then $a\,b = |a|\ |b|$

The associative property of grouping factors can be extended to find the product of three or more signed numbers. Thus,

$$\begin{aligned}(-3)(-4)(-5) &= -3[(-4)(-5)]\\ &= -3[20]\\ &= -60\end{aligned}$$

and

$$\begin{aligned}(-2)(-3)(-4)(-5) &= [(-2)(-3)][(-4)(-5)]\\ &= [+6][+20]\\ &= +120\end{aligned}$$

It should now be clear that the product of two or more signed numbers is positive if there is an even number of negative factors and is negative if there is an odd number of negative factors.

5-10 DIVISION OF SIGNED NUMBERS In Chap. 1 it was pointed out that division is the inverse of multiplication. Hence the rules for division of signed numbers will follow those for multiplication. For example, we have shown that $(-4)(+3) = -12$. It then must follow that $\frac{-12}{3} = -4$ and $\frac{-12}{-4} = +3$. We are now ready to state the following general rule:

PROCEDURE *The quotient of two numbers with like signs is a positive number. The quotient of two numbers with unlike signs is a negative number.*

Exercise 5-4

Perform the indicated operations:

1. $(+6)(-5)$
2. $(-3)(+7)$
3. $(-7)(-9)$
4. $10(-2)$
5. $(-2)(-2)$
6. $(-2)(-2)(-2)$
7. $(-\frac{1}{2})(-\frac{1}{2})$
8. $(-\frac{1}{2})(-\frac{1}{2})(-\frac{1}{2})$
9. $(-\frac{2}{3})(-\frac{2}{3})$
10. $(\frac{2}{3})(-\frac{2}{3})(\frac{2}{3})$
11. $(-1)(-4)(+5)$
12. $(-2)(+3)(-4)$
13. $(-7)(+\frac{1}{2})(+8)$
14. $(-12)(-2)(-3)$
15. $(-\frac{5}{7})(+\frac{14}{15})(+\frac{9}{8})$
16. $(-1)(-1)(-2)(-2)$
17. $(-5)(-2)(+4)(+3)$
18. $(7)(-1)(-3)(-2)$
19. $(8)(-\frac{1}{4})(-\frac{1}{2})(-5)$
20. $-8(-14 + 10)$
21. $-6(27 - 15)$
22. $\frac{-84}{-12}$
23. $\frac{210}{-14}$
24. $(-16) \div (12)$
25. $(19) \div (-4)$
26. $\frac{(-5)(-2)}{(+4)(+15)}$
27. $\frac{(7)(-12)}{(-1)(-14)}$
28. $\frac{(-1)(-24)(+7)}{(-3)(-14)}$
29. $\frac{(-8)(+5)(-3)}{(-6)}$
30. $[(-4)(-4)(-4)] \div [(-2)(-2)]$
31. $[(-4)(-4)] \div [(-2)(-2)(-2)]$
32. $(-\frac{8}{27}) \div (\frac{4}{9}) \div (-\frac{2}{3})$
33. $(3\frac{1}{5}) \div (-1\frac{1}{6})(\frac{4}{15})$
34. $(-5)[(-3) + (+7)]$
35. $(-4)(-7) + (-4)(+6)$
36. $(-2)[(-3) - (+4) + (-5)]$
37. $(-3)[12 - (+7) + (-3)]$

38. $(-1)[-5-(-4)+(-3)]$ 39. $(-5)[15-10-3+4]$

In Probs. 40 to 45, replace the question mark by an appropriate numeral.

40. $(-2)(?)=+18$
41. $(+4)(?)=-24$
42. $(?)(-7)=+7$
43. $(-2)(?)(+3)=+24$
44. $(?)(-2)(-7)=-28$
45. $(-6)(-\frac{1}{2})(?)=-18$

5-11 OPERATIONS WITH ZERO We have noted the value of the symbol 0 in our number system. Sometimes students refer to zero as "nothing." This is an unfortunate mistake. The word "nothing" refers to a nonexistence. Zero does exist. Zero is a number which indicates the absence of any quantity. Zero also has a positional value as has any of the other nine digits in the Hindu-Arabic number system. Thus, the zero in the number 304 indicates the absence of any tens and also is used to place the 3 and 4 in the proper hundreds and ones position.

The number zero has some unique properties and obeys certain special laws which we shall list. The letter N will be used to represent any real number in the following.

1 If zero is added to a number, the sum is equal to the original number. For this reason it is often called the *additive identity element.*

ILLUSTRATIONS $3+0=3$ $N+0=N$
$0+0=0$ $0+N=N$

2 If zero is subtracted from a number, the difference is equal to the original number.

ILLUSTRATIONS $5-0=5$ $N-0=N$

3 A number subtracted from itself is zero.

ILLUSTRATIONS $7-7=0$ $N-N=0$

4 If a number is subtracted from zero, the difference is the negative of that number.

ILLUSTRATIONS $0-6=-6$ $0-N=-N$

5 The product of zero and any quantity is zero.

ILLUSTRATIONS $0\times 4=0$ $0\times N=0$
$4\times 0=0$ $0\times 0=0$

6 If the product of two numbers is equal to zero, one of the numbers must be equal to zero.

ILLUSTRATIONS If $a \cdot b = 0$, then $a = 0$ or $b = 0$ (or both). If $(x - a)(x - b) = 0$, then $x - a = 0$ or $x - b = 0$ (or both).

7 If zero is divided by any number except zero, the quotient is zero.

ILLUSTRATIONS $0 \div 5 = 0$

We have stated that to divide a number a by a number b is to find a number q, called *quotient,* such that b multiplied by q gives a. That is, $a/b = q$ if, and only if, $a = b \cdot q$. If $b = 0$ and $a \neq 0$, there is no number q which satisfies the condition $a = 0 \cdot q$. Hence, it is impossible to divide a number by zero. We say that if $a \neq 0$, then $a \div 0$ has no meaning. If $a = 0$ and $b = 0$, we find there are many numbers q which satisfy the condition $0 = 0 \cdot q$. Thus, $\frac{0}{0}$ does not represent a *single* real number. It is said to be *indeterminate.* Thus, we must exclude division by zero in our operations.

Exercise 5-5

Indicate which of the following are *true* and which are *false:*

1. $(+7) + (-7) = 0$
2. $(+3) - (+3) = 0$
3. $(-5) - (-5) = 0$
4. $0 - (+8) = +8$
5. $8 + 0 = 0 + 8$
6. $(6)(0) = 6$
7. $(0)(-7) < -7$
8. $-3 > 0$
9. $-(-7) = (-1)(-7)$
10. $|5| + |-5| = 0$
11. $0 - 5 = 5 - 0$
12. $[-(-3)](-4) = 12$
13. $\frac{0}{-4} = \frac{0}{4}$
14. $\frac{0}{27} = \frac{0}{1}$
15. $6(0) = 0(6)$
16. $|-6| < -|-6|$
17. $1 + 1 = 1 \cdot 1$
18. $0 \cdot 1 = \frac{0}{1}$
19. $0 \cdot 0 > 0$
20. $(+2)(-3)(0) = 0$
21. $0(5 - 3) = 2$
22. $|0| = 0$
23. $-\frac{5}{4} < 0$
24. $0 \cdot (-1) < 0 + 0$
25. $\frac{0}{1} = \frac{1}{0}$
26. $\frac{0}{-1} > -2$
27. $-\frac{2}{3} > 0$
28. $-5 > 0$
29. If $3x = 0$, then $x = 0$
30. $\frac{0}{0} = \frac{1}{0}$
31. $(ab) \cdot 0 = a(b \cdot 0)$
32. $0(a + b) = 0$
33. $a(0 + b) = ab$
34. $0 < 0 + 0$
35. $0(5 + 3) = 0 \cdot 5 + 0 \cdot 3$
36. If $a + b = 0$, then $b = -a$

CHAPTER

6

THE SLIDE RULE

6-1 USES OF THE SLIDE RULE The slide rule is an instrument which is designed to save time and labor in the common mathematical processes of multiplication, division, raising to powers, taking roots, finding trigonometric and logarithmic functions, and combinations of these processes. Addition and subtraction are two processes for which the slide rule cannot be used.

Every technician should be able to use the slide rule in solving problems and checking his computations. Much valuable time can be saved and tedious work can often be avoided with the help of this instrument.

There are many sizes, types, and costs of slide rules, but they all operate on the theory and properties of logarithms. It is beyond the scope of this text to discuss the various types of rules and the theory on which they are based. We shall limit our discussion in this chapter to the use of the C, D, A, K, CF, DF, CI, and DI scales.

6-2 PARTS OF THE SLIDE RULE The slide rule has three main components (see Fig. 6-1): the stock (or body), the slide, and the indicator (or cursor).

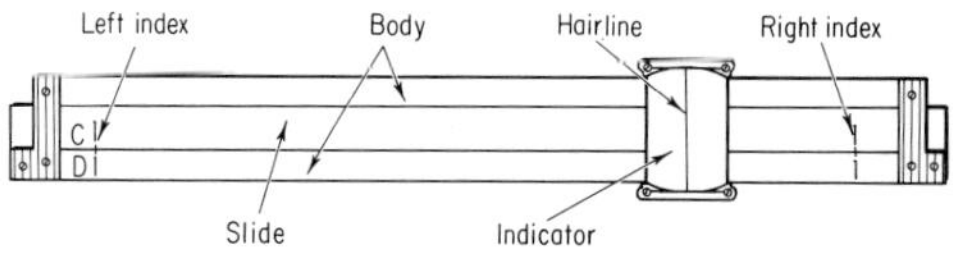

FIG. 6-1.

The *stock* (or body) is the frame of the rule and consists of an upper and lower rectangular bar usually held together by means of metal or plastic strips at the ends.

The middle sliding part of the rule is called the *slide.* The edges of the slide have upper and lower tongues which fit grooves on the inner edges of the bars of the stock.

The transparent runner which slides along the face of the rule is referred to as the *indicator* (or cursor). The indicator has an engraved vertical hairline on each side. This hairline serves as a perpendicular for use in aligning numbers on two scales which are to be used together.

The primary number 1 on the left end of either the C or D scale is called the *left index* of that scale. The primary number 1 on the other end of the scale is called the *right index* of that scale.

A number on one scale is said to be *opposite* a number on another scale if the indicator can be moved so the hairline can cover both numbers at the same time. Each number is said to be *opposite* the other.

The slide rule is said to be *closed* when the left index of the C scale (of the slide) is opposite the left index of the D scale (of the stock).

When the hairline of an adjusted slide rule is placed over the left (or right) index of the D scale on one side of the rule, it will cover the left (or right) indexes of all the scales on both sides of the rule. If the rule is out of adjustment, it usually can be adjusted by loosing screws on the stock and/or indicator and shifting the scales and/or the hairlines. The screws on the stock should be tight enough to prohibit the scales from becoming unadjusted and yet should be loose enough to let the slide be moved easily with some slight effort.

6-3 THE C AND D SCALES Every slide rule has a C and a D scale. The C scale and the D scale are exactly alike in each of the various types of rule. Each of these scales is divided into nine parts by *primary* marks numbered from 1 on the extreme left through the numbers 2, 3, 4, 5, 6, 7, 8, 9, 1 (see Fig. 6-2). The distances between consecutive primary marks decrease toward the right.

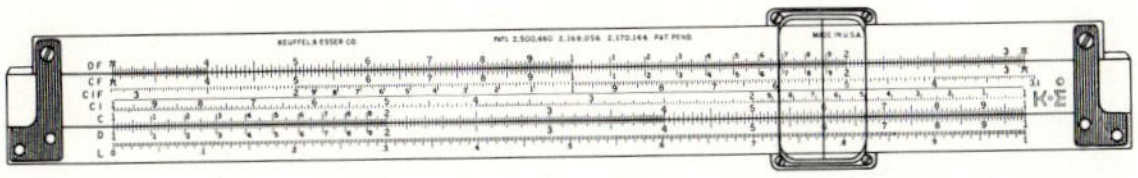

FIG. 6-2.

The space between any two primary marks is divided into ten parts by nine shorter *secondary* marks. The distances between consecutive secondary marks decrease toward the right. With the exception of the secondary marks between the 1 and 2 primary marks, numerals are not printed beside the smaller graduations.

The space between successive secondary graduations at the left end of the rule from primary graduation mark 1 to primary graduation 2 is separated into ten parts by nine tertiary marks. Between primary graduations 2 and 4 each secondary space is divided into five subspaces. Between primary graduation 4 and the right graduation 1 each secondary space is subdivided into two subspaces. None of the tertiary marks is identified by number on the rule.

6-4 READING THE C AND D SCALES The student should note on the C and D scales that between the left primary number 1 and the primary number 2, the space between any two consecutive tertiary marks represents a unit in the third digit of a number. One can estimate the fourth digit of the number. The student should check Fig. 6-3 to familiarize himself with scale readings between primary marks 1 and 2.

Upon examining the portions of the C and D scales between 2 and 4, the student will observe that there are five spaces between any two consecutive

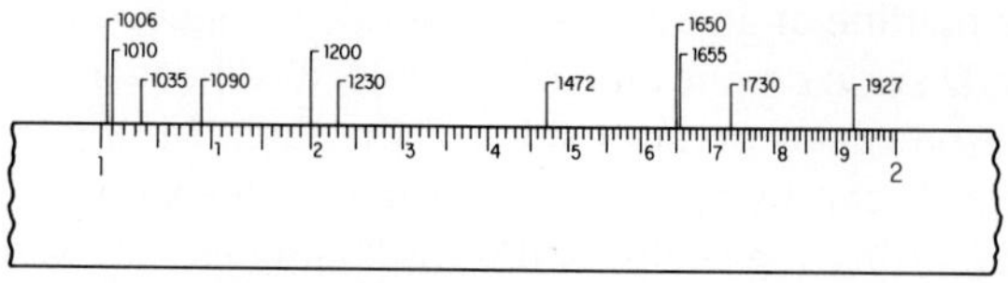

FIG. 6-3. Enlarged section of the D scale from primary mark 1 to primary mark 2.

secondary marks. Hence, each small space between consecutive secondary marks represents two units in the third digit of the number. Numbers represented by points on the scale between the 2 and 4 can be read only to three digits, and even the third digit is an approximation (see Fig. 6-4).

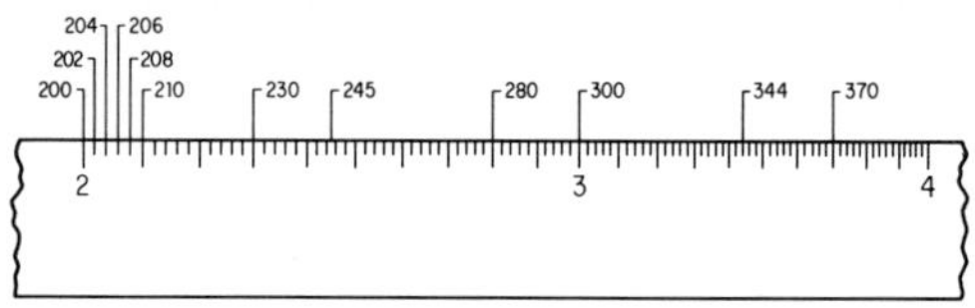

FIG. 6-4. Enlarged section of D scale from primary mark 2 to primary mark 4.

When the student examines the sections of the C and D scales between the primary mark 4 and the right index 1, he will find only two spaces between any two consecutive secondary marks. Hence, each of the spaces between consecutive secondary marks represents five units in the third digit of the number. Such numbers can only be approximated to three significant figures. Figure 6-5 shows numbers on a part of this portion of a rule.

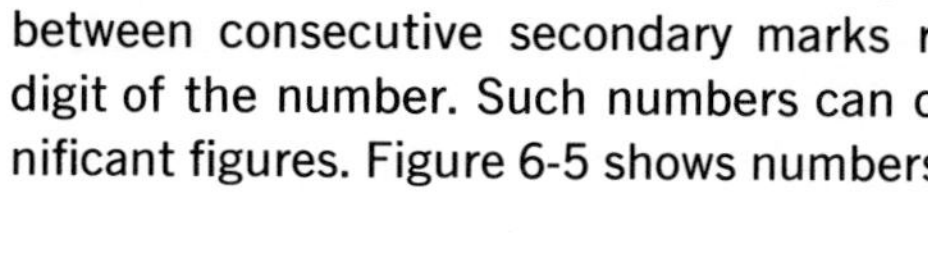

FIG. 6-5. Enlarged section of D scale showing a part to the right of primary mark 4.

The values of the spaces we have discussed for the C and D scales also hold for the following scales: CF, DF, CI, and DI.

We have noted the limited accuracy of the slide rule. Any effort to use the rule for computations involving more than three- or four-place accuracy will prove futile. In this text we will use four digits when the number has 1 as its initial digit and three digits for all other numbers.

On the D scale in Fig. 6-6, the student will find some more sample settings. He should read the numbers associated with the marks lettered *A*, *B*, *C*, . . . , and check his readings with the following: A 127, B 140, C 207, D 285, E 374, F 450, G 1565, H 1237.

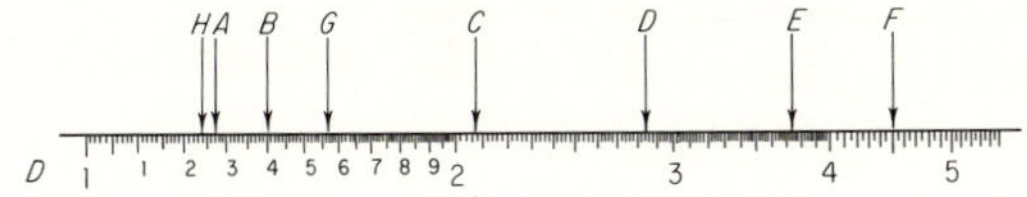

FIG. 6-6.

6-5 LOCATION OF THE DECIMAL POINT The C and D scales on a slide rule give only the sequence of significant digits of a number. The position of the decimal is completely ignored in the scales. When a number is to be located on the D scale, the first significant digit is located by using the primary marks; the second digit is located by using the secondary graduations; the third digit is located by using the tertiary graduations. For example, mark *A* in Fig. 6-6 could represent 0.0127, 0.127, 1.27, 12.7, 127, 1,270, 12,700, etc. The context of the problem will determine for the operator the location of the decimal point.

When the slide rule is used for performing operations such as multiplication and division, the sequence of significant digits in the solution is obtained without regard to the location of the decimal point. The location of the decimal can usually be obtained by "rounding off" numbers and mentally approximating the answer.

For example, if the slide rule is used to multiply 7.83 by 2.59, the sequence of digits obtained in the answer by the rule will be 203. It will be noted that the answer is approximately

$$8 \times 2.5 = 20$$

Thus, the answer will be 20.3.

6-6 DIVISION The operations of multiplication and division can be done on any pair of identical scales such as the C and D, and CF and DF, and the A and B scales. The C and D scales are most often used for this purpose, so we will start with these scales. Since, in general, it is easier to divide than to multiply with a rule, we will discuss division before multiplication.

PROCEDURE *To divide two numbers, set the divisor on the C scale opposite the dividend on the D scale. The quotient is read on the D scale under the index of the C scale* (see *Fig. 6-7*).

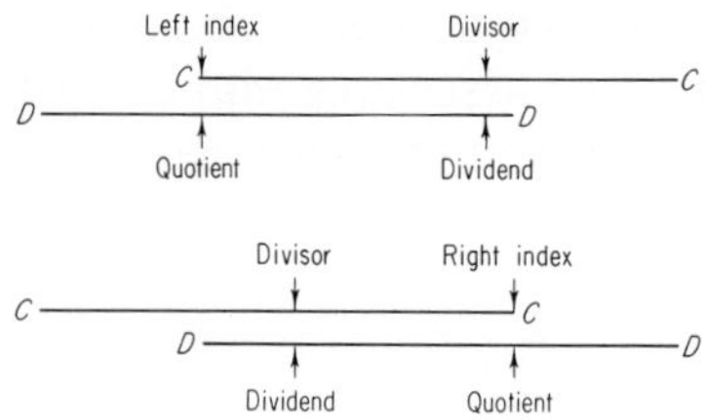

FIG. 6-7.

Example 6-1 Divide 8 ÷ 4.

SOLUTION (See Fig. 6-8.) Move the indicator until the hairline is over 8 on the D scale. Draw 4 of the C scale under the hairline. The quotient (2.00) is read on the D scale under the left index of the C scale.

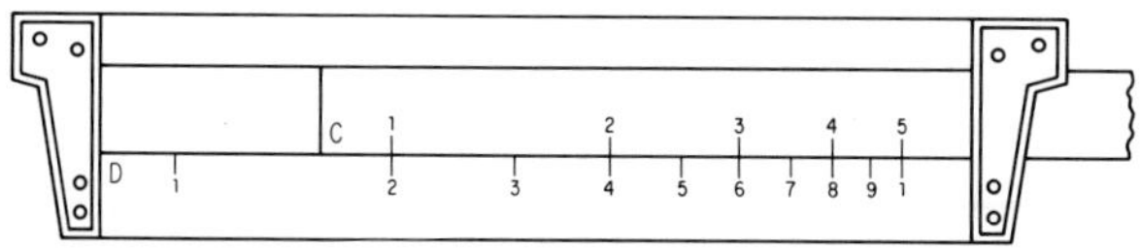

FIG. 6-8. 8 ÷ 4 = 2.

Example 6-2 Divide 2860 ÷ 34.7.

SOLUTION Move the indicator until the hairline is over 286 of the D scale. Move the slide until 347 of the C scale is under the hairline. The quotient is read on the D scale opposite the right index of the C scale. The sequence of digits in the quotient will be 824. Since 2,900 ÷ 30 = 90 (approx.), our quotient will be 82.4.

Exercise 6-1

Perform the indicated operations:

1. 8.34 ÷ 2.76
2. 91.1 ÷ 42.7
3. 37.5 ÷ 0.0444
4. 0.668 ÷ 47.3
5. 0.00397 ÷ 0.0551
6. 0.01462 ÷ 7.89
7. 80,200 ÷ 93.6
8. 192.8 ÷ 13.76
9. 71.30 ÷ 0.01774
10. 1.037 ÷ 100.6
11. 10.05 ÷ 37.0
12. 82,900 ÷ 35.7
13. 94,900 ÷ 725
14. 8.73 ÷ 26,700
15. 1.530 ÷ 0.351
16. 693 ÷ 0.0257
17. 0.01852 ÷ 0.000839
18. 0.00357 ÷ 0.361
19. 40.4 ÷ 718
20. 1.607 ÷ 83.5

6-7 MULTIPLICATION We will use the C and D scales in this section to multiply. The procedure can be summarized as follows:

PROCEDURE *To find the product of two numbers, over one of the factors marked on the D scale set the index of the C scale. Locate the mark on the C scale whose sequence of digits is the same as those of the second factor. Directly below it on the D scale, read the sequence of digits in the answer. When the product falls outside the limit of the D scale, the wrong index of the C scale is being used. In this event "interchange the indexes" of the C scale and proceed as before* (see *Fig.* 6-9).

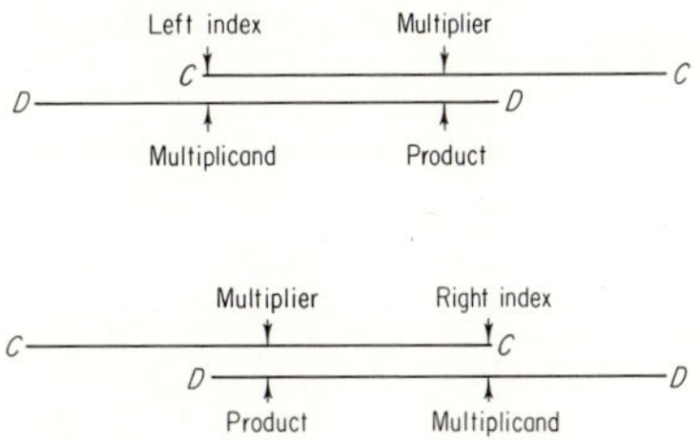

FIG. 6-9.

The following examples illustrate the method of multiplying by using the C and D scales.

Example 6-3 Multiply 2×3.

SOLUTION Set the left index of the C scale on 2 of the D scale. Then push the indicator until the hairline is over the 3 on the C scale. Below the hairline on the D scale read the answer 6. This would also be the procedure to find 200×30, 0.002×0.0003, etc.

Example 6-4 Multiply 15.76×4.28.

SOLUTION To find the product, first disregard the decimal point and proceed as in the above example to get the sequence of digits 675. To locate the decimal point, approximate the answer by noting that $16 \times 4 = 64$. Hence, the answer is 67.5.

Example 6-5 Multiply 0.0372×65.8.

SOLUTION Disregard temporarily the decimal points in the factors. If we proceed to solve this problem as we did the previous two, we shall not be able to read the answer on the D scale opposite 658 on the C scale. We must modify our procedure as follows:

Place the right index of the C scale opposite 658 of the D scale. Push the indicator until the hairline is over 372 on the C scale. Read 245 under the hairline on the D scale. Note that

$0.04 \times 60 = 2.4$

Hence, the answer will be 2.45.

Exercise 6-2

Perform the indicated operations:

1. 2.4×3.2
2. 1.87×5.5
3. 1.264×4.39
4. 4.56×63.8
5. 3.82×724
6. 886×43.2
7. 204×806
8. 1.846×0.00639
9. 955×453
10. 10.95×84.7
11. 71.3×756
12. $48{,}200 \times 0.00013$
13. $0.01016 \times 1{,}008$
14. 207×0.0306
15. 630×0.0000201
16. 0.492×0.00274
17. 0.003×708
18. $917{,}000 \times 0.000372$
19. 0.01947×259
20. 4.76×80.3

6-8 COMBINED MULTIPLICATION AND DIVISION Sometimes the product of three or more numbers is required in solutions to problems. Other times problems call for combinations of multiplication and division. The slide rule is extremely applicable for these computations.

Example 6-6 Multiply $27.49 \times 13.65 \times 5.80$.

SOLUTION Estimate the product as $25 \times 10 \times 6 = 1{,}500$. The true product should be in the neighborhood of 1,500. Remembering that only three significant figures can be placed on the C and D scales for numbers whose first significant digit is not 1, we "round off" 27.49 to 27.5. The 13.65 can be read on the scales, as can 5.80.

Set the left index of the C scale opposite 275 of the D scale. Push the indicator until the hairline is over 1365 on the C scale. (Note that the sequence of digits for this partial product is 375.) Next move the slide until the right index of the C scale is under the hairline (opposite 375 of the D scale). Push the indicator until the hairline is over 58 of the C scale. Below the hairline read the sequence of digits 218 on the D scale. The answer for the product is 2,180.

Example 6-7 Evaluate $\dfrac{81.72 \times 5.34}{62.91}$.

SOLUTION Estimate the answer to be $\dfrac{80 \times 5}{60} = 7$ (approx.). Let us first divide 817 by 629 as follows: Move the hairline over 817 on the D scale. Under the hairline set 629 of the C scale. (Note that the sequence of digits in the partial quotient is 1299.) Next move the indicator until the hairline is over 534 of the C scale. The reading under the hairline on the D scale has 694 as the sequence of digits. The answer then should be 6.94.

Exercise 6-3

Perform the indicated operations.

1. $\dfrac{709 \times 1.124}{0.672}$
2. $\dfrac{326 \times 5.318}{386}$
3. $\dfrac{87.3 \times 2.47}{64.5 \times 12.89}$
4. $\dfrac{4{,}290 \times 7.93}{348 \times 817}$
5. $\dfrac{0.00264}{0.0854 \times 0.27}$
6. $\dfrac{0.0001148}{0.00275 \times 0.0854}$
7. $2{,}641 \times 0.00385 \times 0.01916$
8. $2.37 \times 81.4 \times 0.00886$
9. $0.0000420 \times 805 \times 9.04$
10. $63{,}600 \times 2{,}120 \times 639$
11. $\dfrac{78.4 \times 9.04 \times 1{,}326}{21.7 \times 8.26 \times 7{,}030}$
12. $\dfrac{0.00610 \times 0.0184 \times 67.3}{0.0627 \times 0.810 \times 0.1004}$
13. $\dfrac{23.74 \times 821 \times 0.00608}{83.6 \times 77.70}$
14. $\dfrac{92.87 \times 0.001523 \times 0.07376}{44.05 \times 28.80}$
15. $\dfrac{0.00381 \times 0.000847 \times 0.01166}{0.274 \times 0.00929 \times 0.00001115}$
16. $\dfrac{0.0650 \times 0.6868 \times 55.5}{94{,}700 \times 30{,}004}$
17. $\dfrac{37.2 \times 11.86 \times 995 \times 263}{234 \times 869 \times 9.87}$
18. $\dfrac{0.0762 \times 0.1714 \times 0.417 \times 0.975}{472 \times 0.819}$
19. $\dfrac{0.000389 \times 0.01258}{27.0 \times 0.00182 \times 0.2748}$
20. $\dfrac{8.26 \times 0.0918}{2.14 \times 8.54 \times 918 \times 619}$

6-9 THE CF AND DF SCALES[1] The CF (on the slide) and DF (on the body) scales are identical in graduation with the C and D scales. However, the CF and DF scales are folded in such a way as to place the indexes near the middle and π at the end positions. Because of the position of the graduations, they are called the *folded* scales.

The chief use of the folded scales is to use π as a factor and to avoid the

[1] This section can be omitted. Anything that can be accomplished with the CF and DF scales can also be done with the C and D scales.

necessity of resetting the indexes of the slide in multiplying. It does not increase the capability of the rule, but it does often minimize the number of manipulations in multiplying and dividing. A glance at Fig. 6-10 will show

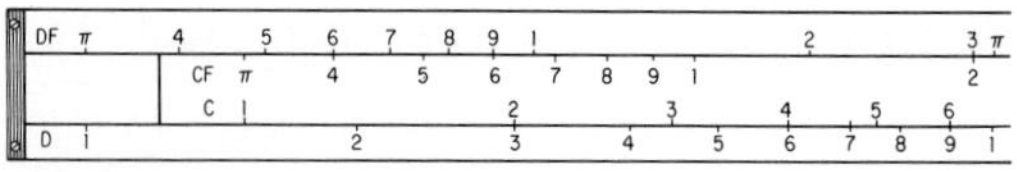

FIG. 6-10.

that, if for any setting of the slide, a number N of the C scale is opposite a number M on the D scale, then that same number N on the CF scale will be opposite the same second number M on the DF scale. Note, for example, that the following ordered pairs can be found on either the C and D scales or on the CF and DF scales:

C or CF	1	2	3	4	5	6
D or DF	1.5	3	4.5	6	7.5	9

Thus, any process which will match two numbers on the C and D scales will at the same time match the same two numbers on the CF and DF scales. This property of the rule is used to shorten the processes of multiplication and division.

Example 6-8 Multiply 4.82×5.37.

SOLUTION If we set the left index of the C scale opposite 4.82 of the D scale and then try to move the hairline over 5.37 of the C scale, we shall realize that this is impossible. However, we can move the hairline over 5.37 of the CF scale and read the product under the hairline on the DF scale. This saves having to interchange indexes. The product will be 25.9.

Example 6-9 Using the folded scales, find $21.6 \div 7.75$.

SOLUTION Set the hairline over 216 of the DF scale. Under the hairline draw 775 of the CF scale. Read the answer on the DF scale opposite the index of the CF scale. The answer will be 2.79. Note the same answer can be read on the D scale opposite the left index of the C scale.

Example 6-10 Find the value of $\dfrac{3.50 \times 96.6}{2.12 \times 57.8}$.

SOLUTION There are many procedures that can be used here. One method follows. Set the hairline over 350 of the D scale. Under the hairline draw 212 of the C scale. Next move the hairline over 966 of the CF scale. Then move the slide until 578 of the CF scale is under the hairline. The sequence of digits in the final answer is found on the DF (or D) scale opposite the index of the CF (or C) scale. The answer is 2.76.

Multiplication and division by π can be simplified by using folded scales.

RULE *Any number on the DF (or CF) scale is π multiplied by its opposite on the D (or C) scale. Any number on the D (or C) scale is its opposite divided by π on the DF (or CF) scale.*

Example 6-11 Find the value of 4π.

SOLUTION Push the hairline over 4 on the D (or C) scale. Under the hairline on the DF (or CF) scale will be the digits for the answer. The product 4π will be 12.57.

Example 6-12 Find the value of $\frac{5}{\pi}$.

SOLUTION Push the hairline over 5 on the DF (or D) scale. Under the hairline on the CF (or C) scale read the answer, 1.592.

Exercise 6-4

Perform the indicated operations.

1. 7π
2. 6.4π
3. 57.8π
4. 0.604π
5. 0.00803π
6. $\frac{6}{\pi}$
7. $\frac{7.5}{\pi}$
8. $\frac{78.4}{\pi}$
9. $\frac{0.508}{\pi}$
10. $\frac{18.27}{\pi}$
11. 2.83×7.54
12. 37.4×95.1
13. 80.4×27.7
14. 5.50×0.0376
15. 0.00839×62.4
16. $\frac{11.37 \times 1.563}{2.74}$
17. $\frac{89.7 \times 72.2}{54.4}$
18. $\frac{42.1 \times 8.55}{253}$
19. $\frac{0.358 \times 0.1106}{0.472}$
20. $\frac{0.0879 \times 5.63}{0.425 \times 1.115}$
21. $\frac{31.7 \times 25.5}{99.2 \times 11.7}$
22. $\frac{2.85\pi}{4.39}$
23. $\frac{29.3 \times 1.156 \times \pi}{843}$
24. $\frac{0.0393 \times 1.506\pi}{0.707}$
25. $\frac{0.1154 \times 0.00246 \times 0.833}{0.276 \times 0.000559}$

6-10 THE CI, DI, CIF SCALES The reciprocal of a number N is 1 divided by the number. Thus $\frac{1}{4}$ is the reciprocal of 4; $\frac{3}{5}$ is the reciprocal of $\frac{5}{3}$.

The CI, CIF, and DI scales are graduated like the C, CF, and D scales but in the reverse direction so that the numbers on the scales increase from right to left. These scales are called the *reciprocal* or inverted scales. The CI and CIF scales usually appear on the same (front) face, while the DI scale appears on the other face of the rule. The scales are usually printed with red numerals to facilitate recognition.

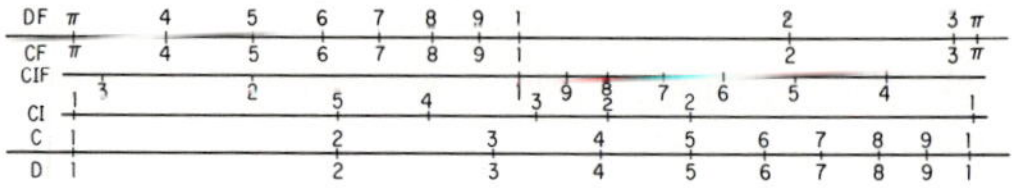

FIG. 6-11.

Figure 6-11 illustrates the relationship between a scale and its inverted counterpart. It will be noted that, if the hairline is set above any number on the C, D or CF scale, the hairline will also be above the reciprocal of that number on the corresponding inverted scale. Note, for example, the following ordered pairs:

CI, CIF, DI	1	0.5 $(=\frac{1}{2})$	0.4 $(=1/2.5)$	0.25 $(=\frac{1}{4})$	0.2 $(=\frac{1}{5})$	0.125 $(=\frac{1}{8})$
C, CF, D	1	2	2.5	4	5	8

The CI, CIF, and DI scales are of considerable value in solving problems that involve continued products.

The student should recognize that the C and D scales are the best to use in performing division, since the answer never falls outside the limits of the D scale. In like manner, he will find that the CI and DI scales are best for multiplication.

Example 6-13 Multiply 5×4.

SOLUTION To multiply 5×4, we consider the equivalent expressions $5 \times \frac{1}{\frac{1}{4}} = 5 \times \frac{1}{0.25} = \frac{5}{0.25}$. The problem reduces to one of dividing 5 by the reciprocal of 4. The reciprocal of 4 can be found on the C scale opposite 4 of the CI scale. The C and D scales are then used to divide. Procedure: First, move the hairline over 5 on the D scale. Then draw the 4 of the CI scale under the hairline. The product (=20) is read on the D scale opposite the left index of the C scale.

Example 6-14 Find the product $2.54 \times 3.68 \times 11.70 \times 5.05 \times 4.29$.

SOLUTION Consider the equivalent expression

$$2.54 \times 3.68 \times 11.70 \times 5.05 \times 4.29 = 2.54 \times \frac{1}{\frac{1}{3.68}} \times 11.70 \times \frac{1}{\frac{1}{5.05}} \times 4.29.$$

Opposite 254 of the D scale set 368 of the CI scale. Move the hairline over 117 of the C scale. Then move the slide until 505 of the CI scale is under the hairline. Finally, move the hairline over 429 of the C scale. The sequence of digits in the answer will be found under the hairline on the D scale. The answer is 2,370.

Example 6-15 Find the product $2.39 \times 0.1875 \times 2.16 \times 0.840$.

SOLUTION Opposite 239 of the D scale set 1875 of the CI scale. Since 216 of the C scale is beyond the D scale, move the hairline over 216 of the CF scale. Next move the slide until 840 of the CIF scale is set under the hairline. The answer, namely, 0.813, is on the DF (or D) scale opposite the index of the CF (or C) scale.

Exercise 6-5

Multiply:

1. $9.18 \times 3.25 \times 1.456$
2. $7.35 \times 0.492 \times 0.364$
3. $0.001184 \times 5.25 \times 7.66$
4. $851 \times 0.00534 \times 0.906$
5. $0.01083 \times 50.2 \times 27.4$
6. $8.27 \times 59{,}100 \times 0.000373$
7. $2.14 \times 5.90 \times 1.104 \times 3.77$
8. $0.499 \times 0.888 \times 663 \times 78.5$
9. $0.00504 \times 40.5 \times 39.6 \times 2.74$
10. $80.70 \times 91{,}200 \times 0.000503 \times 0.0212$
11. $27.1 \times 1.349 \times 3.57 \times 6.14 \times 87.2$
12. $2.83 \times 0.001159 \times 0.938 \times 2.57 \times 5.67$
13. $0.0000404 \times 68{,}000 \times 0.1090 \times 28.4 \times 7.44$
14. $27.2 \times 53.9 \times 0.0804 \times 0.358 \times 22.4$
15. $383 \times 49.2 \times 5.71 \times 0.625 \times 0.00744$
16. $0.208 \times 7.21 \times 5.46 \times 4.33 \times 0.000779$
17. $5.32 \times 8.391 \times 1.467 \times 118.0 \times 0.2357$
18. $60.7 \times 3.92 \times 4{,}550 \times 0.01986 \times 7.358$
19. $42.5 \times 8.06 \times 0.1783 \times 3.33 \times 0.00212 \times 0.663$
20. $0.2761 \times 0.4352 \times 0.008947 \times 50.25 \times 6.139 \times 11.98$

6-11 POWERS AND ROOTS The solutions to many technical problems involve multiplying a number by itself several times. Mathematicians use a convenient bit of symbolism to indicate this process. For example:

$$2 \times 2 \times 2 \times 2 = 2^4$$
$$3 \times 3 \times 3 \times 3 \times 3 = 3^5$$
$$4 \times 4 = 4^2$$
$$10 \times 10 \times 10 = 10^3$$
$$5^3 = 125$$
$$5^1 = 5$$
$$7^1 = 7$$

The number in small type on the upper right of a second number is called an *exponent.* The exponent indicates how many times a second number, called the *base,* is to be used as a factor in repeated multiplication. When no exponent is written, it is understood to imply 1. The numbers 2, 2^2, 2^3, 2^4, . . . are called *powers of* 2. The numbers 3, 3^2, 3^3, 3^4, . . . are *powers of* 3. Thus,

2 or 2^1 is read "two to the first power" or simply "two."

2^2 is read "two to the second power" or "two squared."

2^{10} is read "two to the tenth power."

7^{16} is read "seven to the sixteenth power."

The inverse of raising to a power is called *extracting the root.* When a number is equivalent to the product of two or more equal factors, the root of the number is one of its equal factors. For example, the two equal factors of 16 are 4 and 4 ($4 \times 4 = 16$). Therefore, 4 is a root of 16. In like manner, 2, 2, 2, are the three equal factors of 8. Therefore, 2 is a root of 8.

When a number has two equal factors, either one is called the *square root* of the number. Thus, 4 is the square root of 16.

When there are three equal factors, any of the factors is called the *cube root* of that number. Thus, 2 is the cube root of 8. Likewise, 3 is the fourth root of 81, and so on.

The symbol used for indicating the square root of a number is $\sqrt{\ }$. The symbol $\sqrt[3]{\ }$ is used to indicate the cube root; $\sqrt[4]{\ }$ is used for the fourth root; $\sqrt[5]{\ }$ for the fifth root; and so on. The symbol $\sqrt{\ }$ is called the *radical sign.* The small number at the upper left corner is called the *index.* If no index is shown it is understood to be 2. Thus, $\sqrt{16} = 4$, $\sqrt[3]{8} = 2$, $\sqrt[4]{81} = 3$, $\sqrt[5]{32} = 2$, and so on.

In this book we shall be primarily interested in extracting square roots and cube roots.

6-12 SQUARES AND SQUARE ROOTS Tables are available for finding the squares and square roots of numbers. Consider the following taken from such a table:

(a)	(b)	(c)	(d)
$2^2 = 4$	$11^2 = 121$	$237^2 = 56169$	$0.2^2 = 0.04$
$5^2 = 25$	$48^2 = 2304$	$392^2 = 153664$	$0.5^2 = 0.25$
$8^2 = 64$	$79^2 = 6241$	$581^2 = 337561$	$0.8^2 = 0.64$

(e)	(f)	(g)
$0.11^2 = 0.0121$	$0.237^2 = 0.056169$	$2.3^2 = 5.29$
$0.48^2 = 0.2304$	$0.392^2 = 0.153664$	$2.37^2 = 5.6169$
$0.79^2 = 0.6241$	$0.581^2 = 0.337561$	$23.7^2 = 561.69$

It will be noted in the above that the square of a single-digit number is a one- or two-digit number, the square of a two-digit number is a three- or four-digit number, the square of a three-digit number is a five- or six-digit number, and so on.

If we were to find the square roots of the numbers to the right of the equality signs in the examples just given, we would get the numbers on the left of the equality signs. We would get, for example:

(a) $\sqrt{25} = 5$ (e) $\sqrt{0.0121} = 0.11$

(b) $\sqrt{2{,}304} = 48$ (f) $\sqrt{0.153664} = 0.392$

(c) $\sqrt{56{,}169} = 237$ (g) $\sqrt{5.6169} = 2.37$

(d) $\sqrt{0.04} = 0.2$

We shall now consider illustrative examples of finding square roots by arithmetical processes.

Example 6-16 Find the square root of 288,369.

SOLUTION

```
288369
```

1 Mark off the number in blocks of two, starting at the right, as shown.

```
 5
288369
25
 3
```

2 Find the largest number whose square is equal to or less than the first block, in this case 28. Write that number (5) directly over the first block. Then subtract the square of that number from the first block ($28 - 25 = 3$).

```
        5 ?
      ________
       288369
       25
 10?|  383
```

3 Bring down the next block (83) and annex to the difference (3) to form a new dividend. Multiply the first digit of the root by 2 and place the product (10) to the left of the new dividend (383). The 10 becomes a *trial divisor.* Next ask yourself "What is the largest number (marked ?) that can be placed to the right of 10 and also above the second block of the given number such that the product of the two numbers will be equal to or less than the new dividend (383)?" The number is 3.

```
        5 3
      ________
       288369
       25
 103|  383
       309
        74
```

4 Place the product of these numbers ($3 \times 103 = 309$) under the new dividend. Subtract to get a remainder of 74.

```
        5 3 ?
      ________
       288369
       25
 103|  383
       309
106?|   7469
```

5 Bring down the third block (69) and annex it to the difference (74) to get 7469. Multiply the first two figures of the square root (53) by 2 and place the product to the left of the second new dividend. This number (106) becomes our new trial divisor.

```
        5 3 7
      ________
       288369
       25
 103|  383
       309
1067|   7469
        7469
```

6 Again ask, in step 5, what is the largest number that can be placed to the right of the new trial divisor and above the third block of the given number such that the product of the two numbers will be equal to or less than the new dividend (7469)? This number is 7. Place the product under the new dividend and subtract. In this case there is no remainder. Therefore, our result will be $\sqrt{288{,}369} = 537$.

Example 6-17 Find the square root of 85,849.

SOLUTION The solution to this problem is similar to the solution of Example 6-16. However, as we mark off blocks starting from the right, the last block will have only one digit. We shall proceed as we did in the previous example. The solution follows.

```
        2 9 3
        85849
        4
  49 | 458
        441
583 |   1749
        1749
```

Thus, $\sqrt{85{,}849} = 293$.

Example 6-18 Find the square root of 559.421.

SOLUTION The solution will be the same as in the previous two examples. It will be noted that the blocks are marked from the decimal point, going both to the left and to the right, and that adding zeros to the given number does not alter the size of the number.

```
            2 3 . 6 5 2
            559.421000
            4
     43 |   159
            129
   466 |      3042
              2796
  4725 |        24610
                23625
 47302 |           98500
                   95604
```

Thus, $\sqrt{559.421} = 23.65$ (accurate to four significant digits).

Example 6-19 Find the square root of 0.064839.

SOLUTION Starting at the decimal, we mark off blocks of two in both directions from the decimal point. We then proceed as in the previous examples.

```
          0. 2 5 4 6
          0.06483900
               4
    45 |       248
               225
   504 |         2339
                 2016
  5086 |           32300
                   30516
```

The answer will be closer to 0.255 than to 0.254. Thus, accurate to three significant digits, our answer will be $\sqrt{0.064839} = 0.255$.

Exercise 6-6

Extract the square roots of the following. If the answer is not exact, determine the answer accurate to three significant digits.

1. 841
2. 2,916
3. 21,904
4. 74,529
5. 184,041
6. 337,561
7. 516,961
8. 649,636
9. 9,486.76
10. 83.7225
11. 0.770884
12. 0.585225
13. 0.00378225
14. 0.094864
15. 0.00085849
16. 25.7083
17. 149,739
18. 407,059
19. 0.326083
20. 0.00233209

6-13 SLIDE-RULE SOLUTIONS FOR SQUARES AND SQUARE ROOTS A study of the A and B scales of a slide rule will reveal that they are double scales. Each scale is divided into two identical parts. The A and B scales are comparable to the C and D scales placed end to end and then compressed to the length of one C or D scale. Hence, the A and B scales each have three indexes: a left index, a center index, and a right index.

Multiplication and division can be performed with the A and B scales in the same manner as is done with the C and D scales. The accuracy of the solutions using the A and B scales, however, will not be so high as might be obtained with the C and D scales, since the A and B scales are each half the length of the C and D scales.

The A and B scales are used primarily to square numbers and to extract square roots. It will be noted that, if the rule is closed (left indexes of the A, B, C, and D scales are aligned), any number on the A (or B) scale is the square of the number on the D (or C) scale opposite it (see Fig. 6-12). This affords a ready method for finding the squares and square roots of numbers.

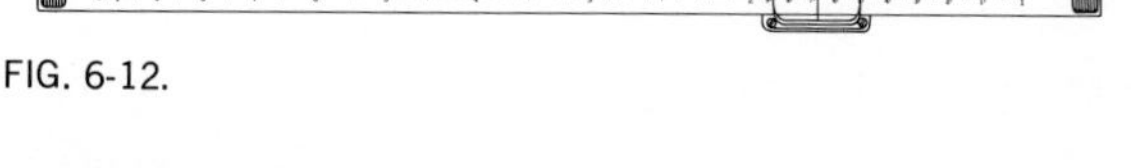

FIG. 6-12.

Example 6-20 Find the square of 2.38.

SOLUTION Move the hairline over the 238 of the D scale. The answer (= 5.66) is on the A scale under the hairline. The answer could also be obtained in similar manner using the C and B scales.

Example 6-21 Find the square root of 546.

SOLUTION In this problem we reverse the procedure of the previous example. However, there are two identical sets of markings on the A (and B) scale. We shall follow the rule: Mark off the number into blocks of two, starting from the decimal, and if there is only one significant digit in the block farthest to the left, use the left half of the A (or B) scale; if the farthest left block contains two significant digits, use the right half of the A (or B) scale. In this instance we shall move the hairline over 546 of the left half of the A (or B) scale and read the answer under the hairline on the D (or C) scale. The result will be $\sqrt{546} = 23.4$.

Example 6-22 Find the square root of 0.00817.

SOLUTION Mark off the digits in blocks of two from the decimal point. The first block containing at least one significant digit has two of them. Hence, use the right half of the A (or B) scale. The result will be $\sqrt{0.008170} = 0.0904$.

Example 6-23 Evaluate $\dfrac{0.274 \times 635 \times \sqrt{2{,}490} \times \sqrt{5.37}}{756 \times \sqrt{1{,}338}}$.

SOLUTION

Multiply 0.274 by 635:	Place hairline over 274 on the D scale and draw 635 on the CI scale under the hairline.
Multiply by $\sqrt{2{,}490}$:	Push the hairline to 2490 of right half of the B scale.
Divide by 756:	Draw 756 of C scale under the hairline.
Multiply by $\sqrt{5.37}$:	Move the hairline to 537 of the left half of the B scale.
Divide by $\sqrt{1{,}338}$:	Draw 1338 of the right half of the B scale under the hairline.
Read the answer:	Opposite the right index of the C scale read the significant digits in the answer. The answer is 0.728.

Exercise 6-7

Using a slide rule, find the squares of the numbers in Probs. 1 to 10.

1. 2.3
2. 5.4
3. 8.7
4. 0.609
5. 42.6
6. 30.8
7. 278
8. 0.542
9. 0.0853
10. 0.0725

Using the slide rule, find the square roots of the numbers in Probs. 11 to 20.

11. 56
12. 7.8
13. 258
14. 600
15. 827
16. 3,840
17. 45,900
18. 0.0537
19. 0.1157
20. 0.00495

Evaluate the following:

21. $24\sqrt{5.27}$
22. $3.58\sqrt{20.6}$
23. $\sqrt{38.5} \div 2.54$
24. $\sqrt{377} \div 8.09$
25. $0.834 \div \sqrt{0.0527}$
26. $83.1 \div \sqrt{5,840}$
27. $\dfrac{\sqrt{60.8}}{4.37 \times \sqrt{5.20}}$
28. $\dfrac{53.4}{8.72 \times \sqrt{12.36}}$
29. $28.9 \times 1.053\sqrt{70,800}$
30. $\dfrac{7.39\sqrt{6.45}}{\sqrt{0.846}}$

6-14 SLIDE-RULE SOLUTIONS FOR CUBES AND CUBE ROOTS The K scale is used to cube a number and to extract cube roots. Examination of the K scale (see Fig. 6-12) will reveal that it is a triple scale. It is divided into three identical parts and can be compared with three C (or D) scales placed end to end and then compressed to the length of one C (or D) scale. The K scale has four indexes which partition it into three identical sections.

If the left indexes of the C, D, and K scales are aligned, any number on the K scale is the cube of the number on the D (or C) scale opposite it. Thus, to raise any number to the third power, the hairline is set over that number on the D scale (or on the C scale if the K scale is on the slide) and the cube is read directly under the hairline on the K scale.

To extract the cube root of a number, the hairline is set over the number on the proper section of the K scale. The cube root is read under the hairline on the D scale (or C scale if the K scale is on the slide).

The rule for selecting the proper section of the K scale is similar to that used in computing square roots: Mark off the digits in the number in blocks of three from the decimal point—in both directions. If the first block at the left of the number contains one significant digit, use the left third of the scale. If the first block contains two significant digits, use the middle third of the scale. If the first block contains three significant digits, use the right third of the K scale.

The student should verify the following cube roots on his slide rule.

$$\sqrt[3]{\overset{\frown}{3},\overset{\frown}{860}} = 15.7 \qquad \sqrt[3]{0.\overset{\frown}{003}\overset{\frown}{860}} = 0.157$$

$$\sqrt[3]{\overset{\frown}{38},\overset{\frown}{600}} = 33.8 \qquad \sqrt[3]{0.\overset{\frown}{038}\overset{\frown}{600}} = 0.338$$

$$\sqrt[3]{\overset{\frown}{386},\overset{\frown}{000}} = 72.8 \qquad \sqrt[3]{0.\overset{\frown}{386}} = 0.728$$

Exercise 6-8

Find the cubes of the numbers in Probs. 1 to 10.

1. 2.8
2. 5.3
3. 0.62
4. 1.44
5. 0.384
6. 0.857
7. 83.2
8. 74.5
9. 0.0884
10. 0.0736

Find the cube roots of the numbers in Probs. 11 to 20.

11. 7.59
12. 28.3
13. 573
14. 84.9
15. 0.680
16. 0.347
17. 0.0359
18. 0.00859
19. 87,200
20. 184,600

Evaluate:

21. $4.2\sqrt[3]{9.27}$
22. $25.7\sqrt[3]{30.5}$
23. $\sqrt[3]{58.6} \div 2.28$
24. $\sqrt[3]{0.256} \div 0.0347$
25. $\sqrt[3]{0.0459} \div 0.808$
26. $\sqrt[3]{0.00852} \div 0.1593$
27. $3.26 \times 5.49\sqrt[3]{0.0803}$
28. $\dfrac{\sqrt[3]{2{,}850}}{\sqrt{57}}$
29. $\dfrac{\sqrt{487}}{\sqrt[3]{52{,}800}}$
30. $\dfrac{5.02\sqrt[3]{0.0836}}{\sqrt{0.651}}$

CHAPTER 7

MEASUREMENT AND DIMENSIONAL RELATIONS

7-1 MEASUREMENT When man started to measure, industrial and technological progress began. With the precision measuring instruments of today it is possible to manufacture parts and subassemblies for complex machines in a locality where raw materials, power, and skilled labor are readily available and assemble the finished product thousands of miles away, months or years later if need be, with full assurance of perfect fit.

Before a quantity can be measured, a *unit of measurement* must be selected. The magnitude of the unit of measurement is quite arbitrary. Consequently, there exist various units of measurements that have been developed throughout the history of civilization.

7-2 FUNDAMENTAL UNITS OF MEASUREMENTS All measurements are relative in the sense that they are comparisons with some standard unit of measurement. Actually there are only five *fundamental units* of measurement. They are units of *length, mass, time, temperature,* and *electric current.* Other measures are called derived *units,* since they can each be expressed in terms of the fundamental units. For example, the unit of *area* (e.g., the square yard) is a product of a length times a length; the unit of *velocity* (e.g., miles per hour) is obtained by dividing a length unit by a time unit; the unit of *density* (e.g., pounds per cubic inch) is derived by dividing a mass unit by the cube of a length unit; and so on. In this chapter we will use only units and derived units of length, mass, and time.

7-3 THE ENGLISH SYSTEM OF MEASUREMENTS The standard unit of length used in English-speaking countries is the *yard.* In Great Britain the yard is legally defined as the distance between two ruled lines scratched on a bronze metal bar while it is at a temperature of 62°F. The metal bar is kept at the British Exchequer. In the United States the standard yard is defined as $\frac{3,600}{3,937}$ meter. In the United States the *foot* is a more commonly used unit. Other units of length are the *inch* and the *mile.*

Mass is defined as the quantity of matter contained in a body and is distinguished from *weight,* which is measured by the pull of gravity on the body. Numerically, mass and weight are equal at sea level. Both mass and weight measurements in the English system are based upon the *pound* as the basic unit. The *ounce,* the *grain,* and the *ton* are also units of mass and weight in the United States and the British Commonwealth of Nations.

The unit of time in the English system is the *second.* It is defined as $\frac{1}{86,400}\left(=\frac{1}{24\text{ hr/day}\times 60\text{ min/hr}\times 60\text{ sec/min}}\right)$ of the mean solar day.

7-4 THE METRIC SYSTEM OF MEASUREMENTS All industrial countries except the United States and Britain use the *metric system* of measurements.

The standard units in the metric system are the *meter,* the *gram,* and the *liter.*

Until 1960 the meter was defined as the distance between the centers of two lines ruled on a platinum-iridium bar at a temperature of 0°C stored at the International Bureau of Weights and Measures at Sèvres, France. By a

TABLE 7-1 UNITS OF MEASUREMENTS

ENGLISH SYSTEM

UNITS OF LENGTH

1 yard = the standard
12 inches (in.) = 1 foot (ft)
3 feet = 1 yard (yd)
$16\frac{1}{2}$ feet = 1 rod
5,280 feet = 1 mile (mi)

UNITS OF MASS OR WEIGHT

1 pound = the standard
16 ounces (oz) = 1 pound (lb)
100 pounds = 1 hundredweight
2,000 pounds = 1 short ton
2,240 pounds = 1 long ton

UNITS OF AREA

144 square inches ($in.^2$) = 1 square foot (ft^2)
9 square feet = 1 square yard (yd^2)
160 square rods = 1 acre
640 acres = 1 square mile (mi^2)

UNITS OF VOLUME

1,728 cubic inches ($in.^3$) = 1 cubic foot (ft^3)
27 cubic feet = 1 cubic yard (yd^3)

DRY MEASURE

2 pints = 1 quart (qt)
8 quarts = 1 peck (pk)
4 pecks = 1 bushel (bu)

LIQUID MEASURE

2 gills = 1 cup
2 cups = 1 pint
2 pints = 1 quart (qt)
4 quarts = 1 gallon (gal)
$31\frac{1}{2}$ gallons = 1 barrel (bbl)
231 cubic inches = 1 gallon

METRIC SYSTEM

UNITS OF LENGTH

1 meter (m) = the standard
1 decimeter = 0.1 meter
1 centimeter (cm) = 0.01 meter
1 millimeter (mm) = 0.001 meter
1 micron = 0.000001 meter
1 kilometer (km) = 1,000 meters

UNITS OF MASS OR WEIGHT

1 kilogram (kg) = the standard
1 gram (g) = 0.001 kilogram
1 centigram = .01 gram
1 milligram (mg) = 0.001 gram

LIQUID MEASURE

1 liter (l) = 1,000 cm^3
1 milliliter (ml) = 0.001 liter
1,000 liters = 1 kiloliter

1960 international agreement the standard meter is now defined as 1,650,763.73 wavelengths of the orange-red line of krypton 86. This definition probably will not mean much to the beginning technology student, but he should recognize the degree of accuracy inherent in the definition.

The metric system is a decimal system. Because of the ease of working with a system based upon the number 10, scientists throughout the world use the metric system of measurement. Even in the United States the metric system is the recognized measuring system of science.

The standard unit of mass in the metric system is the *kilogram.* It originally was defined as the mass of 1,000 cubic centimeters of water at 4°C. The kilogram is now defined as the mass of a platinum-iridium cylinder that is preserved at the International Bureau of Weights and Measures. The volume of 1,000 cubic centimeters is called the *liter.* Subdivisions of the kilogram are the *gram,* the *centigram,* and the *milligram.*

The unit of time in the metric system, as in the English system, is the *second.* For most practical purposes, the second can be defined as $\left(\frac{1}{24\text{ hr/day}} \times \frac{1}{60\text{ min/hr}} \times \frac{1}{60\text{ sec/min}}\right)$ of the mean solar day.

7-5 TABLE OF WEIGHTS AND MEASURES Table 7-1 lists some of the more common units of weights and measures and gives the accepted abbreviation for each.

7-6 CONVERSION TABLES Numbers which express units of measurement are called *denominate numbers.* In using denominate numbers, it is sometimes necessary to change from one unit of measurement to an equivalent unit. Such change of units within a given system of measurement is called a *reduction.* If the change of units is from one system of measurement to another, it is called a *conversion.* Table 7-2 gives some approximate relations between the two systems.

7-7 DIMENSIONAL ANALYSIS In representing a physical quantity mathematically it is necessary to use a combination of a number to describe the magnitude and a name to indicate the unit in which the quantity is measured. Thus 3 ft, 3 hr, and 3 lb describe the same number of different entities—units of length, time, and weight (or mass), respectively. In like manner, 3 in., 3 ft, and 3 mi denote the same number of different units of the same entity—that of length.

Frequently, it becomes necessary to change a given quantity from one unit of measurement to a different unit. For example, it is impossible to add 2 ft, 8 in., and 3 yd until each quantity is expressed in terms of a common unit. Thus,

TABLE 7-2 ENGLISH AND METRIC EQUIVALENTS

MEASURES OF LENGTH	MEASURES OF AREA
1 inch = 2.54 centimeters	1 square inch = 6.452 square centimeters
1 foot = 30.48 centimeters	
1 meter = 39.37 inches	1 square meter = 10.76 square feet
1 mile = 1.609 kilometers	
1 kilometer = 0.621 mile	
MEASURES OF VOLUME	**MEASURES OF WEIGHT**
1 cubic inch = 16.39 cubic centimeters	1 pound = 453.6 grams
1 cubic foot = 28.32 liters	1 ounce = 28.35 grams
1 pint (liquid) = 473.2 cubic centimeters	1 kilogram = 2.205 pounds
1 liter = 1.056 quarts	

$$2\text{ ft} + 8\text{ in.} + 3\text{ yd} = 2\text{ ft} + \tfrac{2}{3}\text{ft} + 9\text{ ft} = 11\tfrac{2}{3}\text{ft}$$

The changing of a unit to an equivalent unit involves a quite simple operation. For example, if it is desired to convert 750 m to miles and it is known that 1 m is equivalent to 100 cm, 2.54 cm is equivalent to 1 in., 12 in. is equivalent to 1 ft, and 5,280 ft is equivalent to 1 mi, we could use the following procedure:

$$\begin{aligned} 750\text{ m} &= 750\,\cancel{\text{m}}\left(\frac{100\,\cancel{\text{cm}}}{1\,\cancel{\text{m}}}\right)\left(\frac{1\,\cancel{\text{in.}}}{2.54\,\cancel{\text{cm}}}\right)\left(\frac{1\,\cancel{\text{ft}}}{12\,\cancel{\text{in.}}}\right)\left(\frac{1\text{ mi}}{5{,}280\,\cancel{\text{ft}}}\right) \\ &= 0.466\text{ mi} \end{aligned}$$

It will be noted that the four fractions in parentheses each equal unity, since the measured values of the numerators are equivalent to the measured value of the corresponding denominators. Essentially, then, we have multiplied 750 m by (1) (1) (1) (1). Each of the units on the right side of the equality sign cancels out except the miles unit.

Example 7-1 Reduce 60 mph to feet per second.

SOLUTION

$$\begin{aligned} 60\text{ mph} &= 60\,\frac{\text{mi}}{\text{hr}} \times \frac{5{,}280\text{ ft}}{1\text{ mi}} \times \frac{1\text{ hr}}{60\text{ min}} \times \frac{1\text{ min}}{60\text{ sec}} \\ &= 88\,\frac{\text{ft}}{\text{sec}} \end{aligned}$$

Example 7-2 Convert 0.0875 ton/ft^3 to kilograms per cubic meter.

SOLUTION Using Table 7-2 we solve as follows:

$$\begin{aligned} 0.0875\,\frac{\text{ton}}{\text{ft}^3} &= 0.0875\,\frac{\text{ton}}{\text{ft}^3}\left(\frac{2{,}000\text{ lb}}{1\text{ ton}}\right)\left(\frac{453.6\text{ g}}{1\text{ lb}}\right)\left(\frac{1\text{ kg}}{1{,}000\text{ g}}\right)\left(\frac{1\text{ ft}}{30.48\text{ cm}} \times \frac{100\text{ cm}}{1\text{ m}}\right)^3 \\ &= 2.80 \times 10^3\,\frac{\text{kg}}{\text{m}^3} \end{aligned}$$

Exercise 7-1

In Probs. 1 to 32, reduce:

1. 28.3 cm to meters
2. 84,072 cm to kilometers
3. 0.00593 cm to millimeters
4. 90,536 mg to kilograms
5. 873.2 cg to grams
6. 45.09 kg to grams
7. 6,115 ml to liters
8. 0.00531 kl to liters
9. 87.5 yd to inches
10. 3,584 yd to miles
11. 2.87 mi to feet
12. 28.7 oz to pounds
13. 2,879 lb to short tons
14. 8.75 short tons to long tons
15. 29.3 pk to bushels
16. 3.46 gal to pints
17. 8 cups to cubic inches
18. 83.9 cm^2 to square meters
19. 5.93 cm^2 to square decimeters
20. 5.41 m^3 to cubic centimeters
21. 83,500 mm^3 to cubic meters
22. 8.45 yd^2 to square feet
23. $\frac{3}{8}$ ft^2 to square inches
24. 8 mi^2 to acres
25. 5 cubic rods to cubic yards
26. 80 cm/sec to meters per minute
27. 4 microns/sec to meters per hour
28. 25 kg/m^2 to grams per square centimeter
29. 875 g/cm^3 to kilograms per cubic meter
30. 20 short tons/hr to pounds per second
31. 70 lb/ft^3 to ounces per cubic inch
32. 14.7 $lb/in.^2$ to short tons per square yard

In Probs. 33 to 45 perform the conversions, using a slide rule when necessary. Leave answers accurate to three significant figures.

33. 1,250 in. to meters
34. 875 gal to liters
35. 2.47 kg to pounds
36. 3.59 km to miles
37. 60.5 ft/sec to centimeters per minute
38. 19.3 ft^3/hr to cubic centimeters per second

39. 0.00522 kg/cm^2 to pounds per square inch
40. 0.0808 km/min to inches per second
41. 1.83 short $tons/yd^3$ to grams per cubic centimeter
42. 3.54 $oz/in.^2$ to grams per square centimeter
43. 25.7 $lb/in.^2$ to milligrams per square millimeter
44. 0.352 $in.^3/oz$ to cubic centimeters per gram
45. 525 mph to kilometers per minute
46. If 8 bars of iron 6 ft long, 3 in. broad, and 1.5 in. thick weigh 735 lb, what will be the weight of 5 bars of the same material if each is 9 ft long, 4 in. broad, and 2.4 in. thick?
47. A skilled craftsman can roll a pound of aluminum into foil so thin that it will make a sheet 6 ft wide and 90 ft long. How thick will this sheet be if aluminum has a density of 168 lb/ft^3? Leave answer in inches.
48. If 10 machines can produce 4,800 castings in 20 days working 8 hr a day, how many days will be required for 15 machines working 6 hr a day to produce 4,320 castings?

SECTION 2

BASIC ALGEBRA FOR TECHNICIANS

CHAPTER 8

BASIC OPERATIONS ON ALGEBRAIC EXPRESSIONS

8-1 THE MEANING OF ALGEBRA In the first seven chapters we have studied the basic concepts and operations of arithmetic. We are now ready to study the language and operations of algebra. Algebra provides a powerful means for expressing mathematical relationships between physical quantities.

Basically, algebra is a generalization of arithmetic. In arithmetic numerical quantities are combined to arrive at a specific numerical answer. For example, if the length of a rectangle is 8 ft and the width is 3 ft, the area of the rectangle is determined by finding the product of 8 ft and 3 ft (to get 24 ft^2). The above example can be generalized by letting l represent the number (any number) of length units, w represent the number (any number) of width units, and A the resultant units of area, and writing

$$A = l \cdot w \qquad \text{or} \qquad A = lw$$

In algebra, when two or more letters representing numbers are written next to each other (with no operation symbol between them), it is understood to mean the product of the numbers representing them.

Algebra essentially is a language in which statements involving numbers can be written in a simple, general way. The algebraic statement $A = lw$ is called a *formula.* It is important to note that in the formula $A = lw$, the letters l and w must represent consistent units of measurement. That is, if l represents lengths of rectangles in feet units, the derived values for A will have meaning only if w also represents width units in feet.

Thus, algebra replaces the numerical units of arithmetic with letters which represent *any* arithmetic numbers. While the 24 ft^2 is a valid answer for the specific dimensions of 8 ft and 3 ft, the algebraic expression $A = lw$ is valid for all lengths and widths.

8-2 ALGEBRAIC SYMBOLS Any letter used to represent a number is called a *literal* number. The four fundamental processes, indicated by the symbols $+, -, \times, \div$, have the same meaning in algebra as they have in arithmetic. If a and b represent two numbers, $a + b$ represents their sum; $a - b$, their difference; $a \times b$, their product; $a \div b$, their quotient; $\sqrt{a}$ represents the square root of a; etc.

Since x is commonly used as a symbol for a number, it is not often used in algebra as a symbol for multiplication. To indicate the product of numbers represented by a and b, we write $(a)(b)$, $a \cdot b$, or simply ab.

Literal numbers often cannot be added, subtracted, multiplied, and divided the same way as arithmetic numbers. In arithmetic, if we want to find the sum of, for example, 6 and 2, we can express their sum by one symbol, namely, 8. If we want to add two numbers represented by x and y, we can only indicate the operation, i.e., $x + y$. In like manner, the difference, product, and quotient of the two numbers can be expressed by single numbers,

4, 12, and 3, respectively. The difference between x and y is expressed $x - y$. The product is xy. The quotient is x/y or $x \div y$.

In arithmetic, $8\frac{1}{3}$ means the sum of 8 and $\frac{1}{3}$, while in algebra, xy is used to indicate the product of two numbers represented by x and y. 54 means $5(10) + 4$. This differs from the algebraic meaning of, say, ab.

The raised dot should be avoided when decimals are present. Parentheses as a symbol for multiplication should be used if needed for clarity's sake. For example, we should write 3(0.05) rather than $3 \cdot 0.05$.

8-3 ALGEBRAIC TERMINOLOGY The expression $3xy$ is used to indicate the product of 3 and numbers represented by x and y. The 3 is called the *numerical factor* and x and y are called the *literal factors* of the product. The numerical factor is called the *numerical coefficient* of the remaining literal factors. Thus, 3 is the numerical coefficient of xy; xy is the literal coefficient of 3; y is the coefficient of $3x$; etc. We usually do not write 1 as a coefficient. When the letter x is written to denote a number, it is understood to mean $1x$.

Any combination of arithmetic numbers, literal numbers, and symbols of operation, when written according to the rules of algebra, is called an *algebraic expression.* Thus $2a$, $5yz$, $3x + 4y$, $3h$ are algebraic expressions.

A *term* or *monomial* is an algebraic expression whose parts are not separated by a + or a − sign. An algebraic expression such as $3x + 4y$ is the sum of two terms. The difference of the terms would be expressed as $3x - 4y$ or $4y - 3x$.

An algebraic expression of two terms is called a *binomial.* An algebraic expression of three terms is called a *trinomial.* An algebraic expression of two or more terms is called a *polynomial.* Thus a trinomial is a polynomial of three terms.

Two terms which have the same literal factors are called *similar terms.* Thus, $2x$ and $5x$ are similar terms. Two terms that do not have the same literal factors are called *dissimilar terms.* Thus $5a$ and $7y$ are dissimilar terms.

A *power* of a quantity is the product obtained when that quantity is multiplied by itself a number of times. A short cut indicating this product is achieved by the use of exponents. An *exponent* is a number written to the right and a little above the quantity to indicate how many times the quantity is taken as a factor. Thus

(5)(5) is written 5^2 and is the second power of 5.

$a \cdot a$ is written a^2 and is the second power of a.

$a \cdot a \cdot a$ is written a^3 and is the third power of a.

$a \cdot a \cdot a \cdot a$ is written a^4 and is the fourth power of a.

The number to which an exponent is affixed is called the *base.* In the previous examples 5 and a were bases.

a^2 is generally read "a squared."

a^3 is generally read "a cubed."

a^4 is generally read "a to the fourth power."

$6a^2b^3$ is read "six a squared b cubed."

$2a^2 + 3ab^2 - 4b^3$ is read "two a squared plus three a b squared minus four b cubed."

If the power of a quantity is not indicated, it is understood to be 1. Thus, we usually write x rather than x^1 and $5ab^2$ rather than $5a^1b^2$.

8-4 EVALUATION OF ALGEBRAIC EXPRESSIONS A very important process in algebra is finding the numerical value of algebraic expressions when the numerical values of the literal factors are known. Whenever we evaluate an algebraic expression, we substitute numbers for letters. Hence, the process is often called *substitution.*

Example 8-1 If $a = 1$, $b = 2$, $c = 3$, find the value of $2a - 3b + 5c$.

SOLUTION

$$2a - 3b + 5c =$$
$$2(1) - 3(2) + 5(3) =$$
$$2 - 6 + 15 = 11$$

Example 8-2 If $x = 2$, $y = -1$, $z = 3$, find the value of $3xy - 5yz - 7xz$.

SOLUTION

$$3xy - 5yz - 7xz =$$
$$3(2)(-1) - 5(-1)(3) - 7(2)(3) =$$
$$-6 + 15 - 42 = -33$$

Example 8-3 If $u = 2$, $v = 3$, $w = -1$, find the value of $3u^2 - 2uv + 4u^2w^3$.

SOLUTION

$$3u^2 - 2uv + 4u^2w^3 =$$
$$3(2)^2 - 2(2)(3) + 4(2)^2(-1)^3 =$$
$$12 - 12 - 16 = -16$$

Exercise 8-1

1. Given the algebraic expression $6x^2y + 5xyz + 9ab - 3x^2y$.
   (a) What is the third term?
   (b) What is the numerical coefficient of the first term?
   (c) What is the coefficient of x in the second term?
   (d) Which terms are similar?

Express the following as algebraic expressions:

2. The sum of 3, s, and t

3. Subtract x from y
4. The difference between a and b
5. 8 more than d
6. The product of 5 and y
7. 6 less than c
8. h increased by 1
9. y decreased by z
10. Four more than three times x
11. The sum of r and s divided by their product
12. The quotient of $15a$ divided by $13b$
13. The product of u and w divided by their sum
14. Six times the sum of x and y
15. The sum of r and s decreased by the sum of h and k
16. Eight times the difference between x and y
17. The product of $2x$ and $3y$

Evaluate the following if $x = 1, y = -1, z = 2, w = -2, k = 3$:

18. $x + y + z$
19. $5xy$
20. $\dfrac{x + y}{z}$
21. $3z - 2$
22. x^2y
23. $5y^2z$
24. $3w^2 - 2yz$
25. $\dfrac{2w + 5k}{7}$
26. $\dfrac{2y - 5w}{4}$
27. $xy + yz + zw$
28. $3(y + z)$
29. $(x + z)^2$
30. $x(y + z)$
31. $5x^2y^2z^2$
32. $(2k - y)^2$
33. $(x + y)(z + k)$
34. $(x + z)(w + k)$
35. $x^2 - y^2$
36. $z^2 + k^2$
37. $5xk - 2yw$
38. $\dfrac{1}{x} - \dfrac{2}{y}$
39. $\dfrac{1}{x} + \dfrac{2}{y}$
40. $7k - (2x + 3z)$
41. $5k + 4(3y - 2z)$
42. The sum of two numbers is 18. If n represents the lesser number, what expression represents (a) the greater number? (b) three times the lesser number?
43. The sum of two numbers is 15. If n represents the greater number, what expression represents (a) the lesser number? (b) four times the lesser number?
44. The difference of two numbers is 20. If n represents the lesser number, what expression represents (a) the greater number? (b) twice the lesser number?

45. The difference of two numbers is 12. If n represents the greater number, what expression represents (a) the lesser number? (b) five times the lesser number?
46. The product of two numbers is 36. If n represents the lesser number, what expression represents (a) the greater number? (b) the difference between twice the lesser number and three times the greater number?
47. If d represents the number of dimes in a box, what expression represents the value of the coins?
48. If l represents the length of a rectangle in yards, what expression represents the length in feet?
49. If x represents the number of gallons of an alcohol-water mixture in a can and 40 percent of the solution is alcohol, what expression represents (a) the number of gallons of pure alcohol in the mixture? (b) the number of gallons of water?
50. How much pure acid is there in $(12 - 5x)/3$ pt of 75 percent acid solution?

8-5 ADDITION OF SIMILAR TERMS In Chap. 1 we added numbers by adding ones to ones, tens to tens, hundreds to hundreds, etc., and by adding tenths to tenths, hundredths to hundredths, etc. In adding fractions we added only like fractions. These were applications of the *first law of addition:* Only like quantities can be added. We now apply this law to obtain the rule:

PROCEDURE *To add similar terms, add their numerical coefficients and multiply the sum by the common literal factor.*

Example 8-4 Add $5xy + 3xy + 4xy$.

SOLUTION

$$5xy + 3xy + 4xy = 12xy$$

Example 8-5 Add $2h^2k + 7hk - 3hk^2$, $5h^2k + 6hk - 9hk^2$, $4h^2k - 5hk + 6hk^2$.

SOLUTION

$$\begin{array}{rr} & 2h^2k + 7hk - 3hk^2 \\ & 5h^2k + 6hk - 9hk^2 \\ & 4h^2k - 5hk + 6hk^2 \\ \hline \text{Sum:} & 11h^2k + 8hk - 6hk^2 \end{array}$$

Often in adding terms we can take advantage of two more properties of addition. The *law of order:* addition is commutative; that is, $a + b = b + a$. The *law of grouping:* addition is associative; that is, $a + b + c = (a + b) + c = a + (b + c) = (a + c) + b$.

Example 8-6 Simplify $3x + 8x + 17x + 32x$.

SOLUTION

$$3x + 8x + 17x + 32x =$$
$$3x + 17x + 8x + 32x =$$
$$20x + 40x = 60x$$

Example 8-7 Simplify $5x^3 + 2x^2y - 3xy^2 + 4y^3 + 7xy^2 - 8x^2y + 2x^3 - 9y^3$.

SOLUTION

$$5x^3 + 2x^2y - 3xy^2 + 4y^3 + 7xy^2 - 8x^2y + 2x^3 - 9y^3 =$$
$$(5x^3 + 2x^3) + (2x^2y - 8x^2y) + (-3xy^2 + 7xy^2) + (4y^3 - 9y^3) =$$
$$7x^3 - 6x^2y + 4xy^2 - 5y^3$$

8-6 SUBTRACTION OF SIMILAR TERMS Algebraic subtraction follows logically from arithmetic subtraction. You will recall that subtraction was defined as the inverse of addition. For example, we know that $8x - 5x = 3x$ because $3x + 5x = 8x$. This leads directly to the following rule.

PROCEDURE *To subtract two similar terms, find the difference of the coefficients and multiply it by the common literal factor.*

Example 8-8 Subtract $12a^2b - 5ab + 8ab^2$ from $6a^2b + 3ab - 4ab^2$.

SOLUTION

$$\begin{array}{rr} & 6a^2b + 3ab - 4ab^2 \\ & 12a^2b - 5ab + 8ab^2 \\ \hline \text{Difference:} & -6a^2b + 8ab - 12ab^2 \end{array}$$

Exercise 8-2

Add:

1. $\begin{array}{r} 4a \\ \underline{3a} \end{array}$

2. $\begin{array}{r} -5k \\ \underline{+2k} \end{array}$

3. $\begin{array}{r} 6xy \\ \underline{xy} \end{array}$

4. $\begin{array}{r} -3ab \\ \underline{-5ab} \end{array}$

5. $\begin{array}{r} -11b^2c \\ \underline{+\ 3b^2c} \end{array}$

6. $\begin{array}{r} 8cd \\ \underline{-3cd} \end{array}$

7. $\begin{array}{r} -4xy^2 \\ \underline{+3xy^2} \end{array}$

8. $\begin{array}{r} bd \\ \underline{-7bd} \end{array}$

9. $\begin{array}{r} -5m^2n \\ \underline{-2m^2n} \end{array}$

10. $\begin{array}{r} 6r^2t^2 \\ \underline{4r^2t^2} \end{array}$

11. $\begin{array}{r} -3x^2y^3 \\ -2x^2y^3 \\ x^2y^3 \\ \underline{8x^2y^3} \end{array}$

12. $\begin{array}{r} -b^3c^2 \\ 2b^3c^2 \\ b^3c^2 \\ \underline{4b^3c^2} \end{array}$

13. $\begin{array}{r} uv^3 \\ 3uv^3 \\ -8uv^3 \\ \underline{2uv^3} \end{array}$

14. $\begin{array}{r} 9.1xy \\ -2.6xy \\ -5.3xy \\ \underline{-4.1xy} \end{array}$

15. $\begin{array}{r} 2x^2 - 3xy + 7y^2 \\ x^2 + xy - 4y^2 \\ \underline{5x^2 - 2xy - 5y^2} \end{array}$

16. $\begin{array}{r} 11a^3 - 5a^2b + 3ab^2 \qquad \\ -5a^3 + 7a^2b \qquad\qquad - 9b^3 \\ \underline{- 5a^2b - 4ab^2 + 7b^3} \end{array}$

17. Subtract $3x^2 - 2xy + 4y^2$ from $5x^2 - 3xy - 6y^2$.
18. Subtract $11r^2s^2 + 7rs - 5$ from $3r^2s^2 - 4rs + 8$.
19. From $8a^3 + b^3$ take away $2a^3 + 3a^2b - ab^2$.
20. From $5a^3 - 6ab + 2b^3$ take away $8a^3 - 4ab - b^3$.

Complete the following:

21. $7x + (?) = 11x$
22. $3a + (?) = -8a$
23. $24w = (?) + 36w$
24. $25k = 13k + (?)$
25. $13xy - (?) = 2xy$
26. $42rs - (?) = -25rs$
27. $18a^2b = (?) - 27a^2b$
28. $21lm = 25lm - (?)$

Combine the following:

29. $5x - 3x + 7x + 4x$
30. $11ab + 4ab - 10ab - 8ab$
31. $x - x^3 + 5x + 6x^3$
32. $2a - 3 + 11a + 5 - 4a$
33. $d^4 + d^2 + 2d^3 + d^2$
34. $5x^2y - 3xy + 5x + 6xy - x^2y$
35. $a^2b - b^3 - ab^2 + 2a^3 + 5ab^2 + 11a^2b$
36. $3a + b + 2a - 5c - b - 2c + 8a$
37. $x + 2x^2 - 5 + x^3 - 2x - 2x^2 + 2$
38. $(2x^2y - xy^2 + 7xy) + (xy^2 - 5x^2y + 8xy)$
39. $(8a^3 - 2a^2b + 4ab^2 - 6b^3) - (4a^3 - 3a^2b - 2ab^2 - b^3)$
40. $(5x^2 - y^2) - (6x^2 - 3y^2) - (8x^2 + 2y^2)$

8-7 LAWS OF EXPONENTS Algebraic multiplication is subject to the same laws as arithmetic multiplication (see Sec. 1-10), namely:

1 Multiplication is commutative; that is $ab = ba$.
2 Multiplication is associative; that is, $abc = (ab)c = a(bc)$.
3 Multiplication is distributive with respect to addition; that is $a(b + c) = ab + ac$.

We shall now develop some laws for exponents for the product and quotient of two powers. Consider the following:

1 Product of powers

$$\begin{aligned} a^6 \cdot a^2 &= (a \cdot a \cdot a \cdot a \cdot a \cdot a)(a \cdot a) && \text{by definition of exponent} \\ &= a \cdot a \cdot a \cdot a \cdot a \cdot a \cdot a \cdot a && \text{by the law of associativity} \\ &= a^8 && \text{by definition of exponent} \end{aligned}$$

PROCEDURE *The exponent of the product of powers with the same base is the sum of exponents of the factors. In general,*

$$a^m \cdot a^n = a^{m+n} \tag{8-1}$$

2 Quotient of powers

$$\begin{aligned} a^6 \div a^2 &= \frac{a^6}{a^2} \\ &= \frac{a \cdot a \cdot a \cdot a \cdot a \cdot a}{a \cdot a} && \text{by definition of exponent} \\ &= a \cdot a \cdot a \cdot a && \text{by fundamental principle of fraction} \\ &= a^4 && \text{by definition of exponent} \end{aligned}$$

PROCEDURE *The exponent of the quotient of two powers with the same base is the difference of the exponents of the factors. In general,*

$$a^m \div a^n = a^{m-n} \qquad \text{if } m > n, a \neq 0 \tag{8-2}$$

3 Power of a power

$$\begin{aligned} (a^3)^2 &= a^3 \cdot a^3 && \text{by definition of exponent} \\ &= a^6 && \text{by Rule 1 above} \end{aligned}$$

PROCEDURE *The exponent of a power of a power is the product of the exponents of the powers. In general,*

$$(a^m)^n = a^{mn} \tag{8-3}$$

4 Power of a product

$(ab)^2 = (ab)(ab)$ by definition of exponent
$= a \cdot b \cdot a \cdot b$ by meaning of ab
$= a \cdot a \cdot b \cdot b$ by commuting the two middle factors
$= a^2b^2$ by definition of exponent

PROCEDURE *To raise a product of two factors to a given power multiply the exponents of the factors by the exponent of the power to which the product is raised. In general,*

$$(a^mb^n)^k = a^{km}b^{kn} \tag{8-4}$$

5 Power of a quotient

$\left(\frac{a^2}{b^3}\right)^2 = \frac{a^2}{b^3} \cdot \frac{a^2}{b^3}$ by definition of exponent

$= \frac{a^2 \cdot a^2}{b^3 \cdot b^3}$ by rule for finding product of fractions

$= \frac{a^4}{b^6}$ by Rule 1 above

PROCEDURE *To raise a quotient of two quantities to a given power, multiply each of the exponents of the two quantities by the exponent of the power and divide. In general,*

$$\left(\frac{a^m}{b^n}\right)^k = \frac{a^{km}}{b^{kn}} \tag{8-5}$$

8-8 MULTIPLICATION OF MONOMIALS Consider the following problem in multiplication: $(-15x^2yz^3)(3xy^3z^4)$. To find the product, let us apply the laws of multiplication and the laws of exponents.

$$\begin{aligned}(-15x^2yz^3)(3xy^3z^4) &= -15(x^2)(y)(z^3)(3)(x)(y^3)(z^4)\\ &= -15(3)(x^2)(x)(y)(y^3)(z^3)(z^4)\\ &= -45x^3y^4z^7\end{aligned}$$

The solution to this problem illustrates the following rule.

PROCEDURE *To find the product of two or more monomials:* (1) *Determine the sign of the product by applying the rule for signs in multiplication;* (2) *multiply the numerical coefficients of the factors to get the numerical coefficient of the product;* (3) *multiply this coefficient (with proper sign) by the product of the literal factors.*

8-9 DIVISION OF MONOMIALS Consider the following problem in division: $(-15x^2yz^3) \div (3xy^3z^4)$. Here we apply the fundamental principle of fractions and the laws of exponents.

$$
\begin{aligned}
(-15x^2yz^3) \div (3xy^3z^4) &= \frac{-15x^2yz^3}{3xy^3z^4} \\
&= \frac{-5x}{y^2z}\left(\frac{3}{3} \cdot \frac{x}{x} \cdot \frac{y}{y} \cdot \frac{z^3}{z^3}\right) \\
&= \frac{-5x}{y^2z}(1 \cdot 1 \cdot 1 \cdot 1) \\
&= \frac{-5x}{y^2z}(1) \\
&= -\frac{5x}{y^2z}
\end{aligned}
$$

PROCEDURE *To find the quotient of two monomials: (1) Determine the sign of the quotient by applying the rules for signs in division; (2) divide the numerical coefficients to find the numerical coefficient of the quotient; (3) apply the fundamental principle of fractions to reduce the quotient of the literal factors to one involving the smallest positive exponents for each factor; (4) multiply this quotient by the numerical coefficient, with proper sign.*

Exercise 8-3

Simplify:

1. $8 \cdot 8 \cdot 8 \cdot 8 \cdot 8$
2. $x \cdot x \cdot y \cdot y \cdot y$
3. $a \cdot b \cdot b \cdot a \cdot a$
4. $(8a)(8a)(8a)(8a)(8a)$
5. $xy(2xy)$
6. $a^5 \cdot a^3$
7. $s \cdot s \cdot s \cdot s \cdot x \cdot x \cdot x \cdot y \cdot z \cdot z \cdot z$
8. $u^7 \cdot u \cdot u^4$
9. $5^3 \cdot 5^2$
10. $(3x)(4x)$
11. $(-5y)(3y)$
12. $(-4y^2)(5y^3)$
13. $(2ab)(3cd)$
14. $(-2rs)(3rt)(-4st)$
15. $(xy)(xyz)(-yz)$
16. $(-5ab^2c)(-2a^3c^2)(-3bc)$
17. $a \cdot a + b \cdot b$
18. $3x \cdot x \cdot x - x \cdot x \cdot x$
19. $k \cdot k \cdot k + k \cdot k$
20. $9z \cdot z \cdot z - 4z \cdot z$
21. $(rs)^3$
22. $(xz^2)^3$
23. $(2ab)^2$
24. $(-5x^2y^3)^2$
25. $(-3c^2d)^3$
26. $(-2xy^2z^3)^4$
27. $\dfrac{-12x^2y^3}{6xy}$
28. $\dfrac{-18ab^3c^2}{-3abc}$
29. $(-24r^2s^3t^7) \div (3rs^3t^4)$
30. $(-6x^3yz) \div (-6xyz)$
31. $(40c^5d^3e^4) \div (-5c^4d^3e^2)$
32. $(-48d^4km^2) \div (16d^3k^3m^3)$
33. $\dfrac{x^3 \cdot x^4}{x^2}$
34. $\dfrac{3^7 \cdot 3^2 \cdot 3^9}{3^5}$
35. $\dfrac{x^4y^2}{(xy)^2}$

36. $\dfrac{a^3(b^2)^4}{(ab^3)^2}$

37. $\dfrac{(-cD^3)^2}{(-c^2D)^3}$

38. $\dfrac{x^3(-y^2)^3}{(-xy)^3}$

39. $\dfrac{(-x^3)^2(y^2)}{x^2(-y)^2}$

40. $\left[\dfrac{a^2(-bc^4)^3}{b^4(-a^3c^5)^2}\right]^3$

8-10 MULTIPLICATION OF A POLYNOMIAL BY A MONOMIAL When a polynomial is multiplied by a monomial, use is made of the associative and distributive laws of addition. Thus,

$$\begin{aligned} a(x+y+z) &= a[(x+y)+z] && \text{by the associative property of addition} \\ &= a(x+y)+az && \text{by the distributive property} \\ &= ax+ay+az && \text{by the distributive property} \end{aligned}$$

PROCEDURE *To multiply a polynomial by a monomial, multiply each term of the polynomial by the monomial.*

Example 8-9 Multiply $2x^2 - 5xy + 8y^2$ by $3xy^2$.

SOLUTION

$$\begin{aligned} 3xy^2(2x^2 - 5xy + 8y^2) &= 3xy^2(2x^2) + (3xy^2)(-5xy) + (3xy^2)(8y^2) \\ &= 6x^3y^2 - 15x^2y^3 + 24xy^4 \end{aligned}$$

8-11 MULTIPLICATION OF A POLYNOMIAL BY A POLYNOMIAL The distributive, commutative, and associative properties for addition and multiplication can now be used to find the product of two polynomials. Consider the product of two binomials. For example,

$$\begin{aligned} (a+b)(x+y) &= (a+b)x + (a+b)y && \text{by the distributive property} \\ &= x(a+b) + y(a+b) && \text{by the commutative property of multiplication} \\ &= xa + xb + ya + yb && \text{by the distributive property} \\ &= ax + bx + ay + by && \text{by the commutative property of multiplication} \\ &= ax + ay + bx + by && \text{by the associative property of addition} \end{aligned}$$

The above solution can be written in a vertical arrangement as follows:

$$\begin{array}{l} \quad x + y \\ \underline{\quad a + b \quad\quad\quad\quad} \\ ax + ay \\ \underline{\quad\quad\quad + bx + by} \\ ax + ay + bx + by \end{array}$$

It will be noted that four separate multiplications are required to obtain the product. The first step is to multiply each term of the multiplicand by the first term of the multiplier. Next each term of the multiplicand is multiplied by the second term of the multiplier. If there are more than two terms in the multiplier, the process is continued until each of the terms in the multiplicand is multiplied by each term in the multiplier. The partial products are then combined by addition. The sum of like terms is replaced by a single term.

Example 8-10 Multiply $(x^2 + 2xy - 3y^2)$ by $(x + 5y)$.

SOLUTION

$$\begin{array}{llll} x^2 + 2xy & - 3y^2 & & \\ x + 5y & & & \\ \hline x^3 + 2x^2y & - 3xy^2 & & \\ \quad + 5x^2y & + 10xy^2 & - 15y^3 & \\ \hline x^3 + 7x^2y & + 7xy^2 & - 15y^3 & \end{array}$$

In multiplying polynomials, it is desirable to rearrange, if necessary, the terms of the multiplicand and multiplier in a descending order of powers of one of the letters. Consider the next example.

Example 8-11 Multiply $(2x - 3x^2 - 8 + x^3)$ by $(4 + 5x)$.

SOLUTION Rearrange the terms in descending powers of x before multiplying.

$$\begin{array}{lllll} x^3 & - 3x^2 & + 2x & - 8 & \\ 5x & + 4 & & & \\ \hline 5x^4 & - 15x^3 & + 10x^2 & - 40x & \\ & + 4x^3 & - 12x^2 & + 8x & - 32 \\ \hline 5x^4 & - 11x^3 & - 2x^2 & - 32x & - 32 \end{array}$$

Exercise 8-4

Perform the indicated multiplications:

1. $2ab(a^2 - bc + c^2)$
2. $-5ab^2(3a^2 - ab + 4b^2)$
3. $3x^2y(x^2 - 2xy + 3y^2)$
4. $4k(k^3 - 2k^2 + 5k - 7)$
5. $6x^2(5x^2 - 2x + 3)$
6. $-3u^2v^2(u^3 - 4u^2v + 7v^2)$
7. $2x(x - y + 3)(5y)$
8. $3ab(a^2 - 2ab - b^2)$
9. $(x + 3)(x + 4)$
10. $(a + 5)(a - 3)$

11. $(y-9)(y-4)$
12. $(z+7)(z-11)$
13. $(2x+3)(x-4)$
14. $(3a-5)(2a-7)$
15. $(3-3y)(2-y)$
16. $(7+2x)(2x-7)$
17. $(3x^2-y)(2x^2+3y)$
18. $(5m^2-2m)(3m^2-4m)$
19. $(a+4b)(a^2-2ab+b^2)$
20. $(x-2y)(2x^2-3xy+y^2)$
21. $(2k^2-3m^2)(3k^2-km-4m^2)$
22. $(x^2-5y^2)(x^2-xy-3y^2)$
23. $(x^2-2x-3)(x^2+x+4)$
24. $(a^2-3a+6)(a^2+2a-5)$
25. $(x^2+xy+y^2)(x^2-xy+y^2)$
26. $(a^2-ab-2b^2)(2a^2-ab+b^2)$
27. $(2x^2-3xy+4y^2)(5x^2-6xy-y^2)$
28. $(3a^2+ab-2b^2)(2a^2-ab+b^2)$
29. $(3x^2-5x+2x^3-7)(2+3x)$
30. $(2x^2y-3xy^2+x^3-y^3)(-7y+4x)$
31. $(x+2)(x-3)(2x+1)$
32. $(a-6)(a-2)(a+1)$
33. $(2x-1)(x+2)(2x+1)$
34. $(3a-2)(a+3)(a+3)$
35. $(x-2y)(2x-y)(x+y)$
36. $(2a-b)(a-3b)(a+5b)$

8-12 DIVISION OF A POLYNOMIAL BY A MONOMIAL Division is inverse to multiplication. To divide, then, we reverse the process of multiplication.

PROCEDURE *To divide a polynomial by a monomial, divide each of the terms in the dividend by the divisor and add the results.*

Example 8-12 Divide $12ax^3y^3 - 8ax^2y^2 + 6axy$ by $2axy$.

SOLUTION

$$\frac{12ax^3y^3 - 8ax^2y^2 + 6axy}{2axy} = \frac{12ax^3y^3}{2axy} - \frac{8ax^2y^2}{2axy} + \frac{6axy}{2axy} = 6x^2y^2 - 4xy + 3$$

8-13 DIVISION OF A POLYNOMIAL BY A POLYNOMIAL The procedure for dividing one polynomial by another polynomial is similar to the process of long division of arithmetic. The procedure can best be shown by an example.

Example 8-13 Divide $(x^3 + 6x^4 - 3x^2 - 19 - 8x)$ by $(2x - 3)$.

SOLUTION To avoid confusion, the terms in the dividend and divisor should be arranged in descending powers of x. The complete steps of the solution follow.

$$
\begin{array}{r|l}
 & 3x^3 + 5x^2 + 6x + 5 + \dfrac{-4}{2x-3} \\
\hline
2x - 3 & 6x^4 + x^3 - 3x^2 - 8x - 19 \\
 & \underline{6x^4 - 9x^3} \\
 & \quad 10x^3 - 3x^2 \\
 & \quad \underline{10x^3 - 15x^2} \\
 & \qquad 12x^2 - 8x \\
 & \qquad \underline{12x^2 - 18x} \\
 & \qquad\quad 10x - 19 \\
 & \qquad\quad \underline{10x - 15} \\
 & \qquad\qquad -4
\end{array}
$$

1 Divide $6x^4$ by $2x$ to get $3x^3$.
2 Place $3x^3$ as the first term in the quotient.
3 Multiply $(2x - 3)$ by $3x^3$ to get $6x^4 - 9x^3$. Place this quantity under the similar terms in the dividend.
4 Subtract similar terms to get a remainder of $10x^3$.
5 Bring down the next term, $-3x^2$, to get a new dividend, $10x^3 - 3x^2$.
6 Divide $10x^3$ by $2x$ to get $5x^2$.
7 Place $5x^2$ as the second term of the quotient.
8 Multiply $(2x - 3)$ by $5x^2$ to get $10x^3 - 15x^2$. Place this quantity under similar terms in the new dividend.
9 Subtract to get a remainder of $12x^2$.
10 Bring down the next term, $-8x$, to get a second new dividend.
11 Continue the process until the last remainder, -4, is of lower order than the first term of the divisor.

The solution, then, can be written:

$$\frac{6x^4 + x^3 - 3x^2 - 8x - 19}{2x - 3} = 3x^3 + 5x^2 + 6x + 5 + \frac{-4}{2x - 3}$$

Exercise 8-5

Perform the indicated divisions:

1. $(12x^4 - 16x^3 - 8x + 4) \div 4$
2. $(15a^3 - 9a^2 + 12a - 6) \div 3$
3. $(24a^4 - 16a^2 - 12a) \div 4a$
4. $(x^2y - xy^2 - 2xy^3) \div xy$
5. $(14a^3b^3 - 21a^2b^2 - 28ab) \div 7ab$
6. $(30x^3y^4 + 21x^2y^2 - 18x^2y^4 - 15x^2y^5) \div 3x^2y^2$

7. $(a^2 - 7a + 10) \div (a - 2)$
8. $(x^2 + 8x + 15) \div (x + 5)$
9. $(10x^2 + 7xy - 12y^2) \div (2x + 3y)$
10. $(3x^4y^2 + 5x^2yz - 2z^2) \div (x^2y + 2z)$
11. $(4x^2 - 20xy + 25y^2) \div (2x - 5y)$
12. $(5z^2 - 25zy - 30y^2) \div (5z + 5y)$
13. $(15a^2 + 28a - 32) \div (-5a + 4)$
14. $(x^3 + 4x^2 + 7x + 6) \div (x + 2)$
15. $(2a^3 - 3a^2 - 13a + 12) \div (a - 3)$
16. $(8x^3 - 10x^2 - 13x + 15) \div (4x^2 - 5 + x)$
17. $(10x^3 - 7x^2y - 16xy^2 + 12y^3) \div (5x - 6y)$
18. $(18y^3 - 27y^2z + 16yz^2 - 4z^3) \div (6y^2 - 5yz + 2z^2)$
19. $(6a^3 - 13a^2b + 8ab^2 - 3b^3) \div (2a - 3b)$
20. $(16a^4 - 9a^2b^2 - 12ab^3 - 4b^4) \div (4a^2 + 3ab + 2b^2)$
21. $(15k^3 - 29k^2 + 16) \div (3k - 4)$
22. $(a^4 + 2a^2b^2 + b^4) \div (a^2 + 3b^2 + 2ab)$
23. $(64a^3 - 729b^6) \div (4a - 9b^2)$
24. $(x^2 + 4y^2 + z^2 + 4xy + 2xz + 4yz) \div (x + 2y + z)$
25. $(a^2 + 9b^2 + 4c^2 - 6ab + 4ac - 12bc) \div (a - 3b + 2c)$
26. $(3k^4 - 31k^3 + 78k^2 - 74k + 30) \div (k^2 - 8k + 6)$
27. $(14a^4 + 45a^3 + 78a^2 + 45a + 11) \div (2a^2 + 5a + 7)$
28. $(12a^4 - 39a^3b + 4a^2b^2 + 37ab^3 + 10b^4)$
 $\div (4a^2 - 5ab - 2b^2)$
29. $(6x^4 - x^3y + 5x^2y^2 + 34xy^3 - 14y^4)$
 $\div (2x^2 - 3xy + 7y^2)$
30. $(42c^4 - 63c^3d - 79c^2d^2 + 45cd^3 + 40d^4)$
 $\div (6c^2 - 9cd - 7d^2)$

CHAPTER 9

LINEAR EQUATIONS

9-1 IDENTICAL AND CONDITIONAL EQUATIONS An *equation* is an algebraic statement that two algebraic expressions name the same number. Thus, $2x + 3x = 5x$, $4x + 1 = 21$, $5x + 3 = 5x + 7$, and $x^2 - 4 = 3x$ are equations. The symbol $=$ is used to indicate that two quantities have the same value. That part of an equation which is to the left of the equality sign is called the *left side,* or *left member,* of the equation. The part of an equation which is to the right of the equality sign is called the *right side,* or *right member,* of the equation.

There are three kinds of equations found in the study of elementary algebra. They can be illustrated by the equations (1) $2x + 3x = 5x$, (2) $4x + 1 = 21$, and (3) $5x + 3 = 5x + 7$. The first statement is true for any value which may be assigned to the letter x. The second statement is true only if x is replaced by an expression equal to 5. The third statement is true for no values for x. When an equation is true for all values that may be assigned to the letter, the equation is called an *identical equation,* or simply an *identity.* An equation which is true for no or only some particular value or values for the letters it contains is called a *conditional equation.* The first equation is an identity; the second and third are conditional equations.

Sometimes the symbol $\equiv$ is used to indicate an identity. Thus, $a + b \equiv b + a$. However, this is not usually necessary, because the context of the statement will make clear which type is indicated. When the word "equation" is used by itself, we shall mean a conditional equation.

9-2 SOLVING EQUATIONS An algebraic statement (equation) must be either true or false. Any number which, when substituted for the letter in a conditional equation involving one letter, will make its left member equal to its right member is called the *root* of the equation. Another way to state this is to say that the root makes the algebraic statement true. The process of finding the roots of an equation is known as *solving* the equation. The set of roots of an equation is called its *solution.* Thus, solving the equation $4x + 1 = 21$ yields the root 5. Since this equation has only one root, the solution of the equation is 5. There is no number which can be substituted for x in the equation $5x + 3 = 5x + 7$ that will make it true. Hence the equation has no solution and is a false algebraic statement.

The solution of conditional equations forms a vital part of the mathematics that is found in the study of physics, chemistry, engineering, and engineering technology.

To solve equations, we often make use of several important self-evident truths called *axioms.* These axioms deal with operations which always lead to equivalent equations.

Axiom 1 Equality is a reflexive relation; that is, $a = a$.

Axiom 2 Equality is a symmetric relation; i.e., if $a = b$, then $b = a$.

Axiom 3 Equality is a transitive relation; i.e., if $a = b$, and $b = c$, then $a = c$.

Axiom 4 The same quantity may be added to both members of an equation; i.e., if $a = b$, then $a + c = b + c$.

Axiom 5 The same quantity may be subtracted from both members of an equation; i.e., if $a = b$, then $a - c = b - c$.

Axiom 6 Both members of an equation may be multiplied by the same quantity; i.e., if $a = b$ then $ac = bc$.

Axiom 7 Both members of an equation may be divided by the same quantity (except zero); i.e., if $a = b$ and $c \neq 0$, then $a/c = b/c$.

Axiom 8 A quantity may be substituted for its equal in an equation. 7+11 same as 18

Example 9-1 Solve the equation $x - 2 = 5$.

SOLUTION

$$x - 2 = 5$$

$$x - 2 + 2 = 5 + 2 \qquad \text{by addition, Axiom 4}$$

$$x = 7$$

Example 9-2 Solve the equation $x + 4 = 5$.

SOLUTION

$$x + 4 = 5$$

$$x + 4 - 4 = 5 - 4 \qquad \text{by subtraction, Axiom 5}$$

$$x = 1$$

Example 9-3 Solve the equation $\frac{2}{3}x = 8$.

SOLUTION

$$\tfrac{2}{3}x = 8$$

$$\tfrac{3}{2}(\tfrac{2}{3}x) = \tfrac{3}{2}(8) \qquad \text{by multiplication, Axiom 6}$$

$$x = 12$$

Example 9-4 Solve the equation $6x = 18$.

SOLUTION

$$6x = 18$$

$$\frac{6x}{6} = \frac{18}{6} \qquad \text{by division, Axiom 7}$$

$$x = 3$$

Usually, before an equation can be solved, several of the axioms on equalities must be used. However, it is often possible to perform some of the operations "in your head."

Example 9-5 Solve the equation $7x - 5 = 3x + 19$.

SOLUTION

$7x - 5 = 3x + 19$

$7x = 3x + 24$ by axiom of addition

$4x = 24$ by axiom of subtraction

$x = 6$ by axiom of division

Example 9-6 Solve the equation $3x + 8 - \frac{1}{2}x - 15 = 2x + 9 + \frac{1}{4}x - 14$.

SOLUTION

$3x + 8 - \frac{1}{2}x - 15 = 2x + 9 + \frac{1}{4}x - 14$

$\frac{5}{2}x - 7 = \frac{9}{4}x - 5$ by combining similar terms

$\frac{1}{4}x = 2$ by axioms of addition and subtraction

$x = 8$ by axiom of multiplication

9-3 CHECKING SOLUTIONS The practicing scientist, engineer, and technician always should *check* his solution to a problem. An answer to a problem is worthless unless it is "checked out." Therefore, the "check" of a solution is an essential part of the solution.

We have defined the root of an equation as the quantity which, when substituted for the letter in an equation, will reduce it to an identity. Therefore, to check the solution of an equation we must substitute the root in the original equation to see if the process will reduce the original equation to an identity. If it does, the root is said to *satisfy the equation.*

Example 9-7 Check to see if 8 is a root of the equation in Example 9-6.

CHECK

$3x + 8 - \frac{1}{2}x - 15 = 2x + 9 + \frac{1}{4}x - 14$

substituting 8 for x

$3(8) + 8 - \frac{1}{2}(8) - 15 \stackrel{?}{=} 2(8) + 9 + \frac{1}{4}(8) - 14$

$24 + 8 - 4 - 15 \stackrel{?}{=} 16 + 9 + 2 - 14$

$13 = 13$

Since the equation is reduced to an identity, 8 is a *root* of the equation. 8 is said to satisfy the equation.

Exercise 9-1

Solve the equations and check your solutions:

1. $2x - 5 = 11$
2. $3x + 4 = 19$
3. $18 + 2x = 12$
4. $17 - 3x = 5$
5. $5x - 8 = x + 16$
6. $6x + 9 = 3 - 6x$

7. $9x - 12 = 7x - 11$
8. $20 - 3x = 6 - 6x$
9. $8y - 5 = 3y - 25$
10. $9z - 11 = 7z - 11$
11. $5a + 7 - a = 2a + 15$
12. $4k + 3k - 4 = 16 - k + 4$
13. $6x - 43 - 28x = 15x - 6 + 37$
14. $44a - 129 - 13a = 5a - 65 - 6a$
15. $5y - 7 - 4y - 8 + 8y - 15 = 0$
16. $21 - 3w = 15 - 2w - 2 - 3w + 2$
17. $6 - 4v + 3 = 5v - 3 - 3v$
18. $16y - 33 = 3 + 4y - 54 + 10y + 23$
19. $6z + 9 - 7z = 9z + 9 + 2z - 6$
20. $9t - 48 - 4t + 16 = 6t - 18 - 5t - 19 + 5$

9-4 TRANSPOSING The student should have noted in studying the previous examples and in working problems in the foregoing exercise that, when the proper term is *added to* or *subtracted from* both sides of an equation, one term disappears from one side and reappears on the other side of the equation *with its sign changed.*

This fact leads to a short cut in applying the axioms of addition and subtraction. The process is called *transposition* and can be expressed by the following rule.

RULE *If any term in an equation is transposed from one side to the other with its sign changed, an equivalent equation will result.*

Example 9-8 Solve the equation $7x - 8 = 3x + 12$.

SOLUTION

transposing -8 and $3x$,

$$7x - 8 = 3x + 12$$
$$7x - 3x = 12 + 8$$
$$4x = 20$$
$$x = 5$$

CHECK

$$7(5) - 8 \stackrel{?}{=} 3(5) + 12$$
$$35 - 8 \stackrel{?}{=} 15 + 12$$
$$27 = 27$$

9-5 EQUATIONS CONTAINING PARENTHESES Often, for clarity's sake, symbols of grouping are required in writing equations. In solving such equations, it is necessary to first remove the symbols of grouping. The equation can then be solved in the regular manner.

Example 9-9 Solve the equation $7x - 3(2x - 4) = 5(3x - 6)$.

SOLUTION

	$7x - 3(2x - 4) = 5(3x - 6)$
By distributive law:	$7x - 6x + 12 = 15x - 30$
Transpose:	$7x - 6x - 15x = -30 - 12$
Combine similar terms:	$-14x = -42$
Divide by -14:	$x = 3$

CHECK

$$7(3) - 3(2 \cdot 3 - 4) \stackrel{?}{=} 5(3 \cdot 3 - 6)$$
$$7(3) - 3(6 - 4) \stackrel{?}{=} 5(9 - 6)$$
$$7(3) - 3(2) \stackrel{?}{=} 5(3)$$
$$21 - 6 \stackrel{?}{=} 15$$
$$15 = 15$$

Exercise 9-2

Solve each equation. Check your answers.

1. $(x + 5) - (x - 7) + x = 10$
2. $(5x - 3) - (x + 7) = 3x + 4$
3. $8x - 2(x - 3) = 5(2x - 2)$
4. $10x + 2(x - 4) = 3(2x - 1) + 7$
5. $7 - (3 - x) + 2x = 11 - (5x - 1)$
6. $3(y + 1) = 12 + 4(y - 1)$
7. $5a - (a - 6) = 3a - 2(a + 3)$
8. $3 - 7(3 + x) = 5x - 6(x - 2)$
9. $5d - 19 - 4(3d - 2) = 18 + (6 - 2d)$
10. $11 - 6(2x - 3) + 5(3x - 4) = 0$
11. $3x(2x - 5) + 7 = 3x(2x - 6) + 13$
12. $7y(y - 2) - 3y(2y - 5) = y(y - 4) + 20$
13. $(z + 3)(z - 6) = (z - 4)(z + 2)$
14. $(2k - 3)(3k + 2) = 3k(2k - 1)$
15. $3x^2 - (x + 4)(3x - 2) = 2(1 - 5x) + 6$
16. $2a^2 - (2a - 3)(a + 5) = 7(2 - a)$
17. $(2y + 3)^2 + 27 = 4y(y - 1) + 7y$
18. $9(z - 1)(z - 3) - 7(z - 2)(z - 3) = 2(z - 1)(z - 2)$
19. $(2y - 3)(y + 4) - 23 = (y + 7)(2y - 5) - 4y$
20. $(t - 1)^2 - (2t - 3)^2 - 46 = 3t(2 - t)$

9-6 EQUATIONS WITH FRACTIONS An equation with fractional coefficients may be reduced to an equivalent equation with no fractional coefficients by multiplying both members of the equation by the lowest common multiple of the denominators of all the coefficients. It first should be noted that such expressions as $\frac{3(x-2)}{4}$ and $\frac{3}{4}(x-2)$ are identical. In like manner $\frac{x-5}{6}$ and $\frac{1}{6}(x-5)$ are identical.

Example 9-10 Solve the equation $\frac{3x-4}{6}+\frac{3}{4}=\frac{4x+2}{3}$.

SOLUTION

$$\frac{3x-4}{6}+\frac{3}{4}=\frac{4x+2}{3}$$

Multiply by 12: $12\cdot\frac{3x-4}{6}+12\cdot\frac{3}{4}=12\cdot\frac{4x+2}{3}$

Simplify: $2(3x-4)+3\cdot 3=4(4x+2)$

Use distributive law: $6x-8+9=16x+8$

Transpose: $6x-16x=8+8-9$

Combine similar terms: $-10x=7$

Divide by -10: $x=-\frac{7}{10}$

CHECK

$$\frac{3(-\frac{7}{10})-4}{6}+\frac{3}{4}\overset{?}{=}\frac{4(-\frac{7}{10})+2}{3}$$

$$\frac{-\frac{21}{10}-\frac{40}{10}}{6}+\frac{3}{4}\overset{?}{=}\frac{-\frac{28}{10}+\frac{20}{10}}{3}$$

$$-\tfrac{61}{60}+\tfrac{3}{4}\overset{?}{=}-\tfrac{8}{30}$$

$$-\tfrac{61}{60}+\tfrac{45}{60}\overset{?}{=}-\tfrac{16}{60}$$

$$-\tfrac{16}{60}=-\tfrac{16}{60}$$

In solving fractional equations, the student should understand that the fractional line is a symbol of grouping in that the numerator is treated as a whole, i.e., as if the numerator were included in parentheses. Consider the next example.

Example 9-11 Solve the equation

$$\frac{x}{4}-\frac{2x-5}{3}=\frac{x}{2}-\frac{5(x-1)}{12}$$

SOLUTION

$$12\cdot\frac{x}{4}+12(-\tfrac{1}{3})(2x-5)=12\cdot\frac{x}{2}+12(-\tfrac{5}{12})(x-1)$$

$$3x-4(2x-5)=6x-5(x-1)$$

$$3x-8x+20=6x-5x+5$$

$$3x-8x-6x+5x=5-20$$

$$-6x=-15$$

$$x=\tfrac{5}{2}\text{ or }2\tfrac{1}{2}$$

CHECK

$$\frac{\frac{5}{2}}{4}-\frac{2(\frac{5}{2})-5}{3}\stackrel{?}{=}\frac{\frac{5}{2}}{2}-\frac{5(\frac{5}{2}-1)}{12}$$

$$\tfrac{5}{8}-0\stackrel{?}{=}\tfrac{5}{4}-\tfrac{5}{8}$$

$$\tfrac{5}{8}=\tfrac{5}{8}$$

Exercise 9-3

Solve each equation. Check your answers:

1. $\frac{x}{2}+\frac{x}{3}+\frac{x}{4}=\frac{117}{2}$
2. $\frac{5x}{6}-\frac{5}{6}=\frac{2x}{3}+\frac{1}{6}$
3. $\frac{2x}{9}-\frac{x}{4}=\frac{x}{6}-\frac{7}{3}$
4. $\frac{x+2}{3}+\frac{2x-4}{4}=3$
5. $\frac{2x-11}{3}-\frac{3x-10}{4}+2=0$
6. $\frac{2x+5}{7}-3=\frac{3x-2}{5}$
7. $\frac{x+5}{14}+\frac{x-5}{4}=2$
8. $\frac{1}{6x}+\frac{1}{3x}=\frac{1}{4}$
9. $\frac{2}{5x}-\frac{7}{3x}-\frac{4}{15x}=\frac{1}{5}$
10. $\frac{1}{x}-\frac{1}{6}=\frac{2}{3}+\frac{7}{2x}$
11. $\frac{2x-9}{10}+\frac{3}{2}=\frac{x+1}{4}$
12. $\frac{4}{15x}-\frac{1}{6}=\frac{2-x}{5x}$
13. $\frac{2x-3}{5x}+\frac{5}{2}=\frac{4x+5}{2x}-\frac{4}{x}$
14. $\frac{2x-1}{3}-\frac{3x+1}{4}=x-6$
15. $\frac{7y-1}{9}=1+\frac{3y+4}{8}$
16. $\frac{2x-3}{7}-\frac{6x-11}{5}=\frac{6}{35}$
17. $\frac{8a+3}{6}-\frac{7a-1}{4}+\frac{1}{2}=0$
18. $\frac{x}{6}-\frac{3x-4}{2}=\frac{2(2x-5)}{3}$
19. $\frac{2x}{3}-\frac{x+5}{2}=\frac{3x-4}{4}$
20. $\frac{3r}{5}+\frac{4(r-5)}{15}=2-\frac{5(2-r)}{6}$

9-7 LITERAL EQUATIONS A *literal equation* is an equation in which some or all of the known quantities, usually constants, are represented by letters instead of numbers. In literal equations letters appear where numbers usually appear. It is customary in writing literal equations to use letters toward the beginning of the alphabet to represent the known quantities and those toward the end of the alphabet as the unknown quantities.

The solution of a literal equation follows the same steps as those of any other equation. The major difference lies in the nature of the solution. The roots of equations studied to this point have always been numbers. The root of a literal equation is expressed in terms of letters which represent numbers.

Example 9-12 Solve the equation $ay - 5b = c$ for y and check the root.

SOLUTION

$$ay - 5b = c$$

Add $5b$ to both sides: $ay = 5b + c$

Divide both sides by a: $y = \frac{5b + c}{a}$

CHECK

$$a\left(\frac{5b + c}{a}\right) - 5b \stackrel{?}{=} c$$

$$5b + c - 5b \stackrel{?}{=} c$$

$$c = c$$

Example 9-13 Solve for z and check the root if $7(z - 2a) - 4(3z + 11a) = 20$.

SOLUTION

$$7(z - 2a) - 4(3z + 11a) = 20$$

Use the distributive law: $7z - 14a - 12z - 44a = 20$

Collect similar terms: $-5z - 58a = 20$

Add $58a$ to both sides: $-5z = 58a + 20$

Divide both members by -5: $z = \frac{58a + 20}{-5}$

or $z = -\frac{58a + 20}{5}$

CHECK

$$7\left[-\frac{58a + 20}{5} - 2a\right] - 4\left[3\left(-\frac{58a + 20}{5}\right) + 11a\right] \stackrel{?}{=} 20$$

$$7\left[\frac{-58a - 20 - 10a}{5}\right] - 4\left[\frac{-174a - 60 + 55a}{5}\right] \stackrel{?}{=} 20$$

Multiply both members by 5:

$$7[-58a - 20 - 10a] - 4[-174a - 60 + 55a] \stackrel{?}{=} 100$$

Apply the distributive law:

$$-406a - 140 - 70a + 696a + 240 - 220a \stackrel{?}{=} 100$$

Combine similar terms: $100 = 100$

9-8 FORMULAS In experimental and theoretical technology many facts are expressed by equations, called formulas. The student should already be familiar with certain mensuration formulas; e.g., $C = 2\pi R$, for the circumference of a circle; $A = \pi R^2$, for the area of a circle; $A = lw$, for the area of a rectangle; $D = rt$, relating distance, rate, and time; etc. The letters in a formula are usually the first letters of the words whose magnitude they represent.

Example 9-14 Develop a formula for the perimeter of a rectangle whose length is three times its width.

SOLUTION Let w represent the width of the rectangle and let P represent the perimeter. Then the length $= 3w$.

$$P = 2(3w) + 2w$$
$$= 6w + 2w$$
$$P = 8w$$

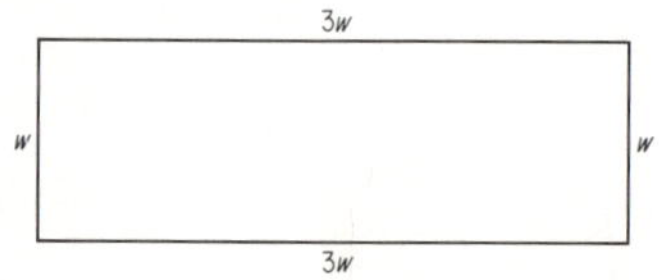

EXAMPLE 9-14.

Example 9-15 Find a formula for the volume V of the block shown in the figure for this example.

SOLUTION

$$V = \text{Volume } 1 - \text{Volume } 2$$
$$= (4a)(3a)(7a) - a(2a)(4a - 1)$$
$$= 84a^3 - (8a^3 - 2a^2)$$
$$= 84a^3 - 8a^3 + 2a^2$$
$$V = 76a^3 + 2a^2$$

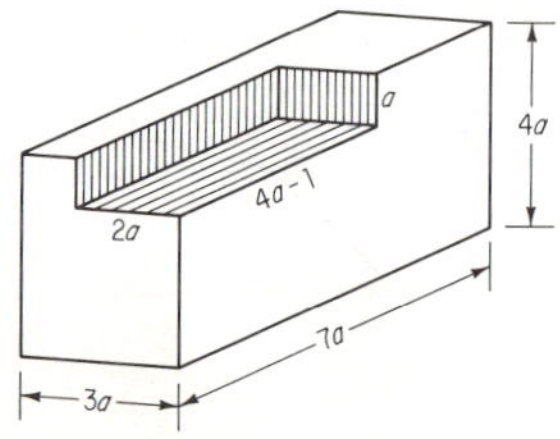

EXAMPLE 9-15.

Often a formula will be changed to an equivalent form in order to "solve for a different unknown." Consider the following examples.

Example 9-16 The formula relating heat H in calories to current I in amperes, resistance R in ohms, and time T in seconds is $H = 0.24I^2RT$. Express T in terms of H, I, and R. (We express this by saying "Solve for T.")

SOLUTION

$$H = 0.24I^2RT$$

By symmetric property of equality: $0.24I^2RT = H$

Divide both members by $0.24I^2R$: $T = \dfrac{H}{0.24I^2R}$

Example 9-17 Solve for r in the formula

$$S = \frac{(a - rL)}{(1 - r)}.$$

SOLUTION

Multiply both members by $(1 - r)$: $S(1 - r) = a - rL$

Apply the distributive law: $S - Sr = a - rL$

Add $rL - S$ to both members: $rL - Sr = a - S$

Apply the distributive law: $r(L - S) = a - S$

Divide both members by $(L - S)$: $r = \dfrac{a - S}{L - S}$

Exercise 9-4

Find the simplified forms for the perimeter P and the area A of each of the following:

1.

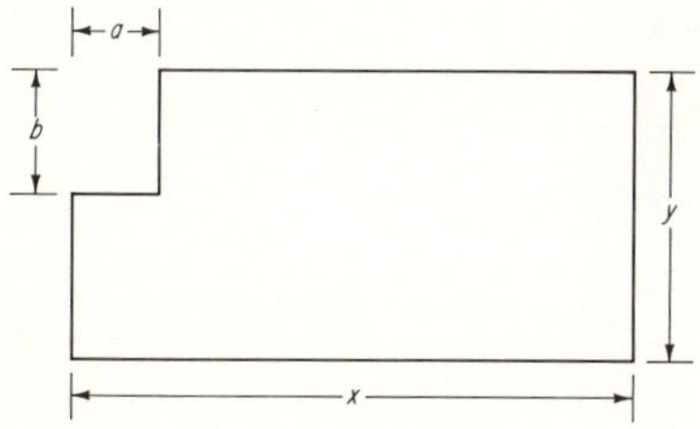

2.

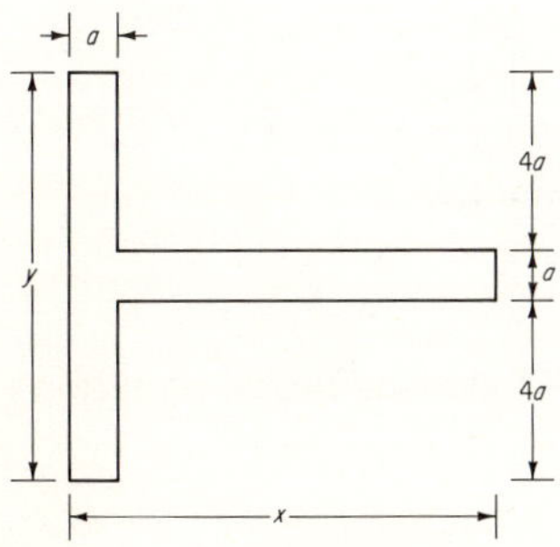

3.

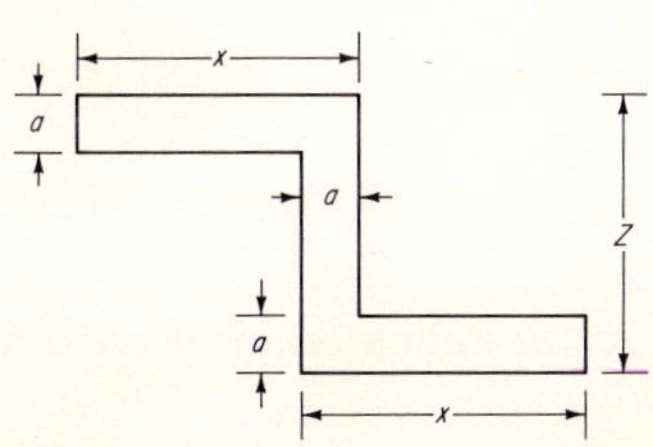

4.

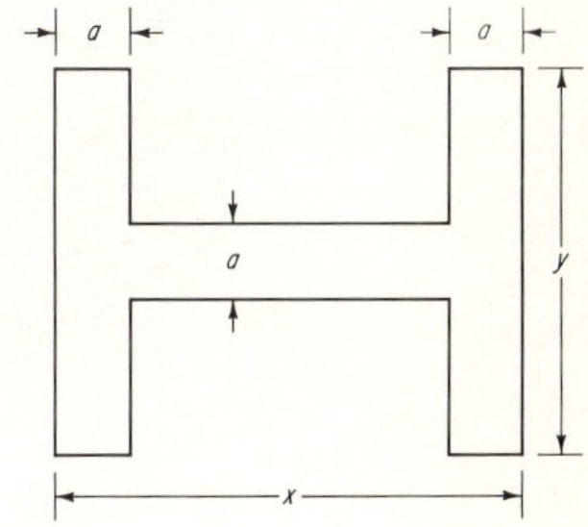

5.

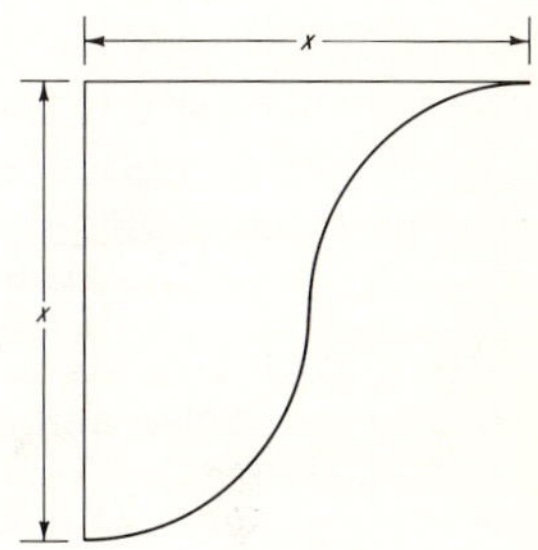

6. Find the shaded area A

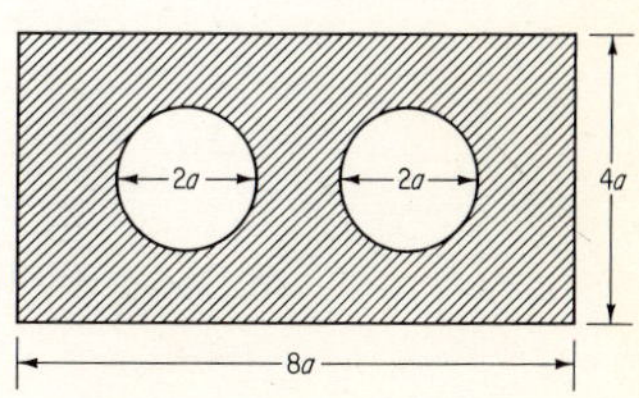

7. Find the shaded area A

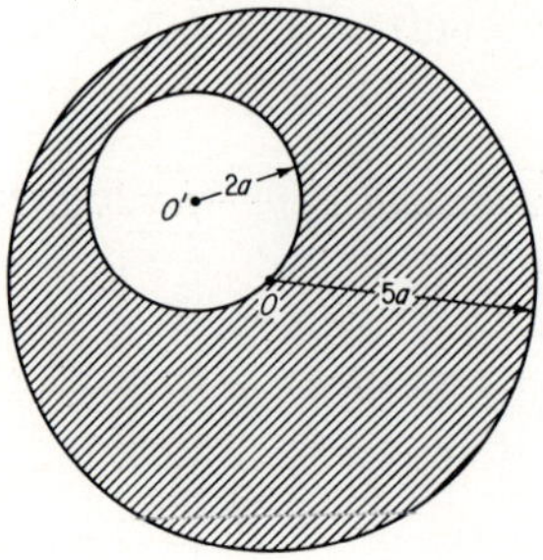

8. A pan is made by cutting out the four shaded areas and folding the remaining sheet along the dotted lines. Find a formula for the volume V of the pan in terms of x, the length of one side of each square cut out.

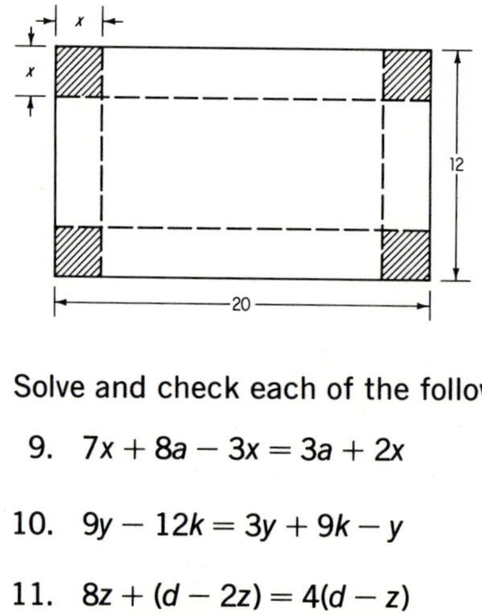

Solve and check each of the following literal equations for x, y, or z:

9. $7x + 8a - 3x = 3a + 2x$

10. $9y - 12k = 3y + 9k - y$

11. $8z + (d - 2z) = 4(d - z)$

12. $ax - b = c - dx$

13. $(y - 3a)(y + 2a) - y(y - 4a) = 9a^2$

14. $(2z - b)(2z + b) - b(z - 8) = 4z(z - 5) - 30b$

15. $\frac{5x}{a} - b = c$

16. $\frac{8y}{3} - \frac{3b}{4} = \frac{5(y - 2b)}{6}$

Solve each formula for the letter indicated after each formula:

17. $V = lwh;\ w$

18. $R = \dfrac{kl}{d^2};\ l$

19. $S = 2\pi r^2 + 2\pi rh;\ h$

20. $y = ax + b;\ x$

21. $V = \frac{1}{3}\pi r^2 h;\ h$

22. $A = P + Prt;\ r$

23. $v^2 = 2gh;\ h$

24. $Fd = Wh;\ h$

25. $M = \frac{1}{6}bh^2;\ b$

26. $Ax + By + C = 0;\ y$

27. $H = \dfrac{PLAN}{33{,}000};\ P$

28. $S = \dfrac{n(a + l)}{2};\ n$

29. $l = a + (n - 1)d;\ d$

30. $F = \frac{9}{5}C + 32;\ C$

31. $A = \frac{1}{2}(a + b);\ a$

32. $\dfrac{p_1V_1}{T_1} = \dfrac{p_2V_2}{T_2};\ V_1$

33. $F = \dfrac{Wv^2}{gR};\ R$

34. $v^2 = v_0^2 + 2gh;\ h$

35. $a = \dfrac{v - v_0}{t};\ v$

36. $s = v_0t + \frac{1}{2}at^2;\ a$

37. $\dfrac{1}{R} = \dfrac{1}{R_1} + \dfrac{1}{R_2};\ R_2$

38. $Ft = mv_2 - mv_1;\ m$

39. Given the formula $S = \dfrac{Mc}{I}$. Solve for I if it is known that $S = 45$ lb/in.², $M = 180$ lb-in., $c = 8$ in.
40. Given the formula $PV = wRT$. Solve for R if it is known that $P = 2{,}160$ lb/ft², $V = 1{,}330$ ft³, $T = 500°F$, and $w = 108$ lb.
41. Given the formula $5FL^4 = 24EIy$. Find the value of y if $L = 100$ in., $E = 30{,}000{,}000$ lb/in.², $F = 18$ lb/in., $I = 15$ in.⁴
42. Given the formula $Q = wc(T_1 - T_2)$. Solve for c if $Q = 14$ Btu, $w = 2$ lb, $T_1 = 105°F$, $T = 65°F$.
43. Given the formula $F = \dfrac{Wv^2}{gr}$. Find the value of r if it is known that $F = 15{,}000$ lb, $W = 30$ tons, $v = 160$ ft/sec, $g = 32$ ft/sec². *Note:* Be certain that your units are consistent.

9-9 SOLVING STATED WORD PROBLEMS The problems to be solved by the engineer, scientist, technician, or whoever uses a mathematical solution are frequently stated in words. A problem usually resolves into finding a number from certain verbal statements about it. Equations are used to solve practical problems, not only in the technical fields, but in all kinds of activity.

The chief difficulty in solving stated problems is that the equations used in solving the problems are not given to us ready-made. We must first make up the equations from the information given in the statement of the problem and from any other information we may have that is pertinent to the problem. After the equations are correctly set up, they can usually be solved with little difficulty.

While there is no unique way to solve stated problems, the following general suggestions should prove helpful.

1 Read the statement of the problem to get the general idea.

2 Carefully reread the problem, noting what is given and what is wanted.

3 Let some letter, such as x, represent one of the unknown quantities. Be certain that the letter represents a number. An explicit statement indicating what the letter stands for should be written; e.g., "Let $x =$ the number of pounds of tin"; "let $r =$ the rate of the plane in feet per second."

4 Express the other unknown quantities using the same letter.

5 From the conditions of the problem, obtain a statement that a certain number can be expressed two ways using the same letter. Indicate the equality of the expression by means of an equation.

6 Solve the equation obtained in step 5 for the unknown letter x. Having x, the other unknowns in step 4 may be determined.

7 Check the results obtained to see whether they fulfill the conditions given in the problem.

It should be clear that the checking must be done against the original worded statement and not against the equation used in solving the problem.

Numerous formulas from mathematics, business, and science can be used to complete step 5 above. In addition to mensuration formulas for geometric figures, other examples are:

The measure of the whole is equal to the sum of the measures of its parts.

Distance equals the product of rate and time.

Density equals the quotient of weight and volume.

Total cost equals the product of the number of items and the cost of each item.

Example 9-18 What are the dimensions of a rectangular plot of land that can be fenced with 540-ft length of fence if the length of the plot is to be twice its width?

SOLUTION

Let $x =$ width of plot, ft

$2x =$ length of plot, ft

The total distance around the plot can be used to arrive at the equation to be solved.

$$x + 2x + x + 2x = 540$$
$$6x = 540$$
$$x = 90$$
$$2x = 180$$

ANSWER The dimensions of the plot are 90 by 180 ft.

CHECK 180 ft is twice 90 ft; the perimeter of the plot is 90 ft + 180 ft + 90 ft + 180 ft = 540 ft.

Example 9-19 Two identical printing presses operating at the same speed printed 25,200 newspapers when the first press operated 120 min and the second press operated 90 min. How many newspapers did each press print?

SOLUTION

Let x = the number of newspapers per minute each press printed

then $120x$ = the number of newspapers the first press printed

and $90x$ = the number of newspapers the second press printed

The problem states that 25,200 newspapers were printed. Thus,

$$120x + 90x = 25{,}200$$
$$210x = 25{,}200$$
$$x = 120$$
$$120x = 14{,}400$$
$$90x = 10{,}800$$

ANSWER The first press printed 14,400 newspapers; the second, 10,800 newspapers.

CHECK

14,400
10,800
25,200 (total production)

Example 9-20 The numerator of a fraction exceeds the denominator by 6. The fraction is equivalent to $\frac{3}{2}$. What is the fraction?

SOLUTION

Let x = the denominator

then $x + 6$ = the numerator

and $\frac{x+6}{x} = \frac{3}{2}$

Multiplying by $2x$, we get

$$2(x + 6) = 3x$$
$$2x + 12 = 3x$$
$$x = 12$$
$$x + 6 = 18$$

ANSWER The fraction is $\frac{18}{12}$.

CHECK

$18 - 12 = 6$

$\frac{18}{12} = \frac{3}{2}$

Example 9-21 A collection of 58 coins consisting of dimes and quarters has a value of \$9.10. Find the number of coins of each kind in the collection.

SOLUTION

Let $x =$ the number of dimes in the collection

Then $58 - x =$ the number of quarters in the collection

and $10x =$ the value, in cents, of the dimes

$25(58 - x) =$ the value, in cents, of the quarters

Expressing the total value of the coins, in cents, as an equation, we get

$$10x + 25(58 - x) = 910$$
$$10x + 1{,}450 - 25x = 910$$
$$-15x = -540$$
$$x = 36$$
$$58 - x = 22$$

ANSWER There are 36 dimes and 22 quarters in the collection.

CHECK

36

22

58 (total number of coins)

$0.10 \times 36 = \$3.60$

$0.25 \times 22 = 5.50$

$\$9.10$ (total value)

Example 9-22 How many pounds each of almonds valued at 40 cents/lb and walnuts valued at 50 cents/lb should be mixed to get a 200-lb mixture which can be sold at 46 cents/lb?

SOLUTION

$$\begin{aligned}
\text{Let } x &= \text{no. of lb of almonds}\\
\text{Then } 200 - x &= \text{no. of lb of walnuts}\\
200 &= \text{no. of lb of nuts in mixture}\\
\text{and}\quad 40x &= \text{value of almonds, in cents}\\
50(200 - x) &= \text{value of walnuts, in cents}\\
46(200) &= \text{value of mixture, in cents}
\end{aligned}$$

Equating expressions for the total value of the ingredients, we get

$$\begin{aligned}
40x + 50(200 - x) &= 46(200)\\
40x + 10{,}000 - 50x &= 9{,}200\\
-10x &= -800\\
x &= 80\\
200 - x &= 120
\end{aligned}$$

ANSWER 80 lb of almonds should be mixed with 120 lb of walnuts.

CHECK

$80 \times \$0.40 = \$32.00$ $\qquad$ $200 \times \$0.46 = \$92.00$

$120 \times \$0.50 = \underline{\ \ 60.00}$

$\$92.00$ (total)

Example 9-23 Two planes take off from airports which are 1,440 mi apart and fly a collision course. The first plane takes off at 2:00 A.M. and averages 750 mph. The second plane takes off 20 min later and averages 1,800 mph. At what time will the planes meet?

SOLUTION

$$\begin{aligned}
\text{Let } t &= \text{time first plane is in flight, hr}\\
\text{Then } t - \tfrac{1}{3} &= \text{time second plane is in flight, hr}
\end{aligned}$$

and $(750 \text{ mph})(t \text{ hr}) = 750t =$ distance first plane covers, mi

$(1{,}800 \text{ mph})(t - \frac{1}{3} \text{ hr}) = 1{,}800(t - \frac{1}{3}) =$ distance second plane covers, mi

$$\begin{aligned}
750t + 1{,}800(t - \tfrac{1}{3}) &= 1{,}440\\
750t + 1{,}800t - 600 &= 1{,}440\\
2{,}550t &= 1{,}440 + 600\\
2{,}550t &= 2{,}040\\
t &= \tfrac{4}{5} \text{ hr or 48 min}
\end{aligned}$$

ANSWER The planes will meet at 2:48 A.M.

CHECK

$$750(\tfrac{4}{5}) = 600 \text{ mi}$$
$$1{,}800(\tfrac{48}{60} - \tfrac{20}{60}) = 840 \text{ mi}$$
$$\overline{1{,}440} \text{ mi (total distance)}$$

Example 9-24 Faucet A can fill a tank in 6 hr; faucet B can fill the tank in 3 hr; outlet C can drain the tank in 4 hr. After the tank is half full, how long will it take to fill the tank if faucets A and B and outlet C are each operating on the tank?

SOLUTION

$\frac{1}{6}$ = part of tank filled per hour by faucet A

$\frac{1}{3}$ = part of tank filled per hour by faucet B

$\frac{1}{4}$ = part of tank drained per hour by outlet C

Let t = hours required to fill tank under above conditions

Then $\dfrac{t}{6}$ = part of full tank delivered by faucet A

$\dfrac{t}{3}$ = part of full tank delivered by faucet B

$\dfrac{t}{4}$ = part of full tank drained by outlet C

and $$\frac{t}{6} + \frac{t}{3} - \frac{t}{4} = \frac{1}{2}$$

$$2t + 4t - 3t = 6$$
$$3t = 6$$
$$t = 2$$

ANSWER The tank will be filled in 2 hr.

CHECK

Faucet A will fill $2(\frac{1}{6}) = \frac{1}{3}$ of the tank in 2 hr

Faucet B will fill $2(\frac{1}{3}) = \frac{2}{3}$ of the tank in 2 hr

Outlet C will drain $2(\frac{1}{4}) = \frac{1}{2}$ of the tank in 2 hr

Together $\frac{1}{3} + \frac{2}{3} - \frac{1}{2} = \frac{1}{2}$ of the tank will be filled in 2 hr

Exercise 9-5

1. A fishing rod and reel together costs \$35. The reel costs \$5 more than the rod. What is the cost of each?
2. The length of a rectangle is 7 in. greater than its width. The distance (called the *perimeter*) around the rectangle is 66 in. Find its length and width.

3. A tract of land bordering a river's edge is to be fenced on one side and both ends. The length of the side of the tract is 25 yd less than twice that of its end. If 87 yd of fence is used, what are the length and width of the tract?
4. Divide 56 into two parts such that the first part is 16 more than the second.
5. The numerator of a fraction is 4 less than the denominator. The reduced fraction is equivalent to $\frac{3}{5}$. What is the fraction?
6. A board 63 in. long is sawed into two pieces such that one piece is 5 in. shorter than twice the length of the other piece. Find the length of the two pieces.
7. A carpenter cuts four pieces with different lengths from a strip of molding 16 ft 9 in. long. Each piece after the first is 3 in. shorter than the preceding piece. How long is each piece?
8. The following three men agreed to raise $140,000 to start a business. Mr. Jones agreed to contribute twice as much money as Mr. Smith, and Mr. Smith agreed to contribute twice as much as Mr. Brown. How much did each agree to contribute?
9. A man is twice as old as his son. Ten years ago he was three times as old as his son. How old is each now?
10. A man is 8 years older than three times his son's age. Four years ago the sum of their ages was 84 years. How old is each now?
11. Mr. Tolle went to work for Company A 5 years ago. Each year he received a $400 raise over his previous year's salary. His total salary for the 5-year period was $28,600. What was his starting salary?
12. A pouch contains a collection of nickels, dimes, and quarters worth $2.95. There are in the pouch twice as many nickels as dimes and eight less quarters than dimes. Find the number of coins of each kind in the pouch.
13. How many gallons each of 8 percent and 20 percent salt solution must be mixed to get 100 gal of 15 percent salt solution?
14. How many pounds of an alloy containing 40 percent nickel must be melted with an alloy containing 60 percent nickel to get 80 lb of an alloy containing 55 percent nickel?
15. How many gallons must be withdrawn from a 100-gal drum full of 75 percent glycerin solution and replaced by water to get the drum full of a 50 percent solution of glycerin?
16. A jet plane is flying 1,500 mph in pursuit of another plane which started $2\frac{1}{2}$ hr earlier and is traveling 600 mph. (a) How long will it take the jet to overtake the slower plane? (b) How far will the jet fly before overtaking the second plane?

17. A man travels from A to B, a distance of 120 mi, at a constant speed of 40 mph. He returns from B to A at a constant speed of 60 mph. What is his average rate for the round trip?
18. Two cars start at 1:00 P.M. going in the same direction. Car A travels at a constant speed of 50 mph. The driver of car B travels 2 hr at a constant speed of 40 mph and then decides to increase his speed to 55 mph until he overtakes car A. At what time does car B overtake car A? How far will car A have traveled before it is overtaken by car B?
19. Machine A can complete a job in 18 hr. Machine B can complete the job in 10 hr. How long will it take both machines working simultaneously to complete the job?
20. A man invests \$50,000, part at 5 percent interest and the rest at 6 percent interest. If his annual income is \$2,680 from his investment, how much has he invested at each rate?
21. A man in a motorboat traveled across a lake at a speed of 20 mph. On his return trip he averaged 12 mph. It took him 4 hr longer on the return trip than on the original crossing. (a) How long did he take for the round trip? (b) What is the round-trip distance?
22. Company A decides to distribute \$68,000 in bonuses to its employees according to their lengths of service with the company as follows: 1 to 5 years, a base bonus; 6 to 10 years, twice the bonus of the 1- to 5-year employee; 11 to 15 years, twice the bonus of the 6- to 10-year employee; 16 to 20 years, twice the bonus of the 11- to 15-year employee. There are 8 employees with 16 to 20 years of service, 30 employees with 11 to 15 years of service, 110 employees with 5 to 10 years of service, 542 employees with 1 to 5 years of service. How much should each employee get?
23. A construction foreman signed a contract to supervise the construction of a motel. He was given a specified time within which to complete the construction. The contract gave him \$35 for each day he worked, plus a bonus of \$50 for each day that the job was completed before the specified time. His crew finished the job in three-fourths of the time allowed. His total salary (wages plus bonus) amounted to \$1,085. What was the specified time limit of the contract?
24. A mechanic was hired to keep the field equipment of a large farm in working condition. He was paid \$30 a day whenever all equipment was in working condition and fined \$50 when any piece could not operate on a given day. At the end of 250 working days he received \$7,180 in wages. How many days was all the equipment in working condition?

9-10 ORDER PROPERTIES OF REAL NUMBERS Real numbers can be compared in terms of their order on the number scale (Fig. 9-1). Thus, we say

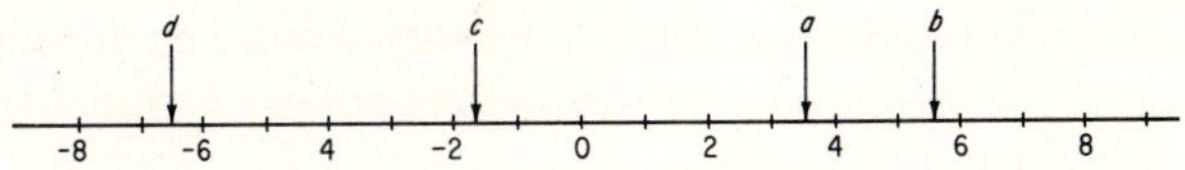

FIG. 9-1.

that "b is greater than a," written $b > a$, if the point on the scale which represents the number b appears to the right of the point which represents the number a. When b is greater than a we also say "a is less than b," written $a < b$. Thus, $b > a$ and $a < b$ have exactly the same meaning.

Two inequalities have the same *sense* if the signs of inequality point in the same direction. Thus, the inequalities $a < b$ and $c < d$ have the same sense. They are said to have opposite senses if the signs of inequality point in opposite directions. Thus, $a < b$ and $c > d$ are inequalities with opposite senses.

The real number system obeys the following order laws.

1 *The trichotomy law* Given any pair a, b of real numbers, exactly one of the following relations holds: $a = b$, $a < b$, and $a > b$.

ILLUSTRATION x cannot be equal to y and at the same time be less than y; or a cannot be less than b and also be greater than b.

2 *The transitive law* For any real numbers a, b, and c, if $a < b$ and $b < c$, then $a < c$; if $a > b$ and $b > c$, then $a > c$.

3 *Addition law of inequality* For any real numbers a, b, and c, if $a < b$, then $a + c < b + c$; if $a > b$, then $a + c > b + c$.

ILLUSTRATION If $x + 3 < 7$, then $(x + 3) + (-3) < 7 + (-3)$ or $x < 4$.

4 *Multiplication law of inequality* For any real numbers a, b, and c, if $a < b$ and $c > 0$, then $ac < bc$; if $a < b$ and $c < 0$, then $ac > bc$.

ILLUSTRATIONS If $\frac{1}{2}a < b$, then $2(\frac{1}{2}a) < 2(b)$ or $a < 2b$. If $-2a < b$, then $-\frac{1}{2}(-2a) > -\frac{1}{2}(b)$ or $a > -\frac{1}{2}b$.

9-11 SOLVING INEQUALITY PROBLEMS Thus far in this chapter we have been solving statements of equalities. Not always are we concerned with the number or numbers which make two algebraic expressions equal. Equally important is the ability to determine the sets of numbers which will make algebraic statements on inequalities true.

The order laws given above can often be applied in solving inequalities. The solution to linear equations involving one variable is at most a single number. Statements involving less-than or greater-than relations often have an infinite number of solutions.

Example 9-25 Solve $4x - 3 < 2x + 5$. Graph the solution on a number scale.

SOLUTION (The solution method is very similar to that used in solving equations, except that now we use the order laws corresponding to the laws of equality given in Sec. 9-2.)

$4x - 3 < 2x + 5$ Given

$4x - 3 + 3 < 2x + 5 + 3$ Using the addition law of inequality

$4x - 2x - 3 + 3 < 2x - 2x + 5 + 3$ Using the addition and associative laws

$2x < 8$ {Combining similar terms

$x < 4$ {Multiplying both terms of the inequality by $\frac{1}{2}$

The graph of the solution is shown.

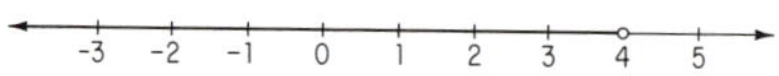

EXAMPLE 9-25.

Exercise 9-6

1. Insert the proper inequality sign ($>$ or $<$) between the following pairs of numbers.
   (a) 8, 7
   (b) 8, -6
   (c) $-4, -\pi$
   (d) $-3, -4$
   (e) $-20, -10$
   (f) $-\sqrt{3}, -2$
   (g) 0.3, 0.2
   (h) 0.5, -0.3
   (i) $-0.7, -0.35$
2. In the following, assume a, b, and c are real numbers. Indicate which statements are always true (mark T) and which are not always true (mark F).
   (a) If $a < 0$, a is a negative number.
   (b) If $0 < b - a$, then $b - a$ is a positive number.
   (c) If $a < b$, then $a - b$ is a positive number.
   (d) The sum of two positive numbers is a positive number.
   (e) If $a > 0$ and $0 < b$, then $ab < 0$.
   (f) The product of two negative numbers is a positive number.
   (g) If $a \neq 0$, then $0 < a^2$.
   (h) If $a < 0$ and $b < 0$, then $0 < ab$.
   (i) $(-a)(-b) = ab$.
   (j) If $0 < a$, then $0 < a^2$.

(k) If $a < b$ and $c < 0$, then $ac < bc$.
(l) If $b < c$, then $a - b > a - c$.
(m) If $a > b$ and $c < 0$, then $ac > bc$.
(n) If $a + b = 0$, then $b = -a$.
(o) If $x + 2 < 7$, then $x < 5$.
(p) If $-2x < 8$, then $x < -4$.
(q) If $x - 3 > 5$, then $8 < x$.
(r) If $5x > 15$, then $3 > x$.
(s) If $-4x > -16$, then $x > 4$.
(t) If $-x < 7$, then $x < -7$.

Solve the following inequalities; graph the solution set on a number scale.

3. $2x - 3 < 7$
4. $2x + 1 > -3$
5. $2x + 1 > x - 3$
6. $3x - 2 < x + 6$
7. $3x + 1 < x - 9$
8. $5x + 2 > 3x - 11$
9. $2x - 3 < 5x + 9$
10. $4x - 7 > -2x + 8$

If you multiply or divide both sides of the equation by a negative number then reverse the direction of the inequality

$4 < 6$
$\div -2 \quad \div -2$
$-2 > -3$

open

$<$ less than
$>$ greater than
$\leq$ less than or equal
$\geq$ greater than or equal
$\neq$ not equal
$=$ equal

solid

CHAPTER 10

SPECIAL PRODUCTS AND FACTORING

10-1 SPECIAL PRODUCTS Multiplication occurs so often in algebra that some products can be found by inspection; that is, it can be done mentally without use of pencil and paper. These products are sometimes called *special products,* because they occur so frequently in algebra that they deserve special attention. They are to algebra what multiplication tables are to arithmetic.

10-2 FACTORING Factoring is a process inverse to multiplication. Skill in factoring is essential in working with fractions and fractional equations and is also useful in many algebraic computations. The student will recall that he had to find prime factors to determine the lowest common denominator of fractions. In this chapter we will study methods of factoring the algebraic expressions that are most often used. Most of them will be inverses of the special products that he will study in this chapter. The student who learns the special products should have no difficulty in factoring.

10-3 FACTORING A MONOMIAL FROM A POLYNOMIAL The distributive property is used to find the product of a monomial and a polynomial.

$$a(x - y + z) = ax - ay + az \tag{10-1}$$

We will now perform the inverse operation. Sometimes the terms of a polynomial can be factored by dividing each term by a common factor.

Example 10-1 Factor $6x^2y^3 - 3xy^2 + 12xy$.

SOLUTION The largest common factor contained in each of the terms is $3xy$. This is one factor of the expression. The other factor is found by dividing each term of the trinomial by $3xy$.

$$6x^2y^3 - 3xy^2 + 12xy = 3xy(2xy^2 - y + 4)$$

10-4 PRODUCT OF THE SUM AND DIFFERENCE OF TWO TERMS Let us multiply the two binomials $2x + 3y$ and $2x - 3y$.

$$\begin{array}{l} 2x + 3y \\ \underline{2x - 3y} \\ 4x^2 + 6xy \\ \underline{ - 6xy - 9y^2} \\ 4x^2 - 9y^2 \end{array}$$

It will be noted that the product contains only two terms. These are the square of the first term $2x$ and the square of the second term $3y$. The product is the difference between the squares.

RULE *The product of the sum of two terms and the difference of the terms is equal to the difference between their squares.*

$$(a + b)(a - b) = a^2 - b^2 \qquad (10\text{-}2)$$

10-5 FACTORING THE DIFFERENCE OF TWO SQUARES To factor an algebraic expression which is the difference between two squares, we need only reverse the rule of Sec. 10-4.

RULE *The factors of the difference of the squares of two terms are the sum and difference of the two terms.*

$$a^2 - b^2 = (a + b)(a - b) \qquad (10\text{-}2a)$$

SOLUTION $36r^2 - 49s^2t^4 = (6r + 7st^2)(6r - 7st^2)$

Exercise 10-1

Find the indicated products by inspection.

1. $2a(3a - 4b)$
2. $5x(6x + 7y)$
3. $-4xy^2(2x^2 - 5xy - 6y^2)$
4. $8r^2s(2rs - r - s)$
5. $(a + 2)(a - 2)$
6. $(x + 5)(x - 5)$
7. $(4 - k)(4 + k)$
8. $(5 - xy)(5 + xy)$
9. $(7x - 6y)(7x + 6y)$
10. $(2a + 5bc)(2a - 5bc)$
11. $(3y + \frac{1}{2})(3y - \frac{1}{2})$
12. $\left(\frac{x}{2} - \frac{y}{3}\right)\left(\frac{x}{2} + \frac{y}{3}\right)$
13. $(a^2b - d)(a^2b + d)$
14. $(8x^2y + 5z)(8x^2y - 5z)$
15. $\left(\frac{k}{5} - 5m\right)\left(\frac{k}{5} + 5m\right)$
16. $\left(\frac{3x}{4} + \frac{2y}{5}\right)\left(\frac{3x}{4} - \frac{2y}{5}\right)$
17. $\left(\frac{2xy}{3} + \frac{1}{4}z\right)\left(\frac{2xy}{3} - \frac{1}{4}z\right)$
18. $(0.1a - 0.3b)(0.1a + 0.3b)$

Factor:

19. $5x + 5y$
20. $4ab - 2b + 6ab^2$
21. $12h - 15k$
22. $9xy - 12y^2 - 3y$
23. $\frac{3}{4}x^2y - \frac{1}{4}xy^2$
24. $7.2w - 12.0w^2$
25. $x^2 - 16$
26. $9a^2 - 16b^2$
27. $36r^2 - 121$
28. $4a^2b^2 - 25$
29. $25x^4 - 16y^2$
30. $16k^2 - 81h^2$
31. $1.21x^2 - 0.64z^2$
32. $1 - 121k^2$
33. $a^2 - \frac{1}{16}$
34. $(x + y)^2 - z^2$
35. $(2x - 3y)^2 - 4z^2$
36. $25 - (3a - 4b)^2$

10-6 SQUARE OF A BINOMIAL Another important product is the *square of a binomial,* sometimes called a *trinomial square.* Consider the square of $(2x + 3y)$, written as $(2x + 3y)(2x + 3y)$ or $(2x + 3y)^2$. If we multiply the two factors by the long method, we get

$$(2x + 3y)^2 = 4x^2 + 12xy + 9y^2$$

In like manner, if we square $2x - 3y$, we get

$$(2x - 3y)^2 = 4x^2 - 12xy + 9y^2$$

For the square of a binomial we can state the following:

RULE *The square of a binomial is equal to the square of the first term, plus twice the product of the two terms, plus the square of the second term.*

$$(a + b)^2 = a^2 + 2ab + b^2 \qquad (10\text{-}3)$$

$$(a - b)^2 = a^2 - 2ab + b^2 \qquad (10\text{-}4)$$

10-7 FACTORING A TRINOMIAL SQUARE It should be noted that not all trinomials are perfect squares. How can we recognize that an algebraic expression is a perfect square? First, it must be a trinomial. Next, the first and third terms must be perfect squares. Finally, the middle term must be twice the product of the square roots of the first and third terms.

Example 10-3 What must be the missing term to make expression $36x^2 + \cdot\cdot\cdot + 25y^2$ a perfect trinomial square?

SOLUTION To find the missing term, we first take the square roots of the given terms, to get $\pm 6x$ and $\pm 5y$. The middle term will then be twice the product of these terms: $2(\pm 6x)(\pm 5y)$. The middle term can be either positive or negative. Hence the answer will be $36x^2 + 60xy + 25y^2$ or $36x^2 - 60xy + 25y^2$.

Example 10-4 What is the missing term to make the expression $4x^2 - 7x + \cdot\cdot\cdot$ a perfect trinomial square?

SOLUTION The middle term $-7x$ is twice the product of the square roots of the end terms of the trinomial. Therefore, $-7x/2$ is the product of the square roots of the two end terms. Since the square root of the first term is $2x$, we ask "$2x$ times what number is equal to $-7x/2$?" The answer is $-7x/2 \div 2x$ or $-\frac{7}{4}$. Therefore, the last term is $(-\frac{7}{4})^2$ or $\frac{49}{16}$.

This process is called "completing the square." Completing the square for $4x^2 - 7x + \cdot\cdot\cdot$ gives $4x^2 - 7x + \frac{49}{16}$.

Example 10-5 Factor $169r^2 - 260rs + 100s^2$.

SOLUTION If this is to be factored as a trinomial square, we must first take the square roots of $169r^2$ and $100s^2$ to get $\pm 13r$ and $\pm 10s$. Noting that the middle term of the trinomial is $-260rs$ will tell us that we must take the positive root of one term and the negative root of the others. Thus

$$169r^2 - 260rs + 100s^2 = (13r - 10s)(13r - 10s) = (13r - 10s)^2$$

$$\text{or } 169r^2 - 260rs + 100s^2 = (-13r + 10s)(-13r + 10s) = (10s - 13r)^2$$

Exercise 10-2

Multiply by inspection:

1. $(x + 3)(x + 3)$
2. $(2a + 3b)(2a + 3b)$
3. $(x - 7y)(x - 7y)$
4. $(5k - m)^2$
5. $(x - 2b)(x + 2b)$
6. $(ax + by)^2$
7. $(4c - \frac{1}{2})^2$
8. $(\frac{1}{2}x + \frac{3}{4}y)^2$
9. $(-6x - 5y)^2$
10. $(2w - 0.1v)(2w - 0.1v)$
11. $(3 + 2xy^2)(3 + 2xy^2)$
12. $(-5 + 2x)(-5 + 2x)$
13. $(r_1 - 3r_2)^2$
14. $(w/3 + 2z/5)^2$

Factor each of the following.

15. $x^2 + 12x + 36$
16. $a^2 - 4ab + 4b^2$
17. $p^2 - 10pq + 25q^2$
18. $9y^2 + 30yz + 25z^2$
19. $16a^2 - 40ab + 25b^2$
20. $r^2 - 4ar + 4a^2$
21. $64 - 16b^2 + b^4$
22. $9x^4y^2 + 42x^2y + 49$
23. $1 - 30rs^2 + 225r^2s^4$
24. $144x^2y^6 - 8xy^3 + \frac{1}{9}$
25. $a^2b^2 - 3abc + \frac{9}{4}c^2$
26. $81k^2 - 9lk + l^2/4$
27. $0.01a^2b^4 + 0.16ab^2c^3 + 0.64c^6$
28. $0.09x^2y^6 - 0.30xy^3z^2 + 0.25z^4$

Complete the square by filling the parentheses so that the expression becomes a trinomial square.

29. $a^2 + (\quad) + 9$
30. $b^2 + 8b + (\quad)$
31. $4x^2 + 4xy + (\quad)$
32. $z^2 - (\quad) + 49y^2$
33. $h^2 - 12h + (\quad)$
34. $9x^2 - (\quad) + 4y^2$
35. $121a^2 + (\quad) + b^2$
36. $9k^2 - 24km + (\quad)$
37. $0.36a^2 + 0.24ab + (\quad)$
38. $0.64 - (\quad) + 0.25m^2$
39. $c^2/9 - (\quad) + d^2/4$
40. $0.49t^2 + 0.70st + (\quad)$

10-8 PRODUCT OF TWO BINOMIALS. GENERAL CASE We will now consider the general case of the product of two binomials. Suppose the product of $(2x - 3)$ and $(3x + 5)$ is desired. Study the following two solutions.

Long Method

$$\begin{array}{r} 2x - 3 \\ 3x + 5 \\ \hline 6x^2 - 9x \qquad \\ 10x - 15 \\ \hline 6x^2 + \quad x - 15 \end{array}$$

Short Method

$$(2x - \overset{-9x}{3)(3}x + 5) = 6x^2 + x - 15$$

$$10x$$

It will be noted that the first term of the product is obtained by multiplying the first term of the two binomials: $(2x)(3x) = 6x^2$.

The third term of the product is obtained by multiplying the second terms of the binomial: $(-3)(5) = -15$.

The middle term of the product is the algebraic sum of the *cross products:* $(-3)(3x) + (2x)(5) = x$. The first and third terms of the product are called *end products.*

RULE *The product of any two binomials is the sum of the end products and the cross products.*

$$(ax + b)(cx + d) = acx^2 + adx + bcx + bd = acx^2 + (ad + bc)x + bd \qquad (10\text{-}5)$$

The product of two binomials as those above is called a *quadratic trinomial.*

10-9 FACTORING THE QUADRATIC TRINOMIAL The quadratic trinomial can be expressed in the form $px^2 + qx + r$. It is one of the most difficult to factor. Comparing $px^2 + qx + r$ with the form $acx^2 + (ad + bc)x + bd$ of the previous example will show that $ac = p$, $bd = r$, and $ad + bc = q$. Our task in factoring will be to reverse the steps used in obtaining the product.

Example 10-6 Factor $2x^2 + 5x - 3$.

SOLUTION Write two parentheses next to each other following the trinomial: $2x^2 + 5x - 3 = (\quad)(\quad)$. The first terms in the parentheses must be factors of $2x^2$. It should be clear that the only two factors suitable here are $2x$ and x. So we will write them as the first terms of the binomial factors: $2x^2 + 5x - 3 = (2x \quad)(x \quad)$.

The only possible pairs of factors of the last term (that are integers) are $(-3,1)(3,-1)(1,-3)(-1,3)$. Next we try the various pairs as second terms of the binomials and see if any will yield the middle term $5x$ for the product of the binomials.

$$(2x - 3)(x + 1) \neq 2x^2 + 5x - 3$$

$$(2x + 3)(x - 1) \neq 2x^2 + 5x - 3$$

$$(2x + 1)(x - 3) \neq 2x^2 + 5x - 3$$

$$(2x - 1)(x + 3) = 2x^2 + 5x - 3$$

Hence, $2x^2 + 5x - 3 = (2x - 1)(x + 3)$

Example 10-7 Factor $6x^2 - 5x - 4$.

SOLUTION We know that the first terms of our binomials must be factors of $6x^2$. The second terms must be factors of -4. To determine the proper pairs of factors, we know that they must yield not only the first and third terms of the trinomial but must also be so matched as to yield the middle term.

The following are the various arrangements of the sets of factors that will give $6x^2$ and -4 as the first and third terms for the product of the binomials. Only one will give the desired $-5x$ as the middle term. The student should check each one to determine which will give the correct product.

$(6x+2)(x-2)$	$(6x-1)(x+4)$	$(3x+4)(2x-1)$
$(6x-2)(x+2)$	$(6x+1)(x-4)$	$(3x-4)(2x+1)$
$(6x+4)(x-1)$	$(3x+2)(2x-3)$	$(3x-1)(2x+4)$
$(6x-4)(x+1)$	$(3x-2)(2x+2)$	$(3x+1)(2x-4)$

The only pair of factors of the above set which will give the desired product is $(3x-4)(2x+1)$.

With practice the student can learn to readily recognize the correct set of factors for any quadractic trinomial that can be factored.

Exercise 10-3

Multiply by inspection:

1. $(x+6)(x+5)$
2. $(y-2)(y+3)$
3. $(2x+5)(3x+2)$
4. $(3z-5)(2z-7)$
5. $(2a+9)(5a-7)$
6. $(2h+4)(h-3)$
7. $(4k-3)(k+2)$
8. $(6c-5d)(6c-11d)$
9. $(5x-6)(5x-6)$
10. $(12z+1)(z-12)$
11. $(2-3x)(3+4x)$
12. $(5+3n)(5-2n)$
13. $(1+9x)(4-9x)$
14. $(4-9k)(3+7k)$
15. $(x-a)(x+2a)$
16. $(2x-3a)(5x+4a)$
17. $(y-2b)(3y-b)$
18. $(5z+7C)(2z-3C)$
19. $(0.3x-7)(0.2x+5)$
20. $(0.2x+0.5)(0.1x-0.4)$

Factor each of the following.

21. x^2+5x+6
22. y^2-2y-3
23. $a^2-3a-10$
24. $z^2+3z-70$
25. $r^2-8r+16$
26. $x^2+10x-24$
27. $10x^2+17x+3$
28. $6k^2-k-1$
29. $2y^2+7y+6$
30. $4x^2+12x+9$
31. $6k^2-16k-6$
32. $2a^2-7a+3$
33. $4x^2-11x+6$
34. $4c^2+16c+15$

35. $16x^2 + 48x + 36$
36. $48d^2 + 119d - 24$
37. $9a^2 - 13ab + 4b^2$
38. $25r^2 - 40rs + 16s^2$
39. $12x^2 - 89xz + 60z^2$
40. $36x^2 - 100xy + 25y^2$

10-10 FACTORING THE SUM AND DIFFERENCE OF TWO CUBES If we divide $a^3 + b^3$ by $a + b$, we get a quotient of $a^2 - ab + b^2$. Also $a^3 - b^3$ is exactly divisible by $a - b$, the quotient being $a^2 + ab + b^2$. Therefore, we know that

$$a^3 + b^3 = (a + b)(a^2 - ab + b^2) \tag{10-6}$$

and

$$a^3 - b^3 = (a - b)(a^2 + ab + b^2) \tag{10-7}$$

RULE *One factor of the sum (difference) of two cubes is the sum (difference) of the numbers, and the other factor is the square of the first number minus (plus) the product of the numbers plus the square of the second number.*

Example 10-8 Factor $x^3 + 27$.

SOLUTION

$$x^3 + 27 = (x + 3)(x^2 - 3x + 9)$$

Example 10-9 Factor $27x^6 - 64y^3$.

SOLUTION

$$\begin{aligned} 27x^6 - 64y^3 &= (3x^2)^3 - (4y)^3 \\ &= [3x^2 - 4y][(3x^2)^2 + (3x^2)(4y) + (4y)^2] \\ &= (3x^2 - 4y)(9y^4 + 12x^2y + 16y^2) \end{aligned}$$

10-11 FACTORING BY METHOD OF GROUPING Sometimes by proper grouping of terms it is possible to reduce an algebraic expression to one of the more common types of factoring. We will illustrate the method by examples.

Example 10-10 Factor $ax + bx + ay + by$.

SOLUTION

$$\begin{aligned} ax + bx + ay + by &= x(a + b) + y(a + b) \\ &= (a + b)(x + y) \end{aligned}$$

Example 10-11 Factor $ax - ay - 2bx + 2by + 3az - 6bz$.

SOLUTION

$$\begin{aligned} ax - ay - 2bx + 2by + 3az - 6bz &= (ax - ay + 3az) - (2bx - 2by + 6bz) \\ &= a(x - y + 3z) - 2b(x - y + 3z) \\ &= (x - y + 3z)(a - 2b) \end{aligned}$$

Example 10-12 Factor $x^2 - 2xy + y^2 - z^2$.

SOLUTION

$$x^2 - 2xy + y^2 - z^2 = (x^2 - 2xy + y^2) - (z^2)$$
$$= (x - y)^2 - (z)^2$$
$$= [(x - y) + (z)][(x - y) - (z)]$$
$$= (x - y + z)(x - y - z)$$

Example 10-13 Factor $a^2 - b^2 + 5a - 5b$.

SOLUTION

$$a^2 - b^2 + 5a - 5b = (a + b)(a - b) + 5(a - b)$$
$$= (a - b)(a + b + 5)$$

10-12 COMPOSITE FACTORING Not all problems of factoring fall into a single type. Some are composites of several types. In these cases two or more steps are needed to complete the factoring.

Example 10-14 Factor $2xy^2 - 8xy - 10x$.

SOLUTION

$$2xy^2 - 8xy - 10x = 2x(y^2 - 4y - 5)$$
$$= 2x(y - 5)(y + 1)$$

Example 10-15 Factor $8x^6 - 19x^3y^3 - 27y^6$.

SOLUTION

$$8x^6 - 19x^3y^3 - 27y^6 = (x^3 + y^3)(8x^3 - 27y^3)$$
$$= (x + y)(x^2 - xy + y^2)(2x - 3y)(4x^2 + 6xy + 9y^2)$$

Exercise 10-4

Factor completely:

1. $x^3 + 27$
2. $y^3 - 8$
3. $8a^3 + b^3$
4. $27k^3 - m^3$
5. $8a^3 + 64y^3$
6. $27y^3 - 8z^3$
7. $2xy^3 + 16x$
8. $16a^2x^3 - 54a^2y^3$
9. $5r^2 - 125$
10. $18x^3y - 98xy^3$
11. $x^4 - y^4$
12. $x^4 - 5x^2 + 4$
13. $2y + ay + 2x + ax$
14. $bx + by - bz - 3cx - 3cy + 3cz$
15. $6ax - 12ay - 10bx + 20by$
16. $6x^2 - 10xy + 35yz - 21xz$
17. $a^6 - b^6$
18. $x^2 - 2xy + y^2 - 4z^2$
19. $x^3 - x^2 + x - 1$
20. $4a^2 + 9b^2 - 14a + 12ab - 21b$

CHAPTER 11 ALGEBRAIC FRACTIONS

11-1 ALGEBRAIC FRACTIONS An algebraic fraction is the indicated quotient of two algebraic expressions. Thus, if $a - b$ and $c + d$ are algebraic expressions in which $c + d \neq 0$, then $\dfrac{a-b}{c+d}$ is an algebraic fraction. The expression $a - b$ is the *numerator* and $c + d$ is the *denominator* of the fraction. The fraction is undefined if $c = -d$, since division by zero has no meaning.

11-2 EQUIVALENT FRACTIONS A fraction is in its lowest terms if the numerator and denominator contain no common factors other than 1. Since operations on algebraic fractions follow the same laws as on arithmetic fractions, we can state the following fundamental principle.

PRINCIPLE *The value of a fraction is not changed if its numerator and denominator are both multiplied or divided by the same nonzero number.*

$$\frac{a}{b} = \frac{ac}{bc} \qquad \text{if } c \neq 0 \tag{11-1}$$

Fractions whose values are the same are called *equivalent fractions.*

Examples

11-1a $\dfrac{ax}{ay} = \dfrac{x}{y}$

11-1b $\dfrac{ax + 5a}{ay} = \dfrac{a(x + 5)}{ay} = \dfrac{x + 5}{y}$

11-1c $\dfrac{a^2 + ab}{ab + ac} = \dfrac{a(a + b)}{a(b + c)} = \dfrac{a + b}{b + c}$

11-1d $\dfrac{10x^2 - 19x - 15}{5x^3 + 8x^2 + 3x} = \dfrac{10x^2 - 19x - 15}{x(5x^2 + 8x + 3)} = \dfrac{(5x + 3)(2x - 5)}{x(5x + 3)(x + 1)} = \dfrac{2x - 5}{x^2 + x}$

11-1e $\dfrac{a - 3}{a + 2} = \dfrac{(a - 3)(a - 1)}{(a + 2)(a - 1)} = \dfrac{a^2 - 4a + 3}{a^2 + a - 2}$

CAUTION *The following are examples of invalid applications of the fundamental principle of fractions.*

---

1 $\dfrac{a + b}{a + c} = \dfrac{b}{c}$ (Invalid!)

2 $\dfrac{a + b}{bc} = \dfrac{a}{c}$ (Invalid!)

3 $\dfrac{a^2b + ab^3}{ab^2 - a^2b^2} = \dfrac{a + b}{b - a}$ (Invalid!)

11-3 SIGNS OF A FRACTION In any fraction three signs are involved: namely, the sign of the fraction, the sign of the numerator, and the sign of the denominator. For example, $\frac{x}{y}$ is really $+\frac{+x}{+y}$. The following rule is used in changing the signs of a fraction.

RULE *When two, and only two, of the three signs of a fraction are changed, the value of the fraction is unchanged.*

Examples

11-2a $\frac{x}{y} = \frac{+x}{+y} = \frac{-x}{-y} = -\frac{-x}{+y} = -\frac{+x}{-y}$

11-2b $\frac{a-b}{5} = -\frac{a-b}{-5} = -\frac{-(a-b)}{5}$ or $-\frac{b-a}{5}$

The preceding rule of signs and fractions can be expanded to the following:

RULE *When the signs of an even number of factors (not terms) in the numerator, or in the denominator, or in both are changed, the sign of the fraction is unchanged. When the signs of an odd number of factors is changed, the sign of the fraction is changed.*

Examples

11-3a $\frac{(2a-b)(3a+b)(a-b)}{(b-a)(b-2a)(a+3b)} = \frac{(2a-b)(3a+b)(a-b)}{(a-b)(2a-b)(a+3b)} = \frac{3a+b}{a+3b}$

11-3b $\frac{y^2-x^2}{x^2-2xy+y^2} = \frac{(y-x)(y+x)}{(x-y)(x-y)} = \frac{(x-y)(x+y)}{(x-y)(y-x)} = \frac{y+x}{y-x}$

Exercise 11-1

Reduce each of the following fractions to its lowest terms.

1. $\frac{12x^2y^3}{15xy^2}$
2. $\frac{18a^3b^2}{36a^2b}$
3. $\frac{48x^6y^5z^3}{84x^4y^3z^3}$
4. $-\frac{108a^5y^7z^2}{-72a^5y^6z^5}$
5. $\frac{-4rs^2t^3}{-12r^3s^2t}$
6. $\frac{x-a}{ax-a^2}$
7. $\frac{15x^6y^4}{12x^2y^3-21x^3y^2}$
8. $\frac{a^2+5a+6}{a^2+6a+9}$
9. $\frac{2x^2-8}{x^2-4x-12}$
10. $\frac{x-y}{x^2-2xy+y^2}$

11. $\dfrac{3a - 3b}{a^2 - b^2}$

12. $\dfrac{-4a}{4a^2 - 16a}$

13. $\dfrac{(-2x + 3y)(y - 5x)(x - 4y)}{(3y - 2x)(5x - y)}$

14. $\dfrac{(a - 2b)^2}{4b^2 - a^2}$

15. $\dfrac{x^3y^2 - xy^4}{x^3y + 2x^2y^2 + xy^3}$

16. $\dfrac{-a^2 - a + 12}{a^2 - a - 6}$

17. $\dfrac{x^2 - 4}{8 - x^3}$

18. $\dfrac{a^3 - 8b^3}{a^2 - 4b^2}$

19. $\dfrac{12a^2 - 28ab + 15b^2}{12b^2 + ab - 6a^2}$

20. $\dfrac{ax + 5x + ay + 5y}{a^2 + 3a - 10}$

21. $\dfrac{ab + bx + ax + x^2}{ac + cx - ay - xy}$

22. $\dfrac{2ax - 3ay - 8bx + 12by}{-2ay - 12bx + 3ax + 8by}$

23. $\dfrac{x^2 - 6xy + 9y^2}{(3y - x)^2}$

24. $\dfrac{27a^3 - 64b^3}{9a^2 + 12ab + 16b^2}$

11-4 MULTIPLICATION OF FRACTIONS The product of two fractions is the fraction whose numerator is the product of the numerators of the given fractions and whose denominator is the product of the denominators of the given fractions. Thus,

$$\frac{a}{b} \cdot \frac{c}{d} = \frac{ac}{bd} \qquad (11\text{-}2)$$

Generally, the product should be reduced to its lowest terms. It should be apparent that it is simpler to divide common factors into the numerators and denominators *before* forming their products. Final numerators and denominators are usually left in their factored form.

Example 11-4 Multiply:

$$\frac{8ab^2}{21x^2y} \cdot \frac{9xy^8}{28b^2c^3} \cdot \frac{15c^4y^2}{2ab}$$

SOLUTION

$$\frac{\overset{\overset{1}{\cancel{2a}}}{\cancel{8ab^2}}}{\underset{7x}{\cancel{21x^2y}}} \cdot \frac{\overset{3y^7}{\cancel{9xy^8}}}{\underset{\underset{7}{\cancel{7c^3}}}{\cancel{28b^2c^3}}} \cdot \frac{\overset{15cy^2}{\cancel{15c^4y^2}}}{\underset{b}{\cancel{2ab}}} = \frac{45cy^9}{49bx}$$

Example 11-5 Simplify:

$$\frac{12b - 8a}{2a^3 - 3a^2b + 2ab^2 - 3b^3} \cdot \frac{a^4 - b^4}{a^2 + 2ab + b^2}$$

SOLUTION

$$\frac{12b-8a}{2a^3-3a^2b+2ab^2-3b^3}\cdot\frac{a^4-b^4}{a^2+2ab+b^2}=$$

$$\frac{-(8a-12b)}{a^2(2a-3b)+b^2(2a-3b)}\cdot\frac{(a^2+b^2)(a^2-b^2)}{(a+b)(a+b)}=$$

$$-\frac{4(2a-3b)}{(2a-3b)(a^2+b^2)}\cdot\frac{(a^2+b^2)(a+b)(a-b)}{(a+b)(a+b)}=-\frac{4(a-b)}{a+b}$$

11-5 DIVISION OF FRACTIONS The quotient of two fractions is the fraction obtained by multiplying the dividend by the divisor inverted. The divisor is inverted by interchanging the numerator and the denominator of the divisor.

The reciprocal (also called the *multiplicative inverse*) of a number is 1 divided by that number. The reciprocal of a is $\frac{1}{a}$. The reciprocal of $\frac{a}{b}$ is

$$\frac{1}{a/b}=\frac{1}{1}\cdot\frac{b}{a}=\frac{b}{a} \tag{11-3}$$

Hence, the reciprocal of a fraction is that fraction inverted.

Example 11-6 Simplify:

$$\frac{a^2+ab-2b^2}{a^3-b^3}\div\frac{3a^2+4ab-4b^2}{9a^2-4b^2}$$

SOLUTION

$$\frac{a^2+ab-2b^2}{a^3-b^3}\div\frac{3a^2+4ab-4b^2}{9a^2-4b^2}=\frac{(a-b)(a+2b)}{(a-b)(a^2+ab+b^2)}\cdot\frac{(3a+2b)(3a-2b)}{(a+2b)(3a-2b)}$$

$$=\frac{3a+2b}{a^2+ab+b^2}$$

Example 11-7 Simplify:

$$\frac{3x^2+5xy-2y^2}{x^2-y^2}\div\frac{6x^2+xy-y^2}{x^3-x^2y}\cdot\frac{4x^2-y^2}{x^4+3x^3y+2x^2y^2}$$

SOLUTION

$$\frac{3x^2+5xy-2y^2}{x^2-y^2}\div\frac{6x^2+xy-y^2}{x^3-x^2y}\cdot\frac{4x^2-y^2}{x^4+3x^3y+2x^2y^2}=$$

$$\frac{(3x-y)(x+2y)}{(x+y)(x-y)}\cdot\frac{x^2(x-y)}{(3x-y)(2x+y)}\cdot\frac{(2x+y)(2x-y)}{x^2(x+y)(x+2y)}=\frac{2x-y}{(x+y)(x+y)}$$

Exercise 11-2

Perform the indicated multiplication and division of fractions:

1. $\dfrac{15xy^3}{28ab^2} \cdot \dfrac{14a^2b}{5xy^2}$

2. $\dfrac{7a^3b^2}{8c^2d^3} \cdot \dfrac{6cd}{35ab}$

3. $\dfrac{-5uv^2w}{7u^2vw^2} \div \dfrac{25u}{49v}$

4. $\dfrac{9ab^3c}{25b} \div \dfrac{6abc}{20c}$

5. $\dfrac{ab^2c^3}{x^2y^2z^2} \div \dfrac{abc}{x^3y^3z^3} \div \dfrac{x^3y^4z^5}{a^3b^2c}$

6. $\dfrac{ab^2c^3}{x^2y^2z^2} \div \left(\dfrac{abc}{x^3y^3z^3} \cdot \dfrac{x^3y^4z^5}{a^3b^2c}\right)$

7. $\left(\dfrac{ab^2c^3}{x^2y^2z^2} \div \dfrac{abc}{x^3y^3z^3}\right) \cdot \dfrac{x^3y^4z^5}{a^3b^2c}$

8. $\dfrac{ab^2c^3}{x^2y^2z^2} \div \left(\dfrac{abc}{x^3y^3z^3} \div \dfrac{x^3y^4z^5}{a^3b^2c}\right)$

9. $\dfrac{x^2-5x+6}{x^2-3x+2} \cdot \dfrac{x^2-1}{x^2-2x-3}$

10. $\dfrac{x^2+4x-12}{x^2+4} \cdot \dfrac{x^2-4}{x^2-4x+4}$

11. $\dfrac{a^2-9}{a-6} \cdot \dfrac{a^2-5a-6}{a^2-a-6}$

12. $\dfrac{a^4-16}{(a+2)^4} \cdot \dfrac{a^2-4}{(a-2)^2} \div \dfrac{2a^3+8a}{4+4a+a^2}$

13. $\dfrac{a^2+ab+ac+bc}{ac-bc} \div \dfrac{a^2+ab}{ad-bd}$

14. $\dfrac{a^2+2a-8}{a^2+7a+12} \cdot \dfrac{a^2+2a-3}{a^2-3a+2}$

15. $\dfrac{2a^2+a-36}{3a^2-11a-4} \cdot \dfrac{15a^2-a-2}{2a^2+5a-18}$

16. $\dfrac{9a^3-63a^2}{1-4a^2} \cdot \dfrac{2a^2-5a-3}{18a^2+15a} \cdot \dfrac{12a^2+4a-5}{5a^2-7a-24}$

17. $\dfrac{a^2+b^2}{3a^2b-3ab^2} \div \dfrac{a^2-2ab+b^2}{b} \cdot \dfrac{a^2+ab}{a^4-b^4}$

18. $\dfrac{4x^2-20x+21}{21-4x-x^2} \cdot \dfrac{x-3}{2x^2-5x-7} \div \dfrac{4x^2-12x+9}{x^2+8x+7}$

19. $\dfrac{x^2+x-6}{x^2-6x+8} \div \dfrac{x^2-3x+2}{x^2-x-12} \cdot \dfrac{x^2+x-2}{x^2+6x+9}$

20. $\dfrac{3x^2+4x-4}{4x^2-4x-3} \div \dfrac{2x^2+9x-5}{2x^2+3x+1} \div \dfrac{3x^2+x-2}{4x^2-8x+3}$

21. $\dfrac{c^2-100}{c^2-c-90} \div \dfrac{c^2+10c}{c^3+9c^2} \div \dfrac{2c^2d+3cd}{4c+6}$

22. $\dfrac{27a^3-b^3}{2a+2b} \cdot \dfrac{3a^2-2ab-b^2}{9a^2-b^2} \cdot \dfrac{8}{4a-4b}$

23. $\left(\dfrac{m^4-5m^2n^2+4n^4}{2m^2+mn-n^2} \div \dfrac{6m^3n^2+12mn^4-18m^2n^3}{m^3+n^3}\right) \cdot \dfrac{16mn-8n^2}{2n^2+3mn+m^2}$

24. $\left(\dfrac{x^2-y^2}{6x^2-13xy+6y^2} \div \dfrac{x^3-y^3}{3x^2+13xy-10y^2}\right) \div \dfrac{10y^2-13xy-3x^2}{2x^3-3x^2y+2xy^2-3y^3}$

25. $\left(\dfrac{10xr+12ys-15xs-8yr}{12yr-2xs+8xr-3ys} \div \dfrac{5x^2-2xy-7y^2}{4x^2+12xy+9y^2}\right)\left(\dfrac{7y^2+2xy-5x^2}{4r^2-9s^2}\right)$

11-6 ADDITION AND SUBTRACTION OF FRACTIONS The method of adding and subtracting in algebra is the same as that used in arithmetic. If the fractions have the same denominator, the sum is obtained by placing the algebraic sum of the numerators of the fractions over the common denominator.

$$\frac{a}{b}+\frac{c}{b}=\frac{a+c}{b} \qquad (11\text{-}4)$$

When fractions do not have the same denominators, the first step in addition or subtraction is to change the fractions to respectively equivalent fractions with a common denominator. We use for the common denominator the lowest common multiple of the given denominators, called the *lowest common denominator* (LCD) of the fractions.

The following examples illustrate the method of adding and subtracting fractions.

Example 11-8 Combine into a single fraction $\dfrac{5}{2a+3b}+\dfrac{8}{6a+9b}$.

SOLUTION

$$\frac{5}{2a+3b}+\frac{8}{6a+9b}=\frac{5}{2a+3b}+\frac{8}{3(2a+3b)}$$
$$=\frac{5(3)}{3(2a+3b)}+\frac{8}{3(2a+3b)}$$
$$=\frac{23}{3(2a+3b)}$$

Example 11-9 Combine into a single fraction.

$$\frac{3a}{a^2+2a-3}+\frac{5a-2}{a^2+a-6}-\frac{2a-4}{a^2-3a+2}$$

SOLUTION The factors of the denominators are

$a^2 + 2a - 3 = (a + 3)(a - 1)$

$a^2 + a - 6 = (a + 3)(a - 2)$

$a^2 - 3a + 2 = (a - 1)(a - 2)$

The lowest common denominator is $(a + 3)(a - 2)(a - 1)$. Therefore

$$\frac{3a}{a^2 + 2a - 3} + \frac{5a - 2}{a^2 + a - 6} - \frac{2a - 4}{a^2 - 3a + 2} =$$

$$\frac{3a(a - 2)}{(a + 3)(a - 2)(a - 1)} + \frac{(5a - 2)(a - 1)}{(a + 3)(a - 2)(a - 1)} - \frac{(2a - 4)(a + 3)}{(a + 3)(a - 2)(a - 1)} =$$

$$\frac{3a(a - 2) + (5a - 2)(a - 1) - (2a - 4)(a + 3)}{(a + 3)(a - 2)(a - 1)} =$$

$$\frac{3a^2 - 6a + (5a^2 - 7a + 2) - (2a^2 + 2a - 12)}{(a + 3)(a - 2)(a - 1)} = \frac{6a^2 - 15a + 14}{(a + 3)(a - 2)(a - 1)}$$

The student should check the effect of the minus sign in front of the third fraction. After $(2a - 4)(a + 3)$ was expanded, the sign of each term in the product was changed.

Example 11-10 Combine into a single fraction

$$\frac{2x}{1 + x} - \frac{1 - x}{x} + \frac{3x}{x^2 - 1} + \frac{5}{1 - x}.$$

SOLUTION

$$\frac{2x}{1 + x} - \frac{1 - x}{x} + \frac{3x}{x^2 - 1} + \frac{5}{1 - x} =$$

$$\frac{2x}{x + 1} + \frac{x - 1}{x} + \frac{3x}{(x + 1)(x - 1)} - \frac{5}{x - 1} =$$

$$\frac{2x(x)(x - 1)}{x(x + 1)(x - 1)} + \frac{(x - 1)(x + 1)(x - 1)}{x(x + 1)(x - 1)} + \frac{3x(x)}{x(x + 1)(x - 1)} - \frac{5(x)(x + 1)}{x(x + 1)(x - 1)} =$$

$$\frac{2x^3 - 2x^2 + x^3 - x^2 - x + 1 + 3x^2 - 5x^2 - 5x}{x(x + 1)(x - 1)} = \frac{3x^3 - 5x^2 - 6x + 1}{x(x + 1)(x - 1)}$$

Exercise 11-3

Combine into single fractions and simplify, if possible.

1. $\dfrac{2x}{3} + \dfrac{3x}{4} - \dfrac{x}{6}$

2. $\dfrac{3xy}{7} - \dfrac{5xy}{14} + \dfrac{7xy}{28}$

3. $\dfrac{2x - 3}{5} + \dfrac{3x - 4}{4} + \dfrac{x}{2}$

4. $\dfrac{4a-3}{6}-\dfrac{2a-5}{4}-\dfrac{2a}{3}$

5. $\dfrac{1}{a}+\dfrac{2}{b}+\dfrac{3}{c}$

6. $\dfrac{2}{x}-\dfrac{3}{y}-\dfrac{4}{z}$

7. $\dfrac{3}{4a}-\dfrac{5}{6a}+\dfrac{1}{3a}$

8. $\dfrac{5}{x}-\dfrac{2}{3x^2}-\dfrac{3}{2x^2}+\dfrac{1}{6x}$

9. $\dfrac{a-b}{ab}+\dfrac{b-c}{bc}+\dfrac{c-a}{ac}$

10. $\dfrac{a-b}{ab}-\dfrac{b-c}{bc}+\dfrac{c-a}{ac}$

11. $\dfrac{3}{2x-1}+\dfrac{4}{2x+1}$

12. $\dfrac{6}{5a-10}-\dfrac{5}{3a-6}$

13. $\dfrac{11a-18}{a^2+a-20}+\dfrac{2a+3}{a-4}-\dfrac{4a-5}{a+5}$

14. $\dfrac{2}{3a-6}-\dfrac{3}{2a+1}-\dfrac{1}{6a-12}$

15. $\dfrac{x+y}{y-x}-\dfrac{y-x}{x+y}+\dfrac{x^2}{x^2-y^2}$

16. $\dfrac{1}{a-1}+\dfrac{1}{a-2}+\dfrac{1}{a-3}$

17. $\dfrac{a+b}{a-b}-\dfrac{a^2+b^2}{a^2-b^2}+\dfrac{a-b}{(a+b)^2}$

18. $\dfrac{x-1}{4x^2-11x+6}-\dfrac{x+2}{2x^2-3x-2}-\dfrac{x-3}{3+2x-8x^2}$

19. $\dfrac{1}{(x-y)(x-z)}+\dfrac{1}{(y-x)(z-y)}-\dfrac{1}{(x-z)(y-z)}$

20. $\dfrac{x}{(x-y)(x-z)}+\dfrac{y}{(y-z)(y-x)}+\dfrac{z}{(z-x)(z-y)}$

21. $\dfrac{1}{x-1}+\dfrac{1}{x-2}-\dfrac{1}{x-3}-\dfrac{1}{x-4}$

22. $\dfrac{6b-5a}{(3a-2b)(2a-3b)}-\dfrac{5a-4b}{(4a+5b)(2a-3b)}-\dfrac{4b-3a}{(3a-2b)(4a+5b)}$

SECTION 3

BASIC GEOMETRY FOR TECHNICIANS

CHAPTER 12

ELEMENTS OF GEOMETRY

12-1 USES OF GEOMETRY Geometry had its origin long ago in the measurements by the Babylonians and Egyptians of their lands overswept by floods. The word *geometry* is derived from the words *geo*, meaning "earth," and *metron*, meaning "measure." As early as 2000 B.C. we find the land surveyors of these people reestablishing vanishing landmarks and boundaries by using the principles of geometry.

Geometry is a science that deals with forms made by lines and surfaces. The study of geometry is an essential part of the training of the successful engineer, scientist, technician, architect, and draftsman. The carpenter, machinist, tinsmith, stonecutter, artist, and designer each applies the truths of geometry in his craft. In the next chapters the student will learn about geometric figures such as lines, rays, angles, triangles, circles, parallelograms, cylinders, spheres, and the like.

12-2 POINTS AND LINES What is a point? Everyone has some understanding of the term. Although we can represent a point by marking a small dot on a sheet of paper or on a blackboard, it certainly is not a point. If it were possible to subdivide the marker, then subdivide the smaller dots, and so on indefinitely, we still would not have a point. We would, however, approach a condition which most of us consider to be that of a point. A point is defined as something which has position but no size or dimensions.

We name a point by a capital letter printed beside it, as "point *A*" in Fig. 12-1. Other geometric figures can be defined in terms of sets of points which satisfy certain restricting conditions.

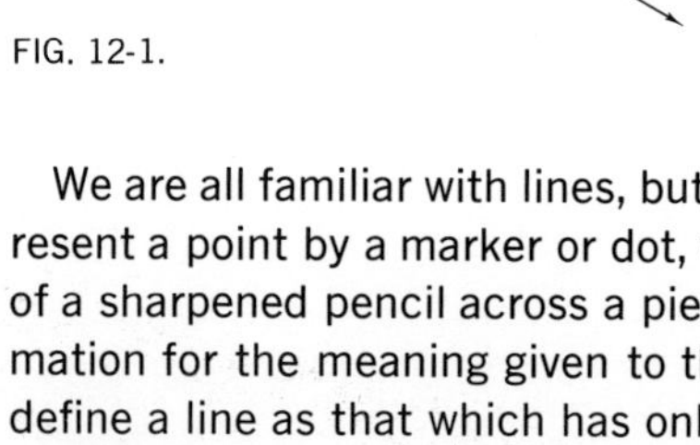

FIG. 12-1.

We are all familiar with lines, but no one has seen one. Just as we can represent a point by a marker or dot, we can represent a line by moving the tip of a sharpened pencil across a piece of paper. This will produce an approximation for the meaning given to the word "line." Some authors attempt to define a line as that which has only one dimension.

While we cannot define the word "line," we recognize it as a *set of points.* The line is named by labeling two points on it with capital letters or by one lowercase letter near it. The straight line in Fig. 12-2 is read "line *AB*" or

FIG. 12-2.

"line *l*." Line *AB is often written* $\overleftrightarrow{AB}$. In this text, unless otherwise stated, when the term *line* is used, the concept *straight* line will be meant.

DEFINITION *A set of points are collinear if and only if they belong to the same line. R, S, and T of Fig. 12-3 are collinear.*

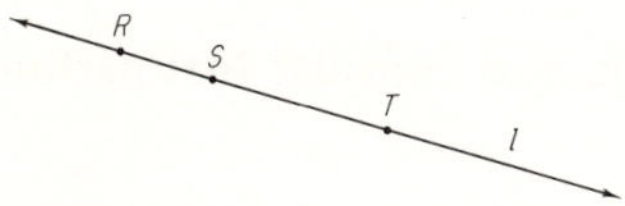

FIG. 12-3.

12-3 SOLIDS AND PLANES A solid is a three-dimensional figure. Common examples of solids are shown in Fig. 12-4.

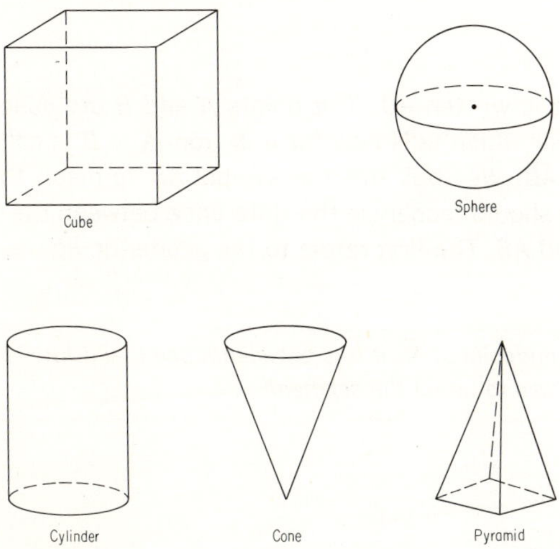

FIG. 12-4.

The geometric solid shown in Fig. 12-5 has six faces which are smooth and flat. These faces are called *plane surfaces* or simply *planes*. A plane surface has two dimensions, length and width. The surface of a blackboard or of a table top are examples of plane surfaces. A plane can be thought of as a set of points.

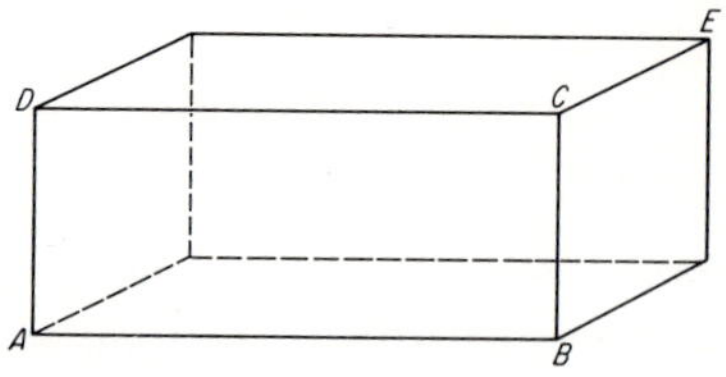

FIG. 12-5.

If two lines cross each other, they are called *intersecting lines.* Two straight lines can intersect in only one point (see Fig. 12-1).

12-4 SEGMENTS. RAYS Let us next consider that part of a line between two points of the line.

DEFINITION *The part of the line between A and B, together with points A and B (see Fig. 12-6), is called the* segment $\overline{AB}$.

FIG. 12-6.

Symbolically it is written $\overline{AB}$. The points A and B are called *end points* of $\overline{AB}$. The number which tells how far it is from A to B is called the *measure* (or *length*) of $\overline{AB}$. We shall use the symbol AB to mean the length of $\overline{AB}$.

The student should recognize the difference between the meanings of the symbols $\overline{AB}$ and AB. The first refers to the geometric figure; the second to a number.

DEFINITION *A point B is the midpoint of* $\overline{AC}$ *if B is* between *A and C and* $AB = BC$ (see *Fig. 12-7*). *The midpoint is said to* bisect *the segment.*

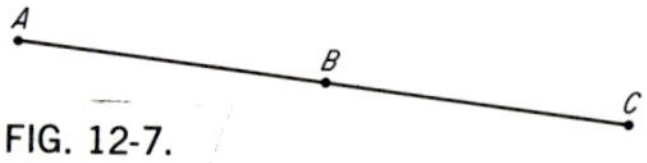

FIG. 12-7.

DEFINITION *A* ray *is the part of a straight line extending from a point on the line indefinitely in only one direction* (see *Fig. 12-8*).

FIG. 12-8.

The symbol for the ray from A through B is $\overrightarrow{AB}$ and is read "ray AB." The point A is called the *end point* of $\overrightarrow{AB}$. $\overrightarrow{AB}$ and $\overrightarrow{AC}$ of Fig. 12-9 are called *opposite rays.*

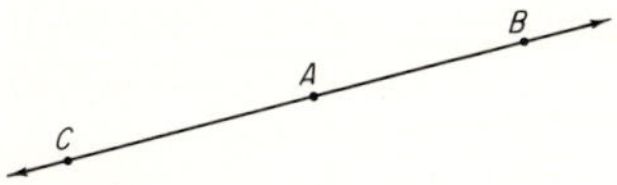

FIG. 12-9.

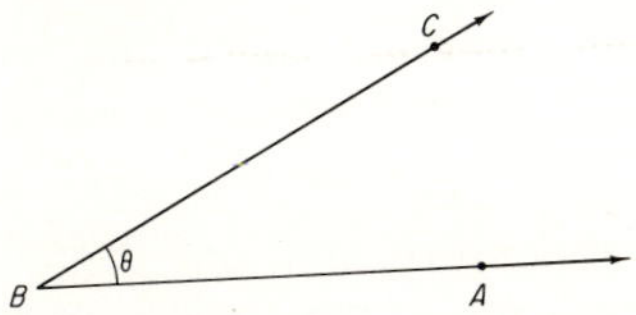

FIG. 12-10.

12-5 ANGLES The figure drawn in Fig. 12-10 is representative of an angle.

DEFINITION *An* angle *is the figure formed by two rays with a common end point. The rays are called the* sides *of the angle, and their common end point is called the* vertex *of the angle.*

The symbol for angle is $\angle$; the plural, $\measuredangle$s. There are three common ways for naming an angle: (1) by three capital letters, the middle letter being the vertex and the other two being points on the sides of the angle, as $\angle ABC$; (2) by a single capital letter at the vertex if it is clear which angle is meant, as $\angle B$; and (3) by a small letter in the interior of the angle. In advanced work in mathematics the small letter used to name the angle is often a Greek letter, as $\angle\theta$.

The student should note that the sides of an angle go out infinitely far in two directions. This is because the sides of an angle are rays, not segments. In Fig. 12-11, $\angle AOD$, $\angle BOE$, and $\angle COF$ all refer to the same angle.

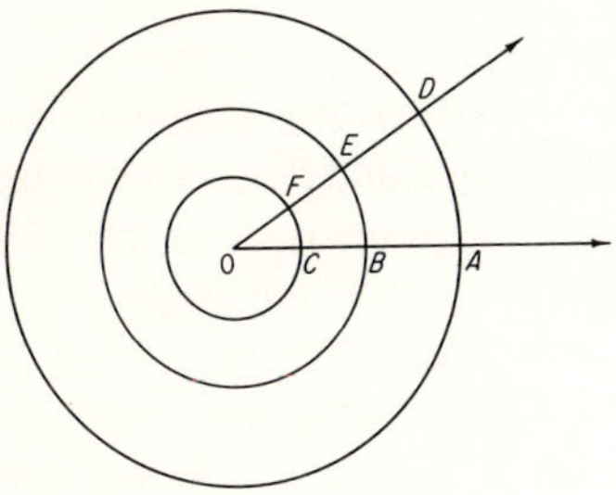

FIG. 12-11.

12-6 MEASURES OF ANGLES Angles are measured in terms of the degree unit. If a circle is divided into 360 equal arcs, and radii are drawn to any two consecutive points of division, the angle formed at the center by these radii has a size of one *degree.* The symbol for degree is °. The degree is quite small. One can gain a rough idea of the size of one degree if he realizes that if in Fig. 12-12 (not drawn to scale), $\overline{BA}$ and $\overline{BC}$ are each 57 in. long and $\overline{AC}$

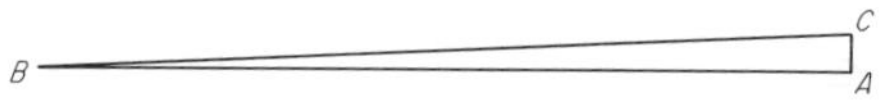

FIG. 12-12.

is 1 in. long, $\angle ABC$ has a size of approximately one degree. If this angle is subdivided into 60 equal angles, each of the smaller angles has a magnitude of one *minute.* The symbol for the minute is ′. If one of these minute angles is subdivided again into 60 equal angles, each of these very small angles will have a size of one *second.* The symbol for the second is ″. Thus 3,600 seconds are equivalent to one degree. Surveyors, engineers, and astronomers commonly measure angles with accuracy to the second.

The number of degrees in an angle can be measured roughly by the student with the aid of a protractor (Fig. 12-13). The student can also draw an angle to a required measure with the use of this instrument.

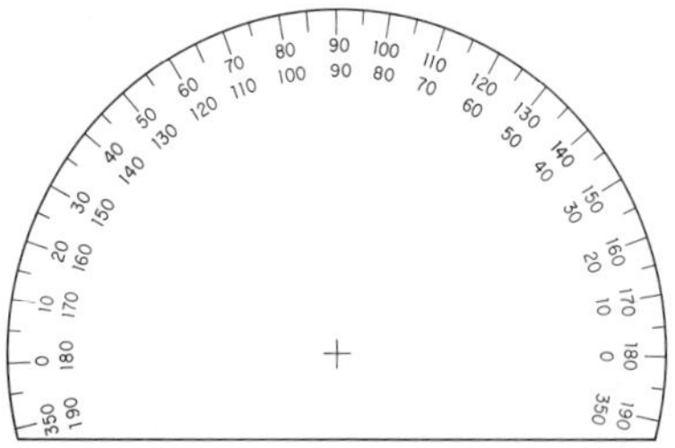

FIG. 12-13.

The United States Armed Services measures angles in terms of mils. If in Fig. 12-12 $\overline{BA}$ and $\overline{BC}$ are 1,000 in. long and $\overline{AC}$ is 1 in. in length, $\angle ABC$ will have a magnitude of one *mil.* A mil is roughly $\frac{1}{20}$ degree. Students of trigonometry and calculus generally use a larger unit angle than the degree. It is called the radian. It is roughly equivalent to 57°.

We shall write the fact that $\angle RST$ has a measure of k degrees in abbreviated form as $m\angle RST = k°$. Thus, in Fig. 12-14, we indicate the angle

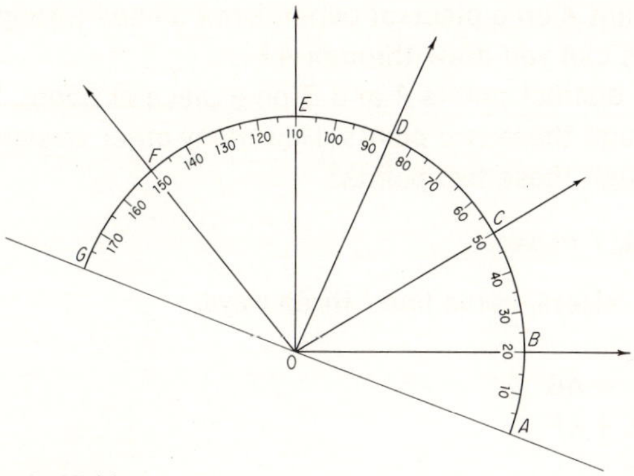

FIG. 12-14.

measures as

$m\angle AOB = 20°$ $m\angle COD = 36°$
$m\angle AOD = 86°$ $m\angle DOF = 64°$
$m\angle AOF = 150°$ $m\angle BOE = 90°$

and so on.

Exercise 12-1

1. Name the three vertices of the figure shown.
2. Which line segments meet at K?

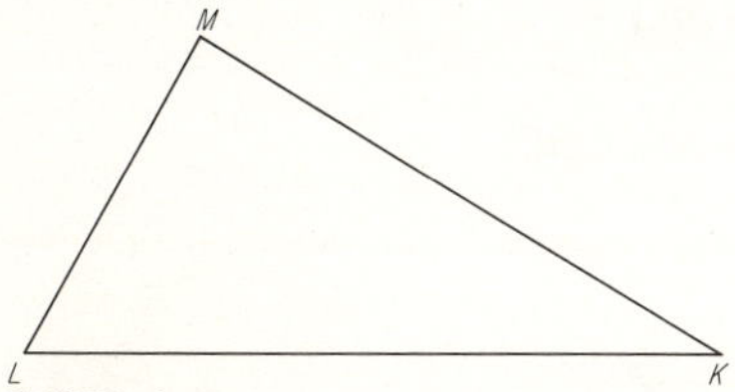

PROBS. 1–3.

3. Which angle has the larger measure, $\angle L$ or $\angle K$? Check your answer by tracing $\angle L$ on a piece of translucent material and superimposing the trace on $\angle K$.
4. Which line segment is the longer? Use tracing paper to check your answer.

PROB. 4.

5. Place a point A on a piece of paper. Draw a line l through A. How many other lines can you draw through A?
6. Place two distinct points A and B on a piece of paper. Draw a straight line l through these two points. How many other straight lines can you draw through these two points?

Complete Probs. 7 to 16.

7. Using two letters, name line l three ways.
8. $AE = AC + \cdots$
9. $OB = \cdots + AB$
10. $\cdots = OE + EF$
11. $OC = OD - \cdots$
12. $AB = \cdots - AO$

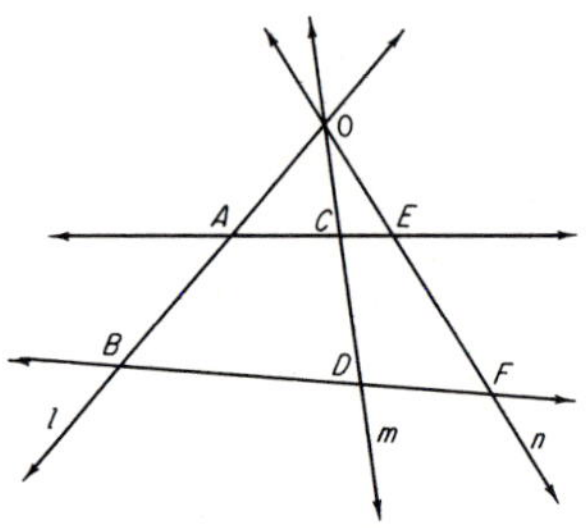

PROBS. 7–12.

13. $m\angle DAB = m\angle DAO + m \cdots$
14. $m\angle DCO = m\angle DCB - \cdots$
15. $m\angle \cdots = m\angle ABO + m\angle OBC$

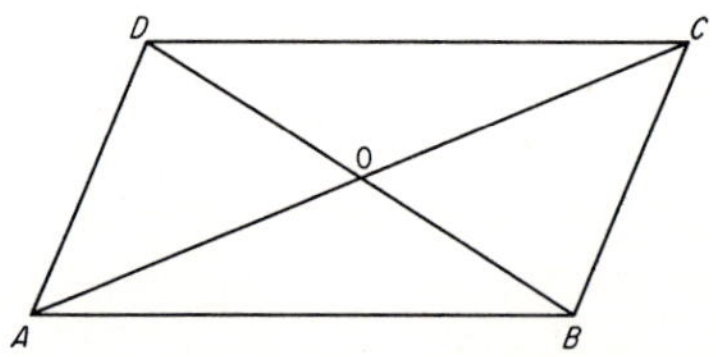

PROBS. 13–17.

16. $m\angle AOB = m\angle AOC - \cdots$
17. Which angle appears to have the same measure as $\angle AOD$? as $\angle DOC$?

In Probs. 18 to 32, draw pictures which illustrate the situations described. If the situations cannot occur, tell why.

18. l and m are two lines intersecting at P.

19. l and m are two lines; P is a point of l; R is a point of l; S is a point of m; $\overleftrightarrow{RS}$ is not the same as $\overleftrightarrow{PR}$.
20. C is not a point on $\overleftrightarrow{AB}$, and A and B are two different points.
21. r and s are two lines that do not intersect.
22. P is not a point of $\overleftrightarrow{KL}$, P is a point of line l, and line l and $\overleftrightarrow{KL}$ do not intersect.
23. R, S, and T are three points; T is the point of intersection of $\overleftrightarrow{RT}$ and $\overleftrightarrow{ST}$.
24. P, Q, R, and S are four distinct points; Q is a point of $\overleftrightarrow{PR}$; and R is a point of $\overleftrightarrow{PS}$.
25. P, Q, R, and S are four distinct points; Q is a point of $\overleftrightarrow{PR}$; and Q is a point of $\overleftrightarrow{PS}$.
26. A, B, and C are three collinear points. C, D, and E are three noncollinear points, and E is a point of $\overleftrightarrow{AB}$.
27. On line AB, C is between A and B.
28. On line AB, B is between A and C.
29. On $\overleftrightarrow{AB}$, A is between B and C.
30. P, Q, and R are three collinear points; P is a point of $\overline{QR}$; and R is not a point of $\overline{PQ}$.
31. l, m, and n are three distinct lines. l and m do not intersect, and m and n do not intersect.
32. l, m, and n are three distinct lines. l and m do not intersect, m and n do not intersect, but l and n do intersect.

Find the value of each of the following:

33. $m\angle AJC$
34. $m\angle CJE$
35. $m\angle HJC$
36. $m\angle DJB$
37. $m\angle BJF$
38. $m\angle CJD + m\angle GJD$

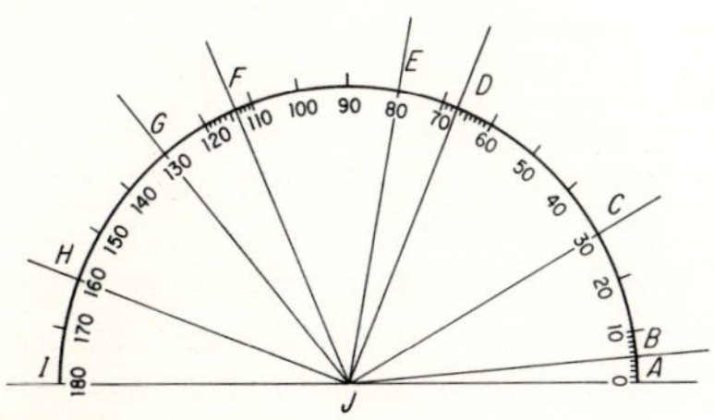

PROBS. 33–40.

39. $m\angle HJC + m\angle FJE$
40. $m\angle HJB - m\angle FJD$

12-7 KINDS OF ANGLES Two angles are said to be adjacent angles if they have the same vertex and a common side between them. In Fig. 12-15,

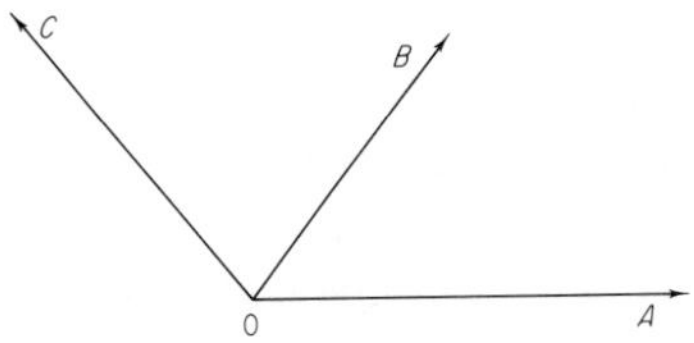

FIG. 12-15.

$\angle AOB$ and $\angle BOC$ are adjacent angles. $\overrightarrow{OB}$ is said to lie in the interior of $\angle AOC$ (except for point O).

The pairs of nonadjacent angles formed when two lines intersect are termed *vertical angles.* In Fig. 12-16 $\angle\alpha$ and $\angle\alpha'$ are vertical angles. Angles β and β' are also vertical angles.

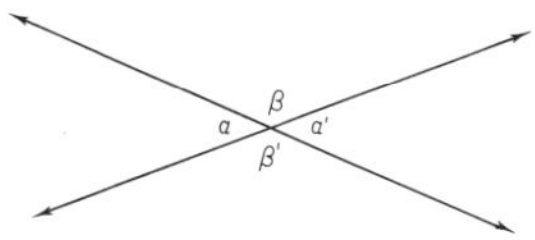

FIG. 12-16.

As the measure of an angle increases from 0 to 180°, the following kinds of angles are formed: acute angle, right angle, obtuse angle, and straight angle (see Fig. 12-17).

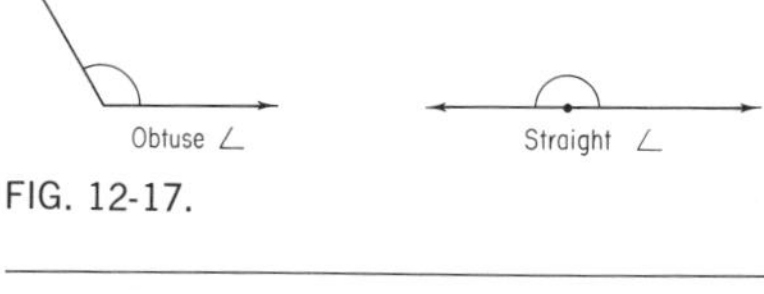

FIG. 12-17.

DEFINITION *An angle is an acute angle if it has a measure less than 90°. An angle is a right angle if it has a measure of 90°. An angle is an obtuse angle if its measure is more than 90° and less than 180°. An angle is a straight angle if its measure is equal to 180°.*

12-8 CONGRUENT ANGLES. CONGRUENT SEGMENTS A common concept in daily life is that of size and comparative sizes. We frequently speak of two things having the same size. The word *congruent* is used in geometry to define what we intuitively speak of as having the same size.

DEFINITION *Angles are congruent if they have the same measure. Segments are congruent if they have the same measure.*

Thus, if we know that $AB = CD$, we can say that $\overline{AB}$ and $\overline{CD}$ are congruent, that $\overline{AB}$ is congruent to $\overline{CD}$, or that $\overline{CD}$ is congruent to $\overline{AB}$. Again, if we know that $m\angle ABC = m\angle RST$, we can say that $\angle ABC$ and $\angle RST$ are congruent angles, $\angle ABC$ is congruent to $\angle RST$, or $\angle RST$ is congruent to $\angle ABC$.

The symbol for "is congruent to" is $\cong$. Thus the following are equivalent statements:

$AB = CD$; $\overline{AB} \cong \overline{CD}$; $\overline{CD} \cong \overline{AB}$.

$m\angle ABC = m\angle RST$; $\angle ABC \cong \angle RST$; $\angle RST \cong \angle ABC$.

DEFINITION *The bisector of an angle is the ray whose end point is the vertex of the angle and which divides the angle into two congruent angles. $\overrightarrow{BD}$ bisects, or is the angle bisector of, $\angle ABC$ if D is in the interior of $\angle ABC$ and if $\angle ABD \cong \angle DBC$ (see Fig. 12-18).*

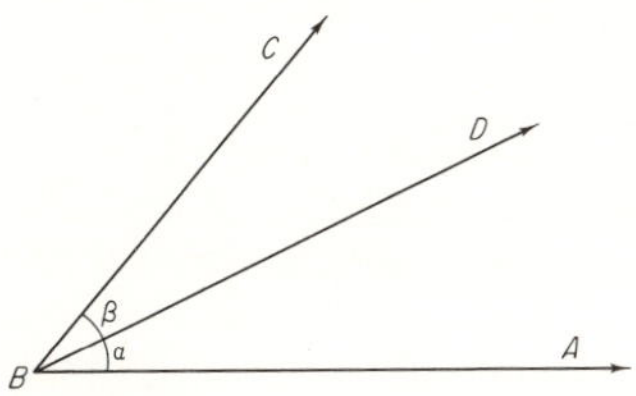

FIG. 12-18.

12-9 PERPENDICULAR LINES AND RIGHT ANGLES Consider the two figures shown in Fig. 12-19. They are examples of representations of right angles and perpendicular lines.

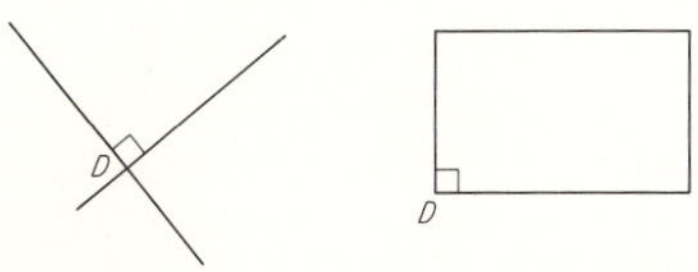

FIG. 12-19.

DEFINITION *Two lines are perpendicular if they intersect to form right angles.*

Since rays and segments are parts of lines, we can have rays, segments, and lines perpendicular to each other.

The symbol for perpendicular is $\perp$. This symbol may also be read "perpendicular to." The right angle of a figure is often designated by placing a square corner mark where the two sides of the angle meet. The *foot* of the perpendicular to a line is the point where the perpendicular meets the line. Thus, D is the foot of the perpendiculars in Fig. 12-19.

12-10 DISTANCE FROM A POINT TO A LINE The *distance* from a point to a line is the measure of the perpendicular line segment from the point to the line. Thus in Fig. 12-20 the measure of $\overline{PM}$ is the distance from point P to $\overleftrightarrow{AB}$. This *perpendicular distance is the shortest distance from a point to a line.*

12-11 COMPLEMENTARY AND SUPPLEMENTARY ANGLES Two angles the sum of whose measures is 90° are termed *complementary angles.* In Fig. 12-21 $\angle\alpha$ and $\angle\beta$ are complementary angles. Each of the angles is the complement of the other. Angle $\angle\alpha$ is the complement of $\angle\beta$; and $\angle\beta$ is the complement of $\angle\alpha$.

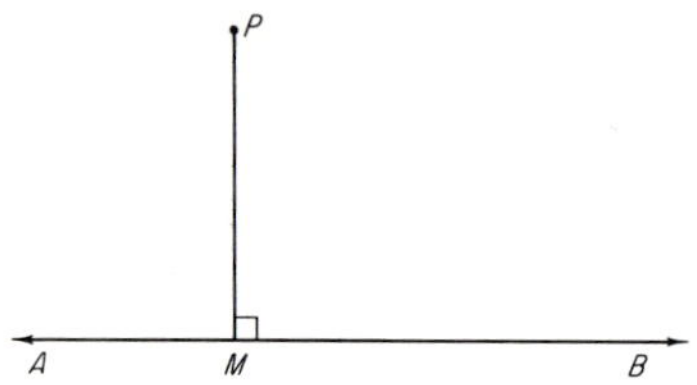

FIG. 12-20.

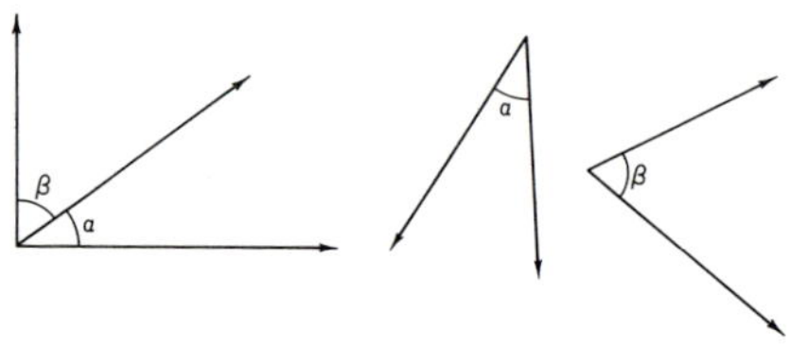

FIG. 12-21.

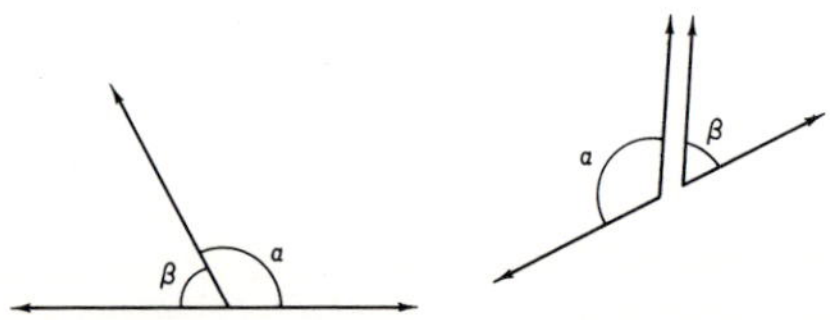

FIG. 12-22.

Supplementary angles are two angles the sum of whose measures is 180°. In Fig. 12-22, $\angle\alpha$ and $\angle\beta$ are supplementary angles. Angle $\angle\alpha$ is the supplement of $\angle\beta$, and $\angle\beta$ is the supplement of $\angle\alpha$.

12-12 TRIANGLES. KINDS OF TRIANGLES If A, B, and C are three noncollinear points, the figure formed by the joining of the three segments $\overline{AB}$, $\overline{BC}$, and $\overline{AC}$ as shown in Fig. 12-23 is called a *triangle.* The symbol for a triangle is $\triangle$ (plural $\triangle$). Thus, Fig. 12-23 is labeled $\triangle ABC$.

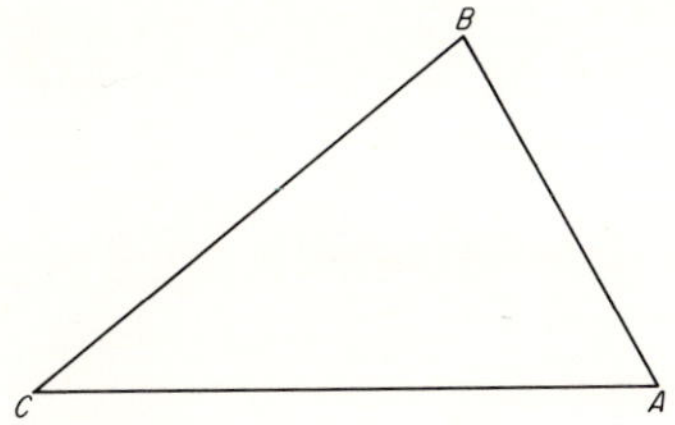

FIG. 12-23.

Each of the noncollinear points is called a *vertex* of the triangle, and each of the line segments is a *side* of the triangle. Each angle which contains two sides of the triangle is an *angle* of the triangle. In Fig. 12-23, A, B, and C are vertices of $\triangle ABC$; $\overline{AB}$, $\overline{BC}$, and $\overline{CA}$ are the sides of $\triangle ABC$; and $\angle A$, $\angle B$, and $\angle C$ are the angles of $\triangle ABC$. Angle C is *opposite* side $\overline{AB}$; AB is *opposite* $\angle C$. The sides AC and BC are said to *include* $\angle C$. $\angle C$ and $\angle A$ *include* side CA.

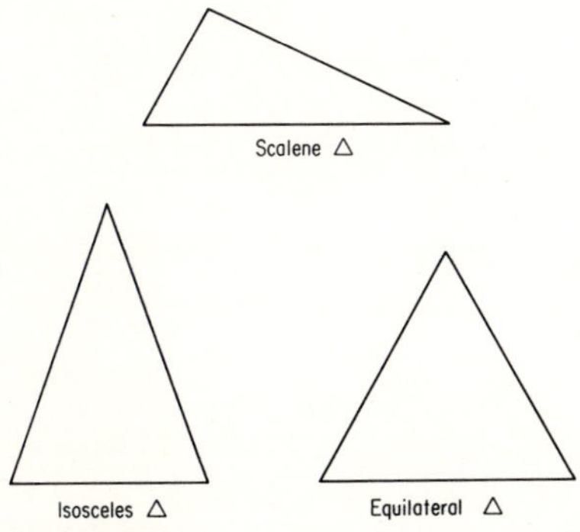

FIG. 12-24.

Triangles may be classified by their sides (Fig. 12-24). A *scalene* triangle is a triangle with no two sides that are congruent. An *isosceles* triangle is a triangle with two congruent sides. An *equilateral* triangle has three congruent sides. The parts of an isosceles triangle are labeled as in Fig. 12-25. In the figure, $AC = BC$. Angle A, opposite $\overline{BC}$, and $\angle B$, opposite $\overline{AC}$, are called the *base angles* of the isosceles triangle. Side AB is the *base* of the triangle. Angle C, opposite the base, is the *vertex angle.*

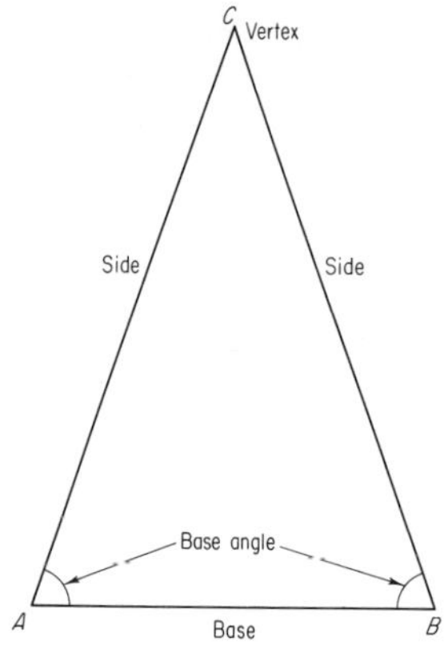

FIG. 12-25.

Triangles may also be classified according to the kind of angles they contain (see Fig. 12-26). An *acute* triangle is one that has three acute angles. An *obtuse* triangle has one obtuse angle. A *right* triangle has one right angle.

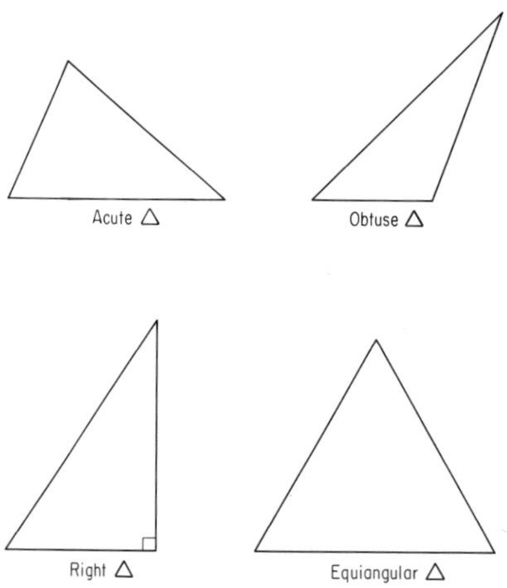

FIG. 12-26.

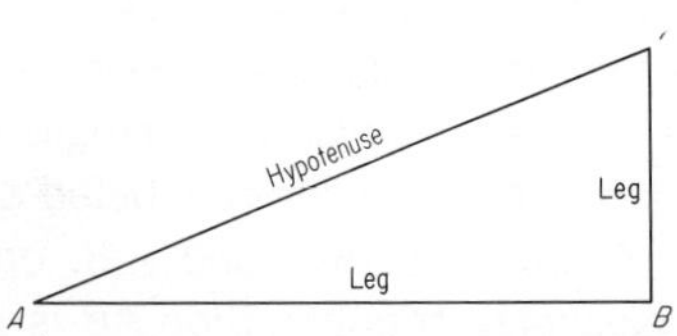

FIG. 12-27.

The sides which form the right angle of the triangle are termed *legs* of the triangle, and the side opposite the right angle is called the *hypotenuse.* In Fig. 12-27, $\overline{AB}$ and $\overline{BC}$ are the legs and $\overline{AC}$ is the hypotenuse of the right triangle. An *equiangular* triangle has three congruent angles.

Exercise 12-2

1. Using a protractor and a ruler, construct a triangle with $AB = 4$ in., $m\angle A = 110°$, and $m\angle B = 25°$. Give two names for this kind of triangle.
2. In the figure, what side is common to $\triangle ADC$ and $\triangle DBC$? What vertices are common to the two triangles?

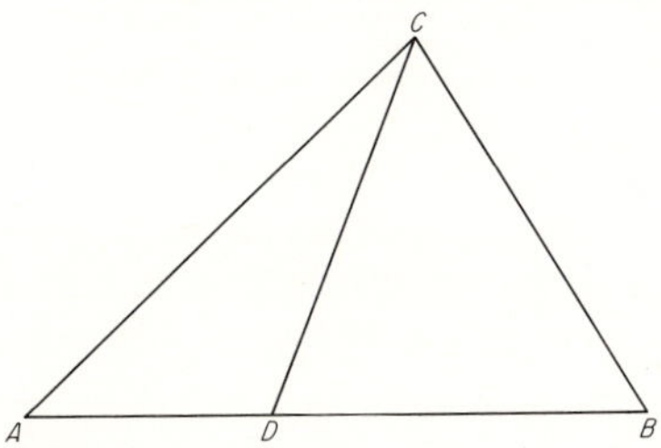

PROB. 2.

In Probs. 3 to 12, state the kind of triangle each figure seems to be (a) according to the sides of the triangles and (b) according to the angles of the triangles.

3. $\triangle KLM$

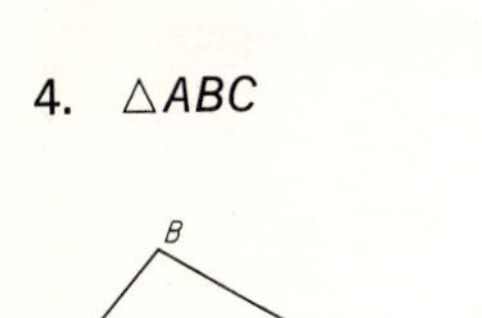

PROB. 3.

4. $\triangle ABC$

PROB. 4.

5. $\triangle DEF$

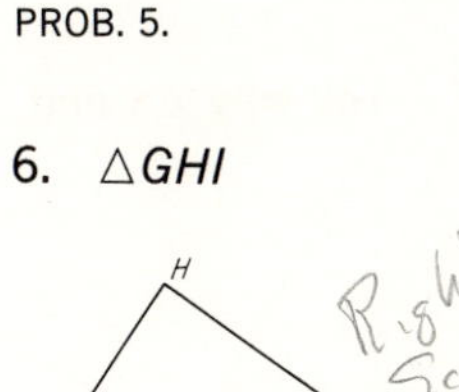

PROB. 5.

6. $\triangle GHI$

PROB. 6.

7. $\triangle JKL$

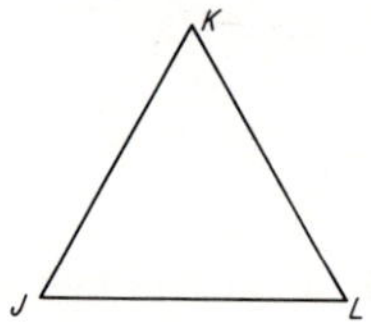

PROB. 7.

8. $\triangle ABC$

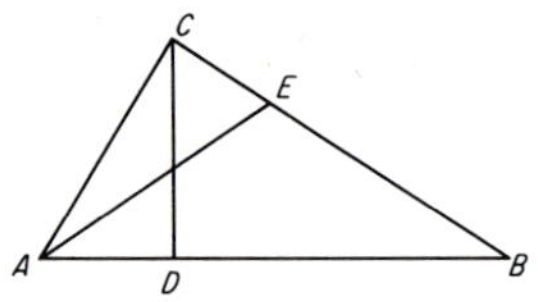

PROBS. 8–12.

9. $\triangle ADC$
10. $\triangle BDC$
11. $\triangle AEC$
12. $\triangle ABE$
13. In the figure for Probs. 8 to 12, indicate two pairs of perpendicular lines.
14.–16. Name a pair of complementary ∠s in each of the diagrams.

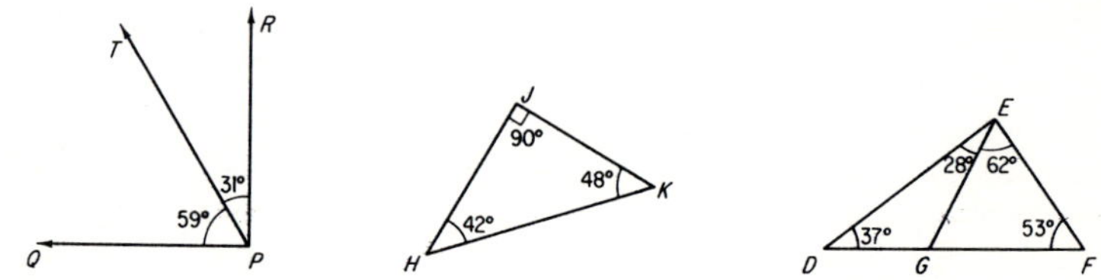

PROBS. 14–16.

17. Tell why $\angle x$ and $\angle y$ are complementary angles.

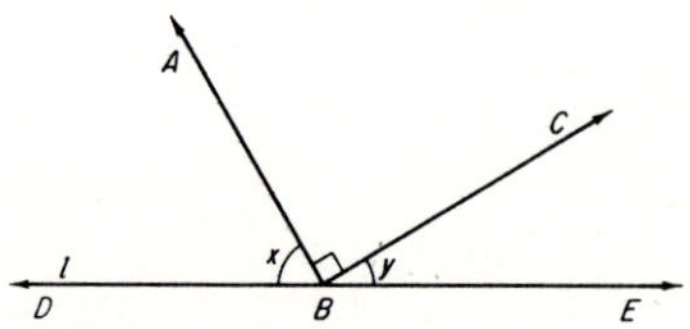

PROB. 17.

18.–20. Name a pair of supplementary angles in each of the diagrams.

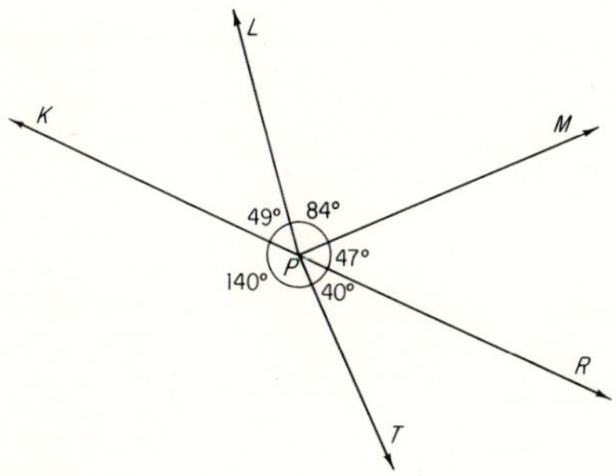

PROB. 18.

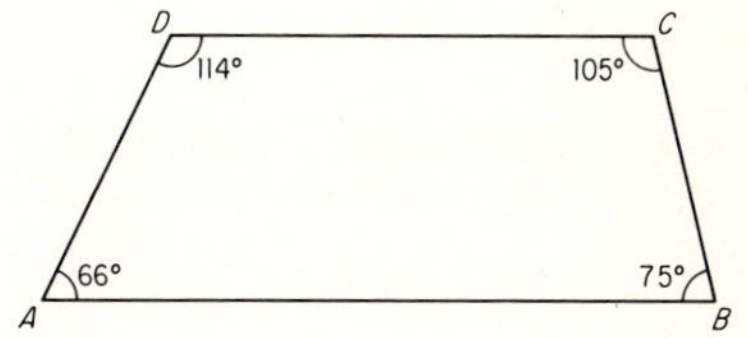

PROB. 20.

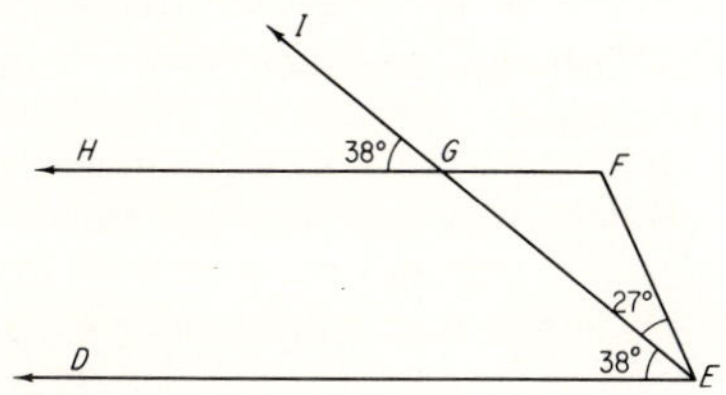

PROB. 19.

21.–23. Name two pairs of adjacent angles in each of the following figures.

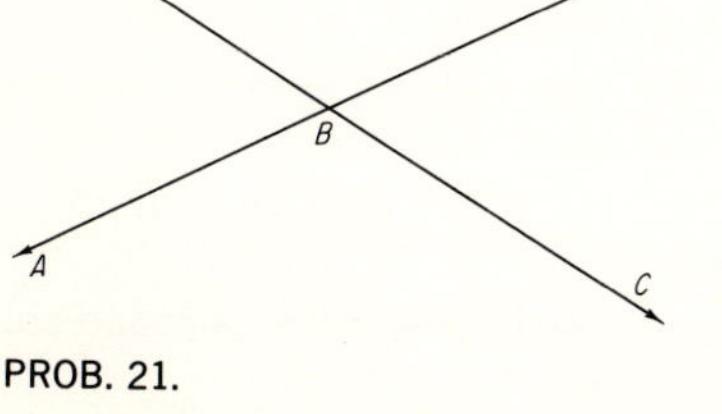

PROB. 21.

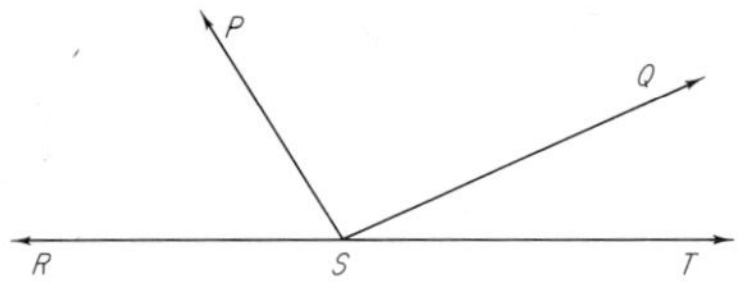

PROB. 22.

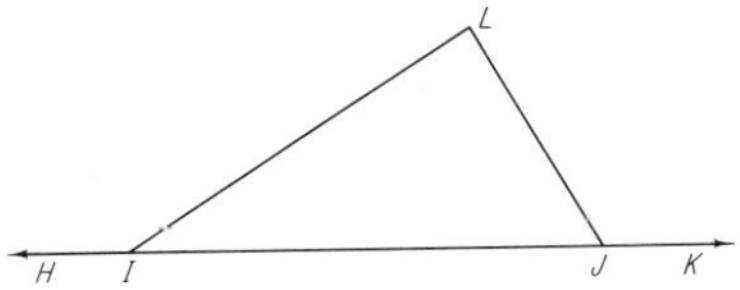

PROB. 23.

24. Find the complement of (a) 30°, (b) 55°, (c) 80°, (d) $\alpha°$.
25. Find the supplement of (a) 30°, (b) 65°, (c) 90°, (d) 135°, (e) $\alpha°$.

12-13 THE METHOD OF GEOMETRY In the study of geometry the student begins with certain basic assumptions which can be agreed upon, even though they cannot be proved. In Chap. 9 we listed a few basic general assumptions called axioms. Let us next list a few fundamental geometric assumptions, called *postulates,* which also are accepted as true without proof.

Postulate 1 Through two given points, one, and only one, straight line can be drawn.
Postulate 2 Two straight lines can intersect in at most one point.
Postulate 3 If B is between A and C on segment AC, then $AC = AB + BC$.
Postulate 4 If D is a point in the interior of $\angle ABC$, then $m\angle ABC = m\angle ABD + m\angle DBC$.
Postulate 5 A line segment has one, and only one, midpoint.
Postulate 6 An angle has one, and only one, bisector.

With these axioms and postulates and a set of definitions, plus a few more postulates, it is possible to *prove* deductively many new statements called theorems. These theorems can then be used to prove new theorems.

Some of the more important theorems proved in geometry follow:

Theorem 12-1 All right angles are congruent.
Theorem 12-2 All straight angles are congruent.
Theorem 12-3 Vertical angles are congruent.
Theorem 12-4 A straight-line segment is the shortest line segment that can be drawn between two points.
Theorem 12-5 At any point on a straight line one, and only one, perpendicular can be drawn to the line.
Theorem 12-6 From any point outside a straight line one, and only one, perpendicular to the line can be drawn.

Theorem 12-7 The perpendicular from a point to a straight line is the shortest path from the point to the straight line.

It will not be our purpose in this text to formally prove any theorems or exercises. We will be more concerned with gaining an understanding of the properties of geometric figures. As a general review of this chapter, the student is urged to answer all the questions of the next exercise.

Exercise 12-3

In Probs. 1 to 15, indicate the one word or number that will make the statement true:

1. The sides of a right angle are ______ to each other.
2. The pairs of nonadjacent angles formed when two lines intersect are called ______.
3. An ______ angle is larger than its supplement.
4. The side opposite the right angle of a triangle is called the ______.
5. A triangle with no two sides congruent is called a ______ triangle.
6. If the sum of the measures of two angles is 180°, the angles are ______.
7. A triangle with two congruent sides is called a(n) ______ triangle.
8. The ______ of an angle divides the angle into two angles with equal measures.
9. Complementary angles are two angles the sum of whose measures is equal to ______.
10. The difference between the measures of the complement and the supplement of an angle is always ______.
11. The sum of the measures of the angles about a point is equal to ______.
12. The angle whose measure equals that of its supplement is a ______ angle.
13. An angle with a measure less than 90° is ______.
14. Angles with the same measures are ______.
15. The only point of a line equally distant from two of its points is the ______ of the segment with these points as its end points.

In Probs. 16 to 30 indicate whether the statement is *always true* (mark T) or *not always true* (mark F).

16. An obtuse angle has a greater measure than a right angle.
17. A straight line has a fixed length.
18. If an obtuse angle is bisected, two acute angles will be formed.
19. The measure of an angle depends on the length of its sides.
20. Complementary angles are angles the sum of whose measures is equal to a right angle.

21. A straight angle is a straight line.
22. All isosceles $\triangle$s are equilateral.
23. When the measure of an acute angle is doubled, an obtuse angle is formed.
24. A ray has two end points.
25. $FG = JK$ and $\overline{FG} \cong \overline{JK}$ are equivalent statements.
26. The sides of an angle are rays.
27. If S is not between R and T, then T is between R and S.
28. A collinear set of points is a line.
29. If G is a point in the interior of $\angle DEF$, then $m\angle DEG + m\angle FEG = m\angle DEF$.
30. If two lines intersect to form vertical angles that are supplementary, the vertical angles are right angles.

31.–38. Given the scaled line with points and their corresponding coordinates as indicated. Complete the following.

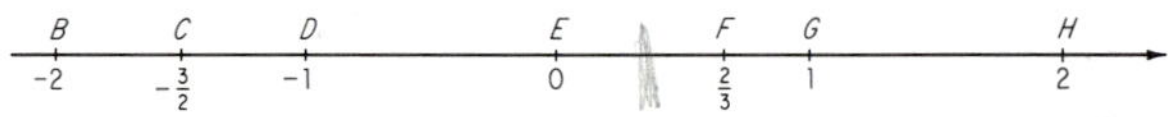

PROBS. 31–38.

31. $DH =$
32. $GB =$
33. $DF =$
34. $HB =$
35. $CG =$
36. $CF =$
37. The coordinate of the midpoint of $\overline{EH}$ is ________.
38. The coordinate of the midpoint of $\overline{DH}$ is ________.
39. If $m\angle A$ is 40°, what is the measure of the complement of $\angle A$?
40. If $m\angle B$ is 110°, what is the measure of the supplement of $\angle B$?

41.–46. Given $m\angle AOC = 40°$; $m\angle COE = 70°$; $\overrightarrow{OB}$ bisects $\angle AOC$ and $\overrightarrow{OD}$ bisects $\angle COE$. Complete the following.

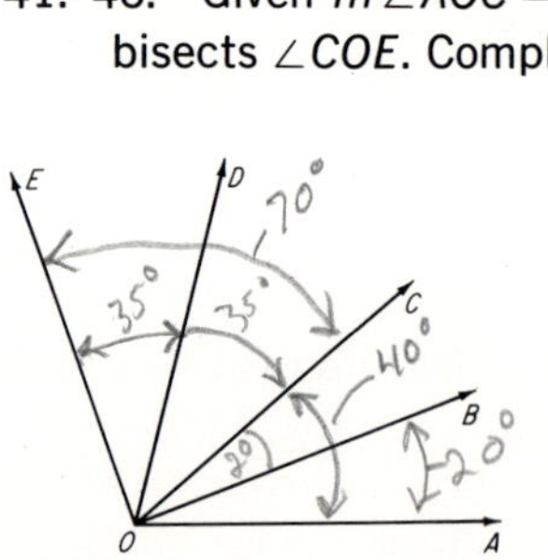

PROBS. 41–46.

41. $m\angle AOB$
42. $m\angle COD$
43. $m\angle BOD$
44. $m\angle EOB$
45. $m\angle AOE$
46. $m\angle BOE$
47. $\overleftrightarrow{AB}$ and $\overleftrightarrow{CD}$ are straight lines intersecting at E. What must the measures of angles α and ϕ be?
48. The measure of angle θ is three times the measure of $\angle\phi$. What is the measure of $\angle\theta$?

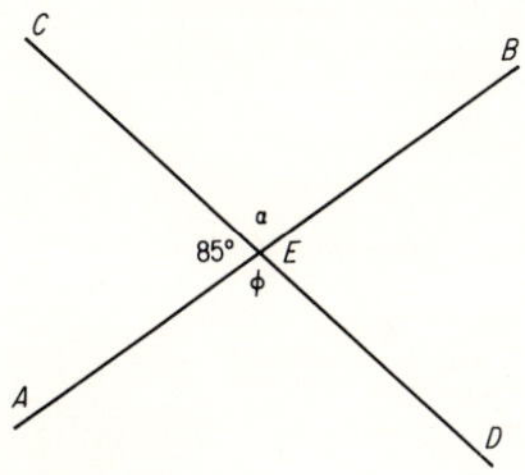

PROB. 47.

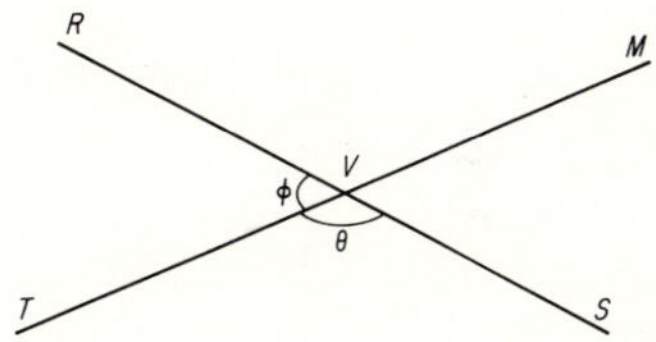

PROB. 48.

49. $\overline{CM}$ bisects $\angle ACB$; $m\angle ACB = 110°$. The $m\angle BCM =$ ______.
50. Angle α is the complement of an angle whose measure is 38°; $\angle\beta$ is the supplement of $\angle\alpha$. The $m\angle\beta =$ ______.

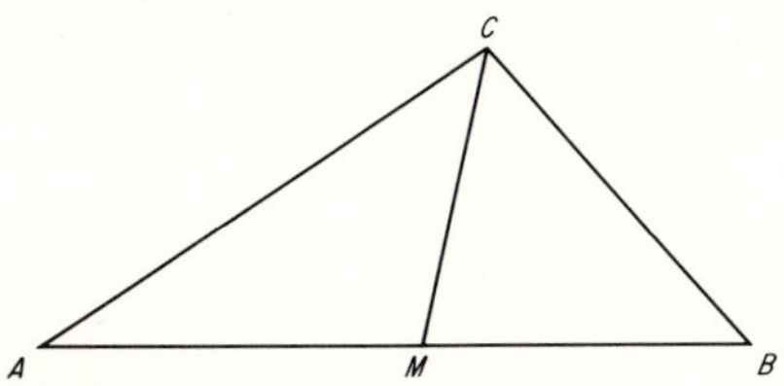

PROB. 49.

CHAPTER 13

CONGRUENCE

13-1 CONGRUENT FIGURES Two figures are *congruent* if they can be made to coincide in all their parts. Congruent figures therefore have the same shape and size. Conversely, two figures whose corresponding parts can be made to coincide are congruent. The coincident parts are called *corresponding parts.*

Congruent figures are common in modern technology. For example, industry today relies a great deal on mass production and assembly-line manufacture. Often each part of a machine or household article is made by precision manufacture to have exactly the same shape and size. These parts are then sent to an assembly plant where the parts can be fitted together to form a complete unit.

The mass production and repair of automobiles, airplanes, television sets, automatic washers, refrigerators, and many other products of modern industry depend on the manufacture of thousands of parts having exactly the same shape and size. It is especially important in repairing a complex machine that the necessary replacement parts match exactly the original parts.

The symbol for congruence is $\cong$. Thus, $\triangle ABC \cong \triangle DEF$ is read "triangle ABC is congruent to triangle DEF." If the triangles are congruent (see Fig. 13-1), we know that $AB = DE$, $BC = EF$, $AC = DF$, $m\angle A = m\angle D$, $m\angle B = m\angle E$, and $m\angle C = m\angle F$. In like manner, if we know that the three corresponding pairs of sides and the three pairs of angles have the same measures, we can state that the triangles are congruent.

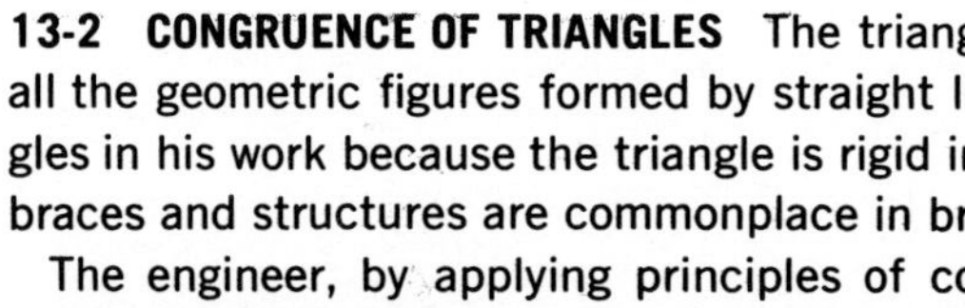

FIG. 13-1.

13-2 CONGRUENCE OF TRIANGLES The triangle is the most widely used of all the geometric figures formed by straight lines. The engineer uses triangles in his work because the triangle is rigid in structural design. Triangular braces and structures are commonplace in bridges, towers, and gates.

The engineer, by applying principles of congruent triangles, is able to duplicate structures. He is able to measure the sides and angles of a triangle and compute its area.

It is not the purpose of this text to develop theorems relating conditions for congruence of triangles. We shall simply list the various methods for establishing congruence and then determine if certain figures are congruent.

While we *define* two triangles as congruent if three pairs of sides and three pairs of angles have equal measures, triangles can be *proved* congruent if fewer corresponding parts are known to have the same measures. Another postulate must first be accepted.

Postulate 7 (the SAS postulate) Two triangles are congruent if two sides and the included angle of one are, respectively, congruent to the two sides and included angle of the other.

Thus, Postulate 7 states that, in Fig. 13-2, if $AB = DE$, $AC = DF$, and $\angle A \cong \angle D$, then $\triangle ABC \cong \triangle DEF$.

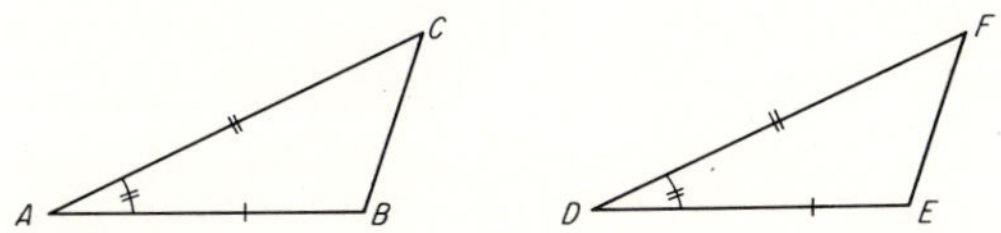

FIG. 13-2.

The student will often find he is aided in checking for congruence of triangles if he designates congruent sides and angles by similar check marks. Thus, in Fig. 13-3, if it is given $RS = QS$, $m\angle RST = m\angle QSP$, and $TS = PS$,

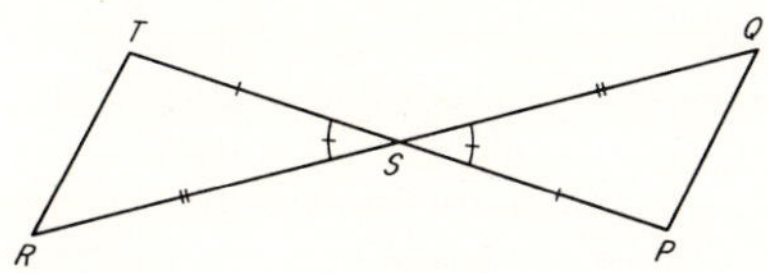

FIG. 13-3.

the student can readily see that the conditions of Postulate 7 are satisfied and that $\triangle RST \cong \triangle QSP$.

It is important that the student recognize, in using Postulate 7 to prove triangles congruent, that the congruent angles must be *between* (formed by) the corresponding sides. If the congruent angles are not between the two known congruent sides, it does not necessarily follow that the correspondence will give a congruence. In $\triangle RST$ and $\triangle KLM$ (Fig. 13-4) note that,

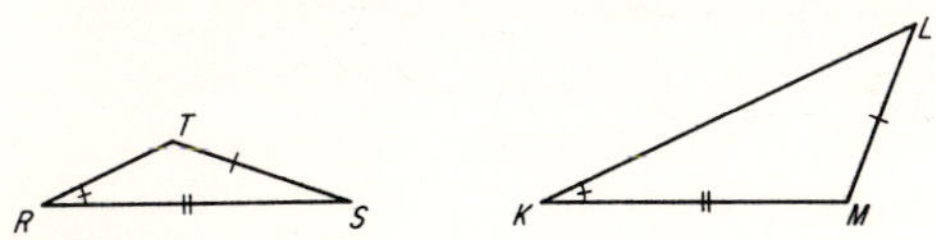

FIG. 13-4.

though $RS = KM$, $ST = ML$, and $\angle R \cong \angle K$, the triangles certainly are not congruent.

Thus, whenever we know or can prove that two sides and the included angle of one triangle are congruent to two sides and the included angle of the other, we can give Postulate 7 as the reason for stating that the two triangles are congruent. We shall abbreviate the words of the postulate with the letters SAS.

Example 13-1

Given: $\overline{CD} \perp \overline{AB}$; $\overline{CD}$ bisects $\overline{AB}$ at D.
Prove: $\triangle ADC \cong \triangle BDC$.

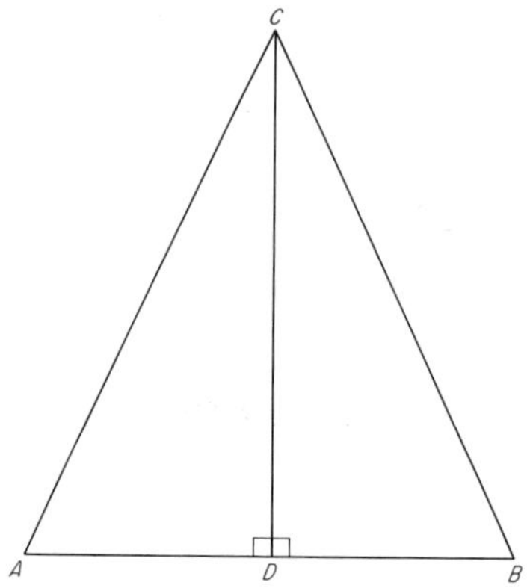

EXAMPLE 13-1.

PROOF

1 $\overline{CD} \perp \overline{AB}$, given.
2 $\angle ADC$ is a right $\angle$ because $\perp$ lines form right angles.
3 $\angle BDC$ is a right $\angle$ because $\perp$ lines form right angles.
4 $\angle ADC \cong \angle BDC$ because right angles are congruent.
5 $\overline{CD}$ bisects $\overline{AB}$, given.
6 $AD = BD$ because a bisector divides a segment into two congruent parts.
7 $CD = CD$, by reflexive property (Axiom 1).
8 Then $\triangle ADC \cong \triangle BDC$, by SAS.

Exercise 13-1

The triangles in Probs. 1 to 15 are marked to indicate congruent sides and angles. Indicate which pairs of triangles can be proved congruent by using the SAS postulate, definitions, axioms, and the other postulates listed so far in this text. If the triangles cannot be proved by the SAS postulate, tell why.

1.

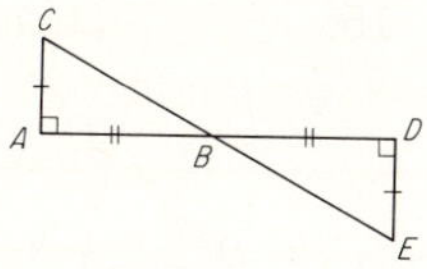

2.

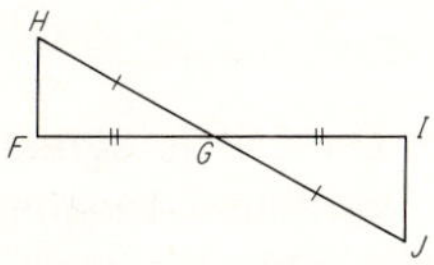

3.

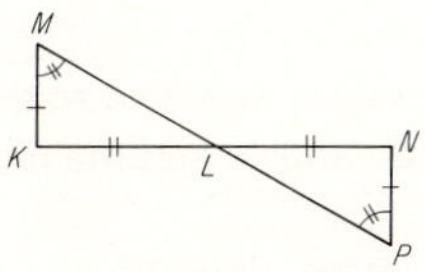

4.

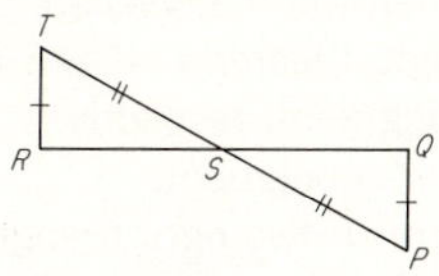

5.

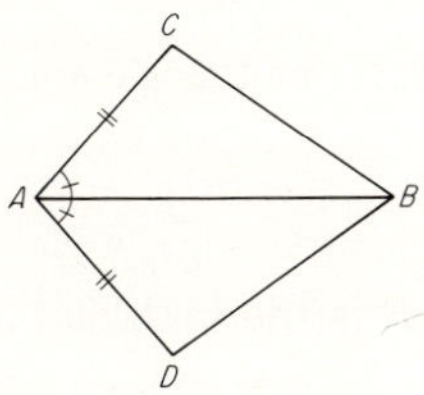

6.

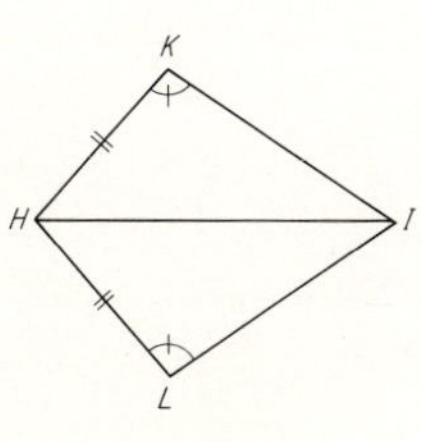

7.

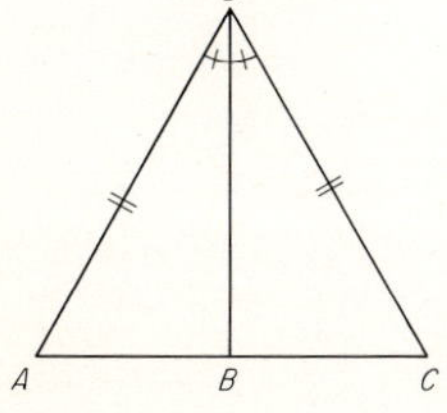

8.

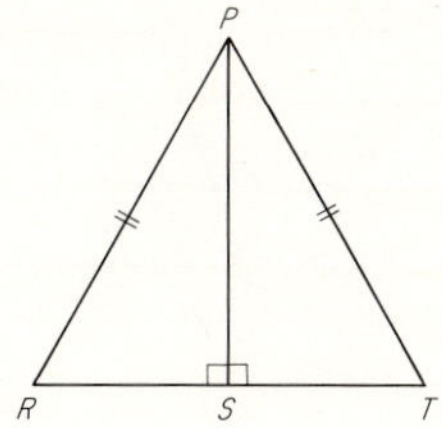

9.

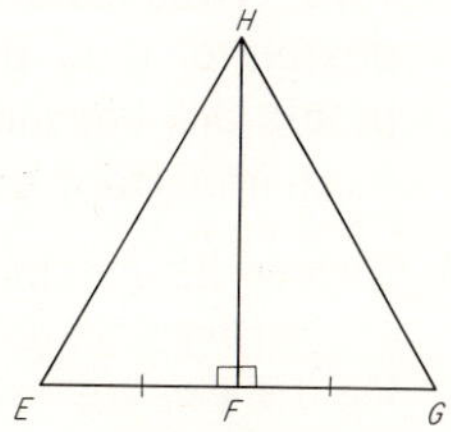

10.

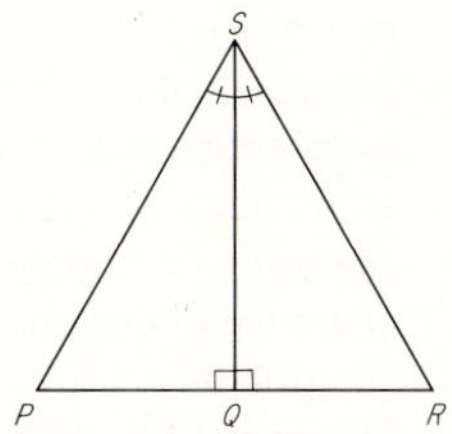

11.

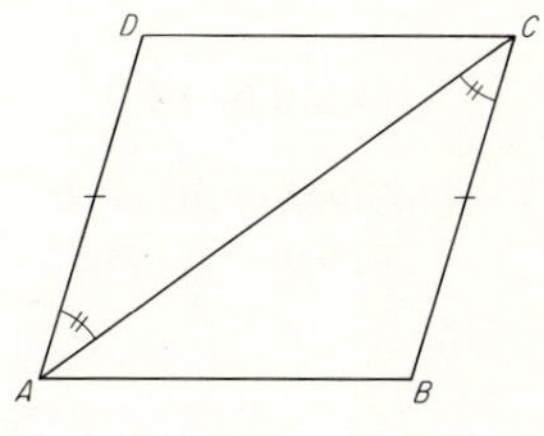

12.

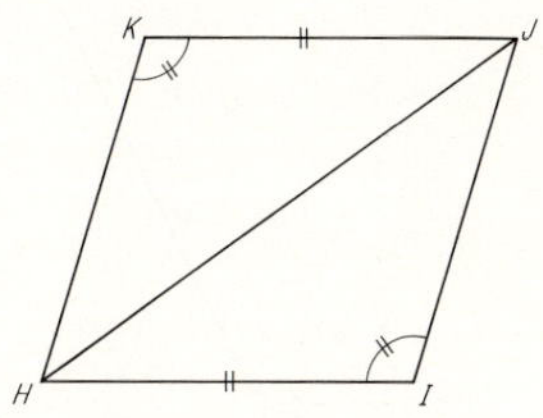

13.

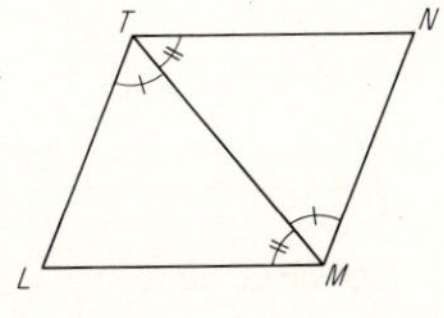

14. 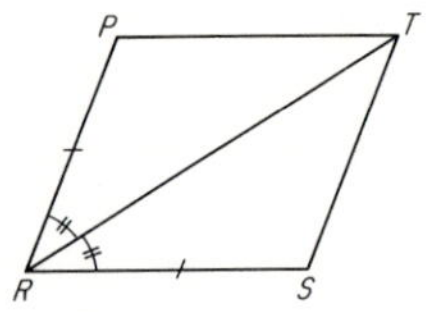

15. 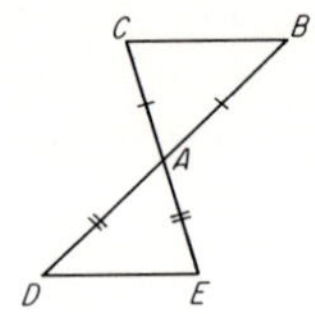

13-3 THEOREMS ON CONGRUENCE Once it is agreed to accept the SAS postulate, it is possible to *prove* additional theorems for triangles. The proofs are beyond the scope of this book. We shall just list some of the more important ones.

Theorem 13-1 (ASA theorem) If two triangles have two angles and the included side of one congruent to the corresponding two angles and the included side of the other, the triangles are congruent.

Theorem 13-2 (SSS theorem) If the three sides of one triangle are congruent, respectively, to the three sides of another, the triangles are congruent.

Theorem 13-3 (hypotenuse–acute-angle theorem) If two right triangles have the hypotenuse and acute angle of one congruent, respectively, to the hypotenuse and acute angle of the other, the triangles are congruent.

Theorem 13-4 (hypotenuse-leg theorem) If two right triangles have the hypotenuse and a leg of one congruent to the hypotenuse and a leg of the other, the triangles are congruent.

Theorem 13-5 The base angles of an isosceles triangle are congruent.

Example 13-2

Given: $AC = BC$; $\angle APM \cong \angle BPN$; P is the midpoint of $\overline{AB}$.
Prove: $\triangle APM \cong \triangle BPN$.

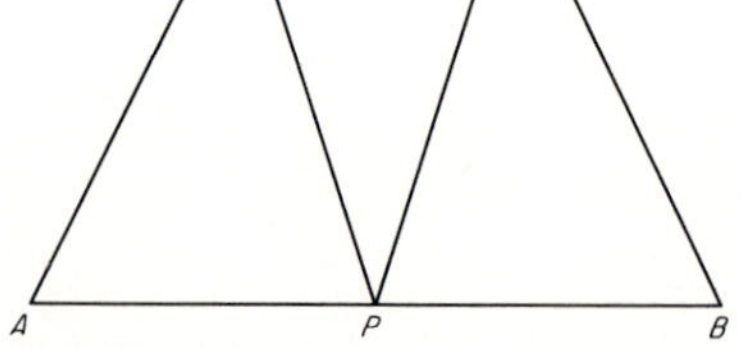

EXAMPLE 13-2.

PROOF

1 $AC = BC$ given.
2 $\angle MAP \cong \angle NBP$ because the base angles of an isosceles $\triangle$ are $\cong$.
3 P is the midpoint of $\overline{AB}$, given.
4 $AP = BP$ because the midpoint of a segment divides it into two congruent segments.
5 $\angle APM \cong \angle BPN$ given.
6 $\triangle APM \cong \triangle BPN$ by ASA.

Exercise 13-2

The triangles in Probs. 1 to 15 are marked to indicate congruent sides and angles. Indicate which pair of sides and angles is given or can be proved congruent, and if possible, indicate the theorem or postulate that can be used to prove pairs of triangles congruent. Indicate which triangles cannot be proved congruent by any of the methods discussed thus far.

1.
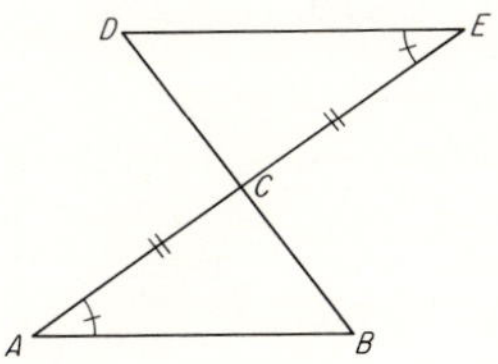

2.
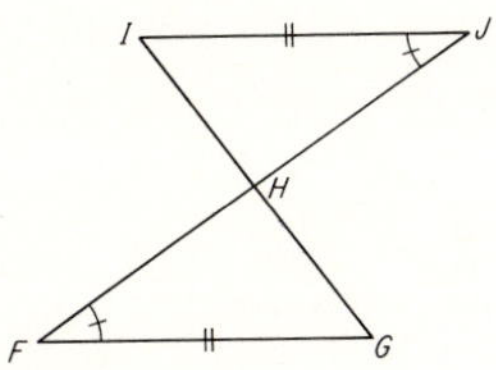

3.
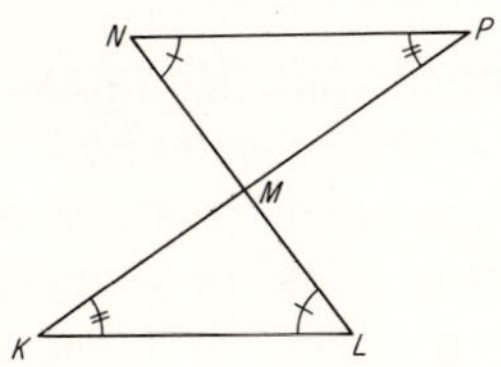

4.
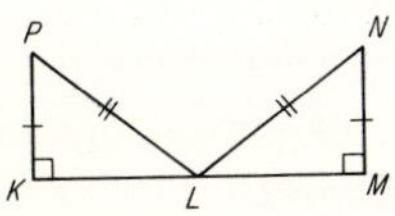

5.
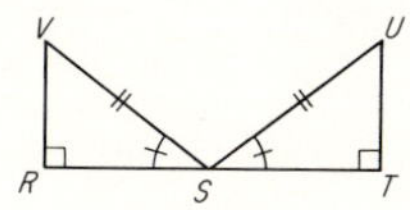

6.
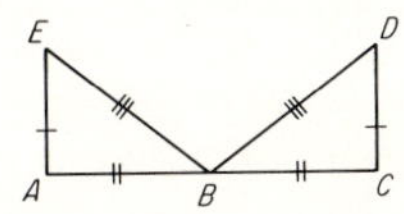

7.
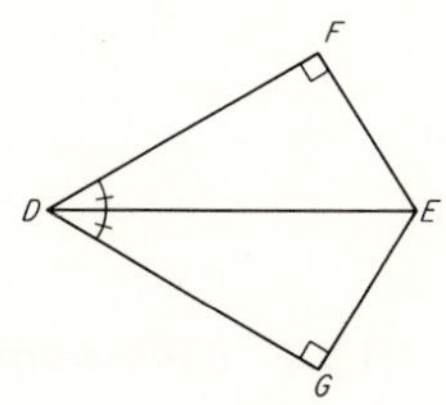

8.
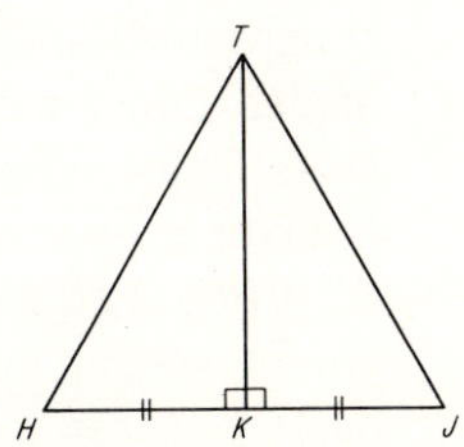

9.

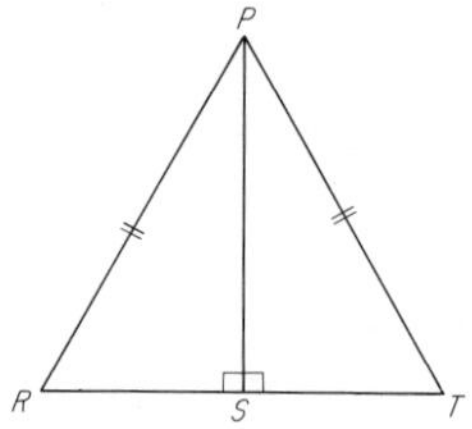

10.

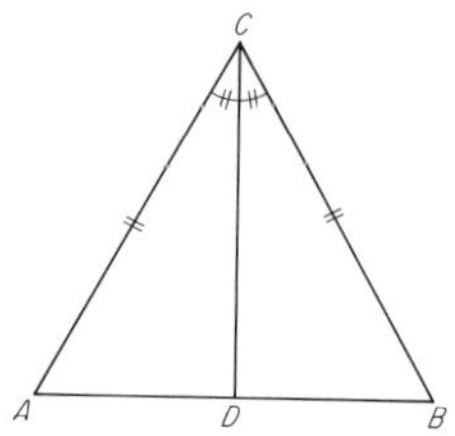

11.

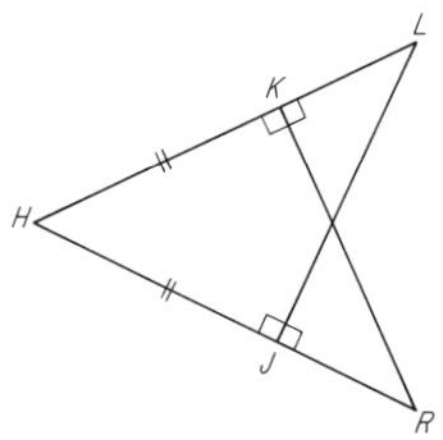

12.

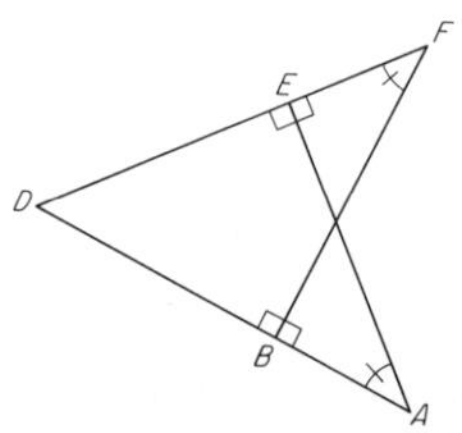

13.

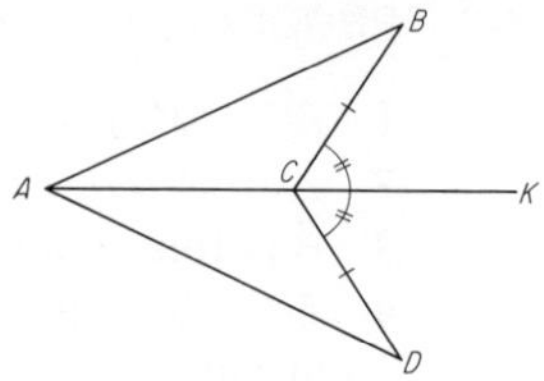

14.

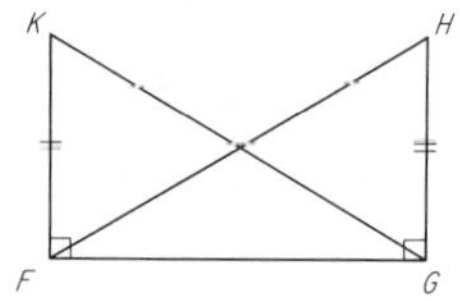

15.

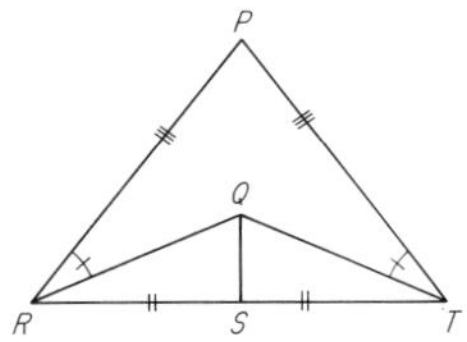

13-4 CORRESPONDING PARTS OF CONGRUENT TRIANGLES Since congruent triangles have the same shape and size, we know that corresponding sides and corresponding angles of congruent triangles are congruent. In two congruent triangles, corresponding sides are found opposite corresponding angles. Conversely, corresponding angles are found opposite corresponding sides. Thus, in Fig. 13-5, $\triangle ABC$ and DEC can be proved congruent by having given or by proving the following corresponding pairs congruent: $\angle E \cong \angle B$; $EC = BC$; $\angle ECD \cong \angle BCA$. Now, since we have proved the triangles congruent, we can state that other pairs of corresponding parts are congruent; namely, $\angle D \cong \angle A$, $ED = BA$, and $DC = AC$.

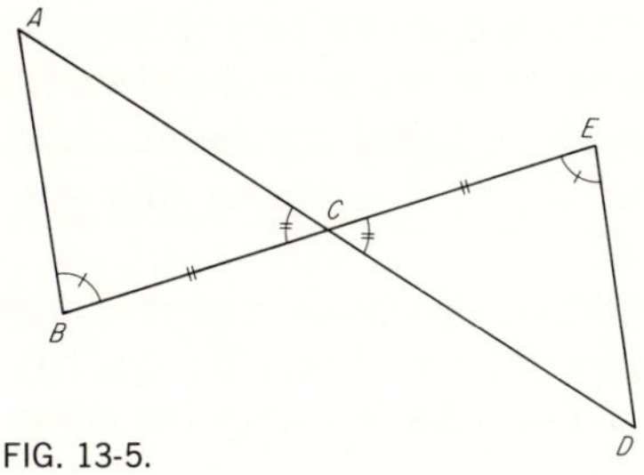

FIG. 13-5.

We now have an additional method of proving segments and angles congruent. If we can show that the segments and angles are corresponding parts of congruent figures, we know that they are congruent.

Example 13-3

Given: $\overrightarrow{AD}$ bisects $\angle CAB$; $\overline{DB} \perp \overline{AB}$; $\overline{DC} \perp \overline{AC}$.
Prove: $DB = DC$.

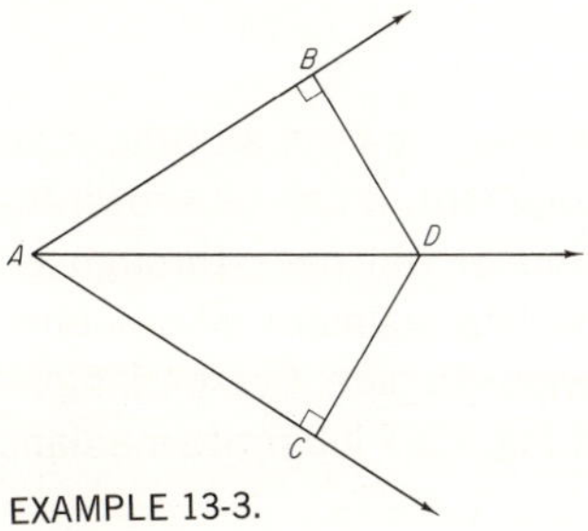

EXAMPLE 13-3.

PROOF

1 $\overrightarrow{AD}$ bisects $\angle CAB$, given.
2 $m\angle DAB = m\angle DAC$ because a bisector divides an angle into two angles with equal measures.
3 $\overline{DB} \perp \overline{AB}$ and $\overline{DC} \perp \overline{AC}$ because it is given.
4 $\angle ABD$ and $\angle ACD$ are right angles because $\perp$ lines form right $\angle$s.
5 $\triangle ABD$ and $\triangle ACD$ are right $\triangle$s by definition of "right triangle."
6 $AD = AD$ by the reflexive property (Axiom 1).
7 $\triangle ABD \cong \triangle ACD$ by hypotenuse–acute-angle theorem.
8 $DB = DC$ because corresponding parts of congruent triangles are congruent.

13-5 SPECIAL LINES AND ANGLES OF A TRIANGLE Several special lines, line segments, and angles have particular significance when associated with triangles.

The *base* of a triangle is the side upon which it appears to rest. Any one of the three sides may be designated as the base of a given triangle. The angle opposite the base of a triangle is called the *vertex angle.* A triangle has three bases and three vertex angles. The angles adjacent to the base are termed *base angles.*

An *altitude* of a triangle is the perpendicular segment drawn from a vertex to the line containing the opposite side. Every triangle has three altitudes. The dotted-line segments of Fig. 13-6 illustrate the three altitudes of an

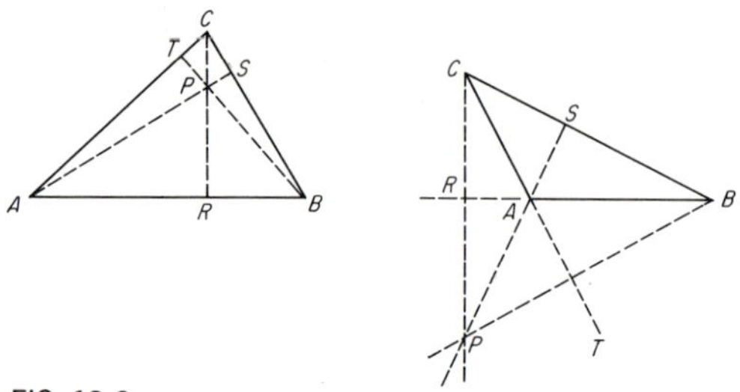

FIG. 13-6.

acute triangle and an obtuse triangle. The term altitude is also often used to refer to the measure of the line segment. It can be shown that the three altitudes (or extension of the altitudes) will pass through a common point.

The *median* of a triangle is a line segment whose end points are any vertex and the midpoint of the opposite side. Every triangle has three medians. The dotted-line segments of Fig. 13-7 illustrate medians of a triangle. It

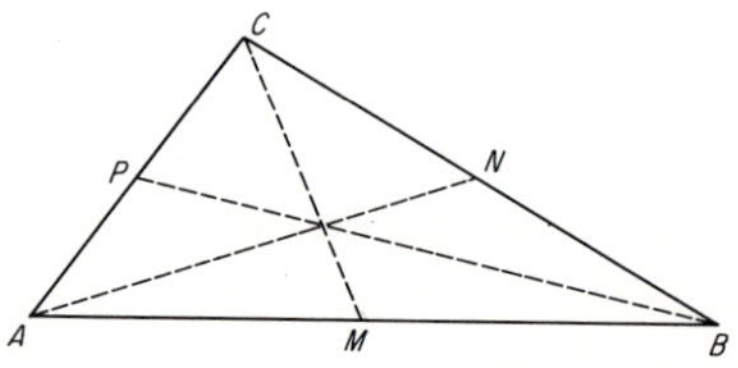

FIG. 13-7.

can be shown that the three medians of a triangle pass through a common point, called the *centroid.* This point is also called the *center of gravity* in the study of mechanics. If a sheet of material is cut into a triangular shape, it will balance on a suspension point through the centroid.

The *angle bisector* of a triangle is a segment which divides an angle of the triangle into two congruent angles. The angle bisector has its end points on a vertex and a side opposite the angle. $\overline{BD}$ is the bisector of $\angle B$ of $\triangle ABC$ in Fig. 13-8. Every triangle has three angle bisectors. It can be shown that the

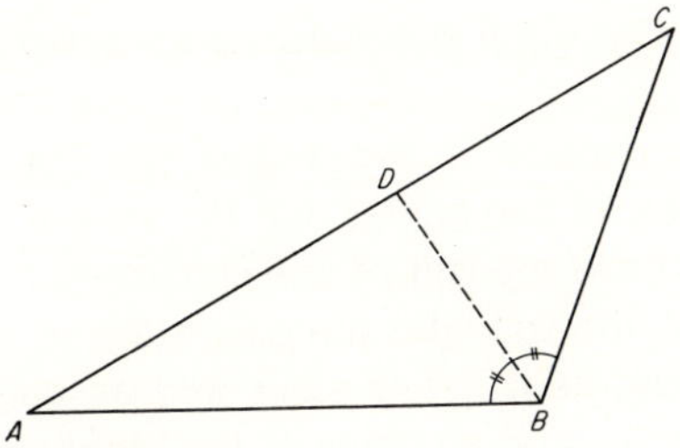

FIG. 13-8.

three angle bisectors meet in a common point which is equidistant from the three sides of the triangle.

In Fig. 13-9, if S is between R and Q, then $\angle QST$ is an *exterior angle* of

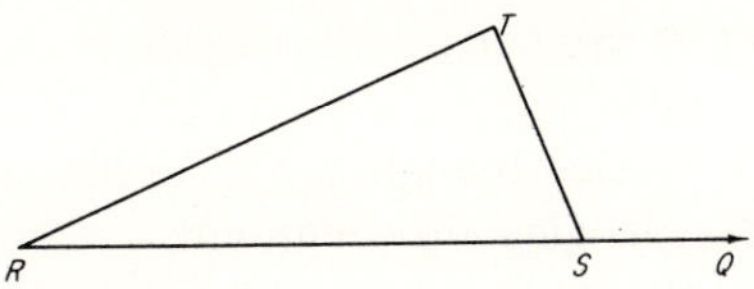

FIG. 13-9.

$\triangle RST$. Every triangle has six exterior angles. These exterior angles form three pairs of vertical angles. $\angle R$ and $\angle T$ are called *nonadjacent interior angles* of $\angle QST$.

Exercise 13-3

In Probs. 1 to 9, indicate the word or words which will make the statement true.

1. An _______ angle of a triangle is an angle formed by one side of the triangle and the prolongation of another side through their common point.
2. A _______ of a triangle is the line segment joining a vertex and the midpoint of the opposite side of the triangle.
3. Corresponding sides of congruent triangles are found opposite the _______ angles of the triangles.
4. An _______ of a triangle is the line segment drawn from a vertex perpendicular to the opposite side.
5. _______ parts of congruent triangles are congruent.
6. The _______ angles of an isosceles triangle are congruent.
7. If the median of a triangle is also an altitude, the triangle is _______.
8. The bisectors of two supplementary adjacent angles form a _______ angle.
9. The side of a right triangle opposite the right angle is called the _______.

In Probs. 10 to 25 indicate whether the statement is *always true* (mark T) or *not always true* (mark F).

10. Two triangles are congruent if two angles and the side of one are congruent respectively to two angles and the side of the other.
11. If two right triangles have the legs of one congruent respectively to the two legs of the other, the triangles are congruent.
12. Two triangles are congruent if two sides and an angle of one are ≅ respectively to two sides and an angle of the other.
13. Two triangles that have ≅ bases and ≅ altitudes are congruent.
14. The bisectors of two adjacent supplementary angles are perpendicular to each other.
15. The bisectors of two angles of a triangle are perpendicular to each other.
16. Two equilateral triangles are congruent if a side of one triangle is ≅ to a side of the other.
17. If the sides of one isosceles triangle are ≅ to the sides of a second isosceles triangle, the triangles are congruent.
18. The altitude of a triangle passes through the midpoint of a side.
19. An exterior angle of a triangle is the supplement of at least one interior angle of the triangle.
20. If two triangles have their corresponding sides congruent, then the corresponding angles are congruent.
21. If two triangles have their corresponding angles congruent, then the corresponding sides are congruent.
22. No two angles of a scalene triangle are congruent.
23. The supplement of an angle is always an obtuse angle.
24. A perpendicular to a line bisects the line.
25. The median to the base of an isosceles triangle is perpendicular to the base.
26. *Given:* $\overline{TM}$ bisects $\angle RTS$; $RT = ST$.
    Prove: $RM = SM$.

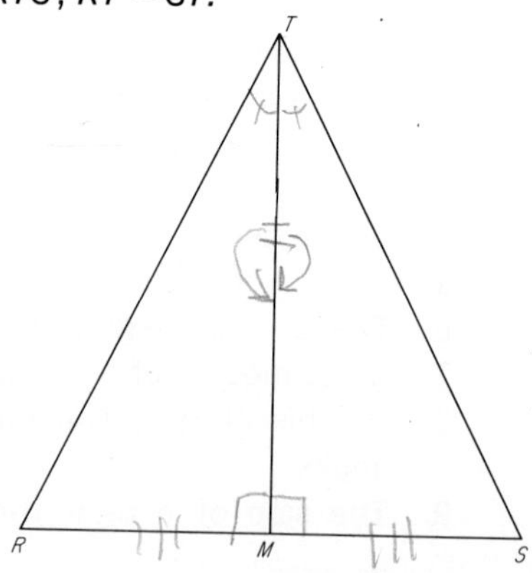

PROB. 26.

27. *Given:* $PB = PC$; $\overline{PB} \perp \overline{AB}$; $\overline{PC} \perp \overline{AC}$.
Prove: $\overrightarrow{AP}$ bisects $\angle BAC$.

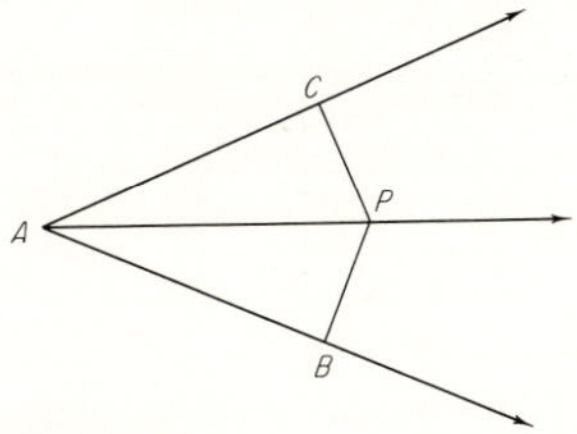

PROB. 27.

28. *Given:* $\overline{PQ} \perp \overline{PS}$; $\overline{RS} \perp \overline{PS}$; T bisects $\overline{PS}$.
Prove: T bisects $\overline{RQ}$.

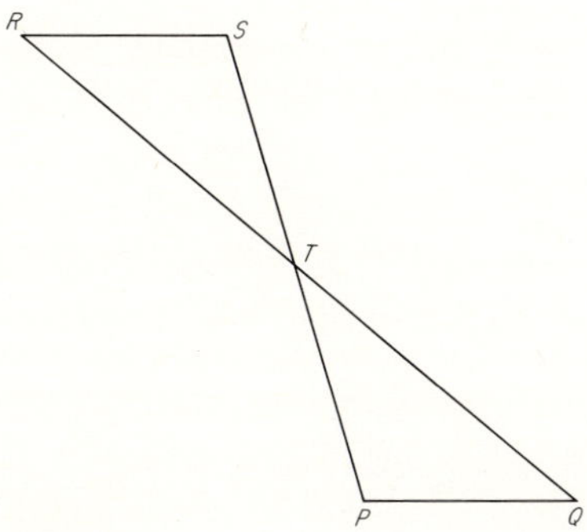

PROB. 28.

29. *Given:* $ST = SR$; $\overline{PT} \perp \overline{ST}$; $\overline{PR} \perp \overline{SR}$.
Prove: $\overrightarrow{SP}$ bisects $\angle RST$.

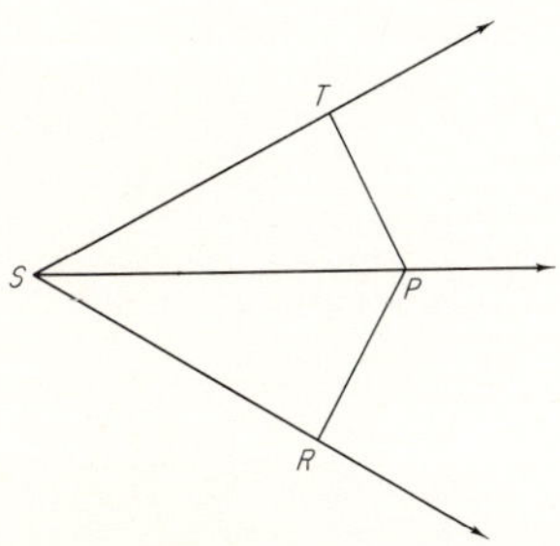

PROB. 29.

CHAPTER 14

PARALLEL LINES AND PARALLELOGRAMS

14-1 PARALLEL LINES ARE COMMONPLACE Two straight lines can intersect in at most one point. If they do not intersect, they are parallel or do not lie in the same plane.

DEFINITION *Two lines which lie in the same plane and do not intersect if extended indefinitely are parallel lines.*

Examples of parallel lines can be found in many places. The opposite edges of this page are parallel. Two railroad tracks, the yardage markers on the football field, and the pickets of a fence are approximations of parallels.

The symbol for "parallel" or "is parallel to" is $\parallel$. Rays and segments are considered parallel if the lines which contain them are parallel. Thus, in Fig. 14-1, the statements $\overleftrightarrow{AC} \parallel \overleftrightarrow{RT}$, $\overline{AB} \parallel \overline{RS}$, $\overrightarrow{AC} \parallel \overrightarrow{RT}$ are each equivalent to the statement $l \parallel m$.

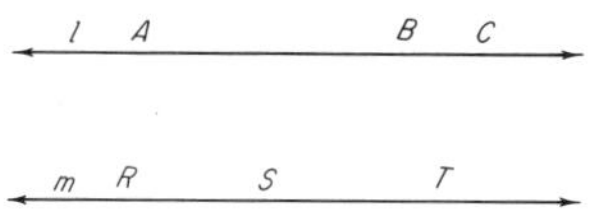

FIG. 14-1.

One of the most important postulates in the study of geometry is the parallel postulate.

Postulate 8 *Parallel postulate* Through a given point not on a given line there is at most one line which can be drawn parallel to the given line.

This postulate tells us that, if, in Fig. 14-2, it is given that $\overleftrightarrow{CD} \parallel \overleftrightarrow{AB}$ and

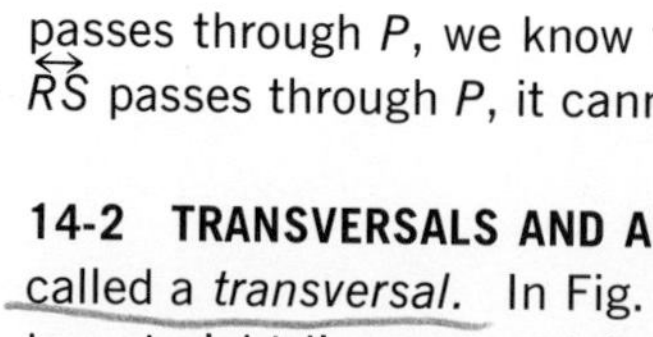

FIG. 14-2.

passes through P, we know that, if $\overleftrightarrow{RS} \parallel \overleftrightarrow{AB}$, it cannot pass through P or, if $\overleftrightarrow{RS}$ passes through P, it cannot be parallel to $\overleftrightarrow{AB}$.

14-2 TRANSVERSALS AND ANGLES A line which cuts two or more lines is called a *transversal.* In Fig. 14-3, $\overleftrightarrow{GK}$ is a transversal of $\overleftrightarrow{AB}$ and $\overleftrightarrow{CD}$. When two straight lines are cut by a transversal, eight angles are formed. These angles have the following names. The pairs $\angle x$, $\angle t$ and $\angle q$, $\angle y$ are called *alternate interior angles.* The pairs $\angle w$, $\angle s$ and $\angle z$, $\angle r$ are *alternate exterior*

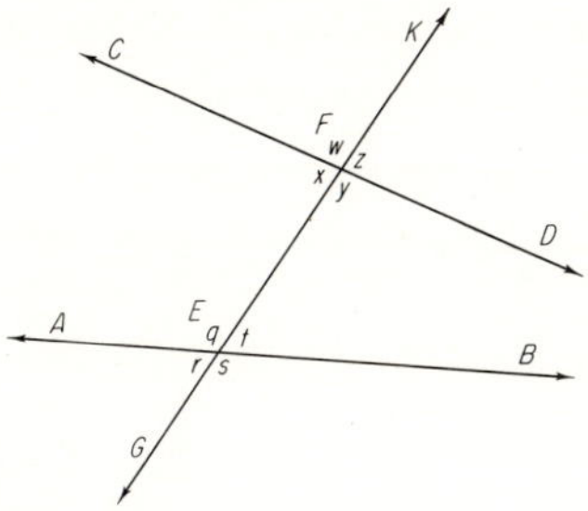

FIG. 14-3.

angles. The pairs $\angle z$ and $\angle t$, $\angle y$ and $\angle s$, $\angle w$ and $\angle q$, and $\angle x$ and $\angle r$ are called *corresponding angles.* A study of the figure will reveal why the terms are used.

14-3 THEOREMS ON PARALLEL AND PERPENDICULAR LINES If Postulate 8 is accepted, it becomes possible to prove many theorems relating parallel and perpendicular lines. We shall now state some of the more important ones. These theorems are often used by the designer, the engineer, the technician, and the craftsman.

Theorem 14-1 Two lines each parallel to a third line are parallel to each other. Thus if, in Fig. 14-4, $l \parallel n$ and $m \parallel n$, then $l \parallel m$.

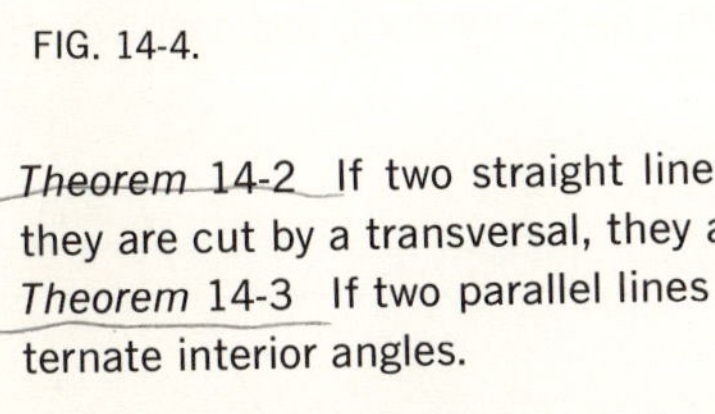

FIG. 14-4.

Theorem 14-2 If two straight lines form congruent alternate interior angles when they are cut by a transversal, they are parallel.

Theorem 14-3 If two parallel lines are cut by a transversal, they form congruent alternate interior angles.

The last two theorems are known as *converse statements*. If, in Fig. 14-5, it is given that transversal t cuts $\overleftrightarrow{AB}$ and $\overleftrightarrow{CD}$ so as to make $m\angle\alpha = m\angle\beta$, we can conclude that $\overleftrightarrow{AB} \parallel \overleftrightarrow{CD}$. Conversely, if we know that $\overleftrightarrow{AB} \parallel \overleftrightarrow{CD}$, we know that $m\angle\alpha = m\angle\beta$.

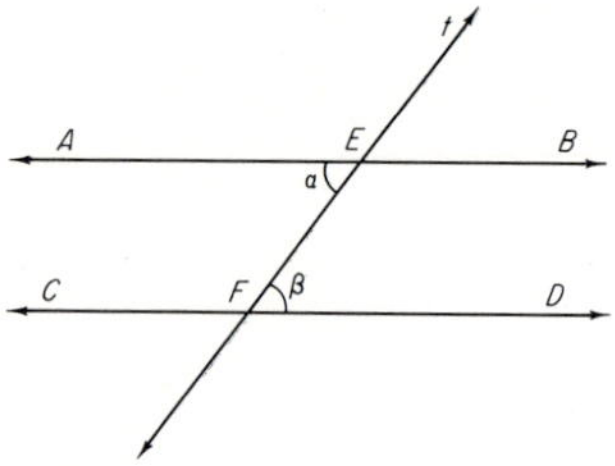

FIG. 14-5.

Theorem 14-4 If two straight lines form congruent corresponding angles when cut by a transversal, the lines are parallel.

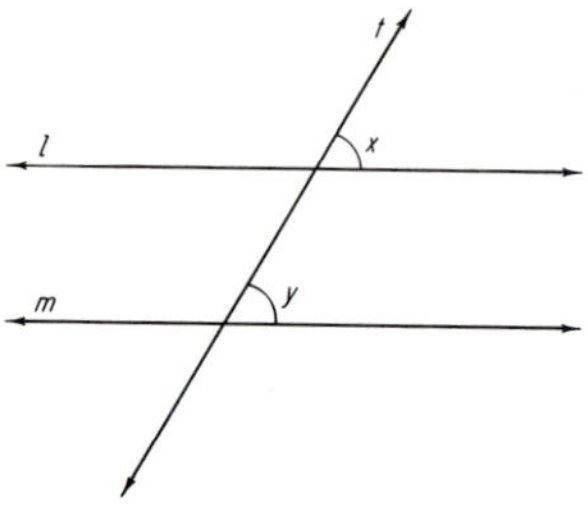

FIG. 14-6.

Theorem 14-5 If two parallel lines are cut by a transversal, they form congruent corresponding angles.

If, in Fig. 14-6, it is known that transversal t cuts lines l and m so that $m\angle x = m\angle y$, we can be certain that $l \parallel m$. Conversely, if it is given that $l \parallel m$, we know that $m\angle x = m\angle y$.

Exercise 14-1

In Probs. 1 to 10, congruent angles are marked with similar hash marks and right angles are indicated with the ⊾ symbol. Indicate whether or not l and m are parallel.

1.

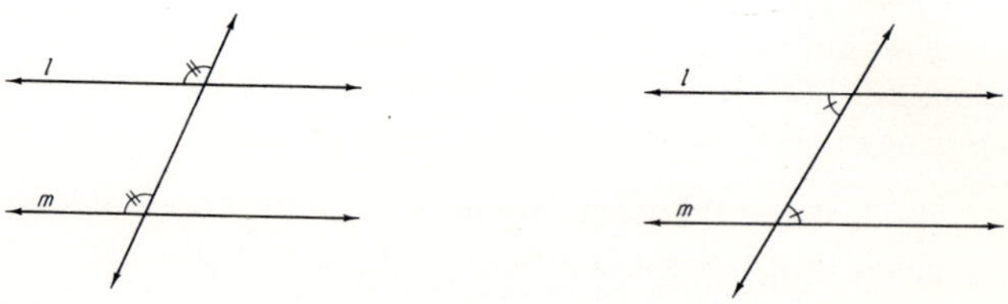

3.

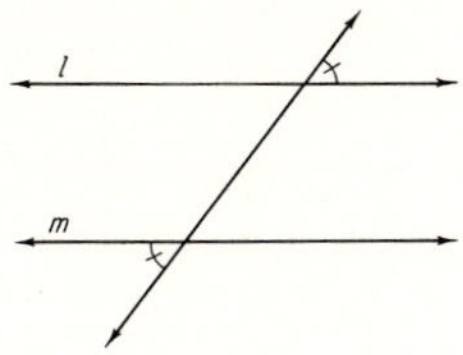

7.

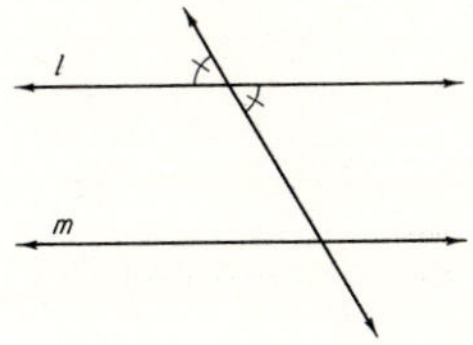

4.

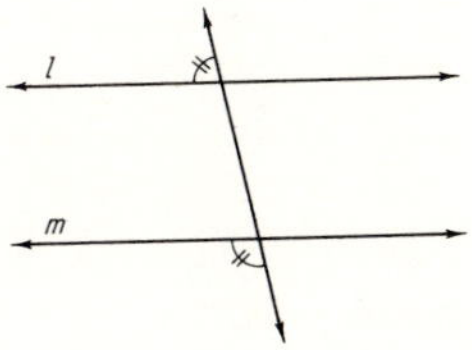

8.

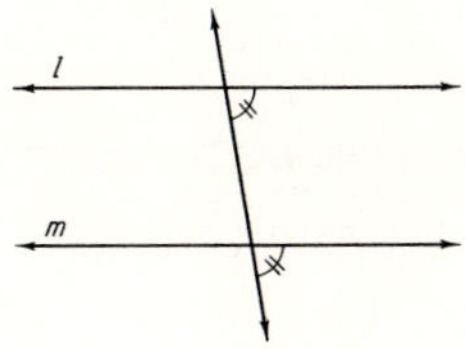

5.

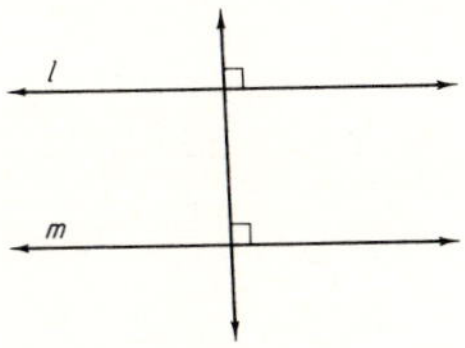

9.

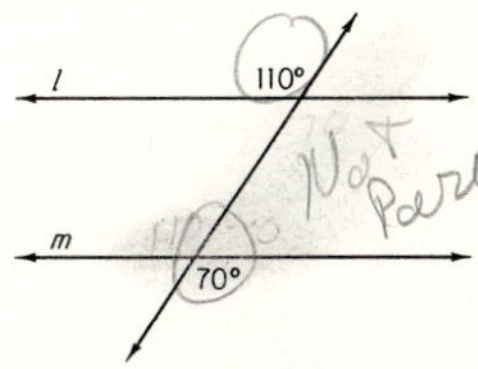

6.

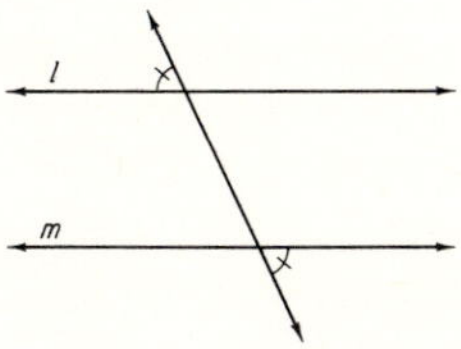

10.

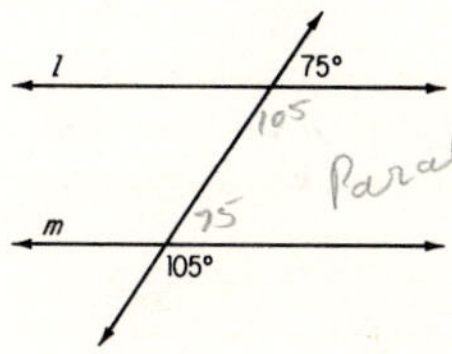

11. *Given:* $\overleftrightarrow{AH} \parallel \overleftrightarrow{EI}$; $\overleftrightarrow{AH} \parallel \overleftrightarrow{CG}$; $m\angle ABD = 55°$; $m\angle BDF = 100°$.
What is the measure of $\angle DFE$?

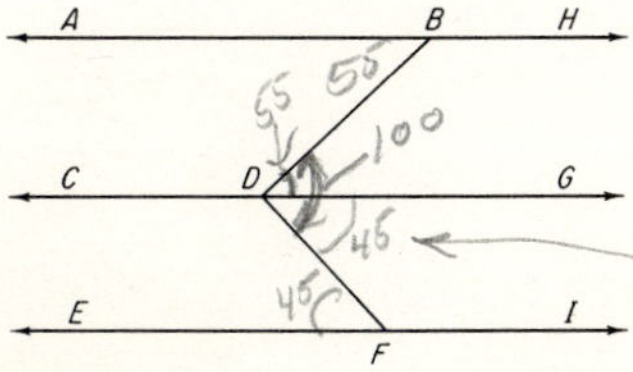

PROB. 11.

12. *Given:* $\overleftrightarrow{AB} \parallel \overleftrightarrow{CE}$; $\overleftrightarrow{BG} \parallel \overleftrightarrow{EF}$; $m\angle B = 75°$
What is the measure of $\angle CEF$?

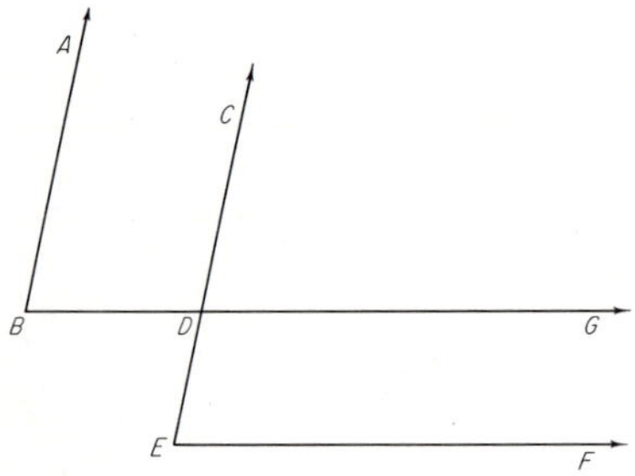

PROB. 12.

13. *Given:* $AC = BC$; $\overrightarrow{CE} \parallel \overrightarrow{AB}$; $m\angle A = 70°$.
What is the measure of $\angle BCD$?

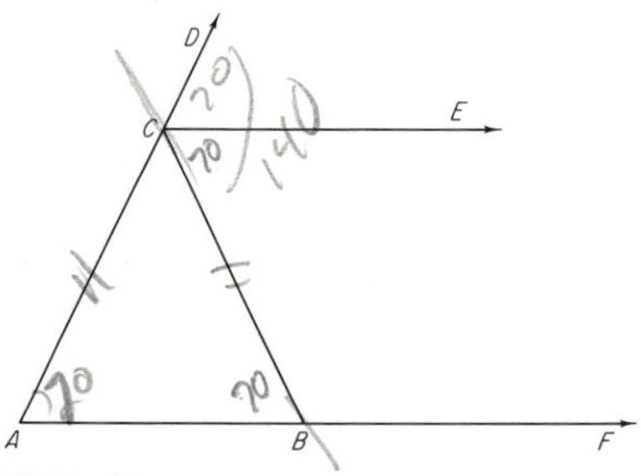

PROB. 13.

14. *Given:* $\overleftrightarrow{AB} \parallel \overleftrightarrow{GH}$; $\overleftrightarrow{EB} \parallel \overleftrightarrow{FH}$; $m\angle EBC = 80°$.
What is the measure of $\angle GHF$?

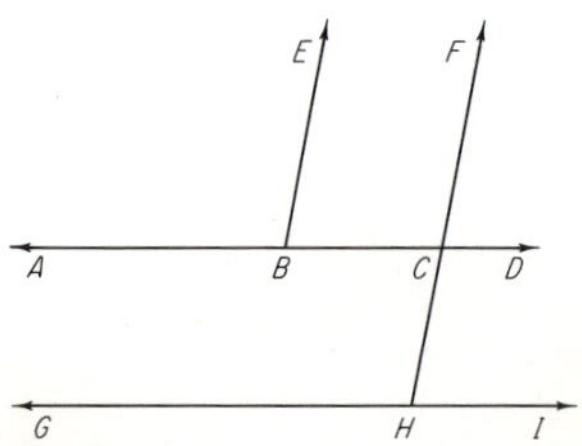

PROB. 14.

15. Indicate which pairs of lines are parallel.

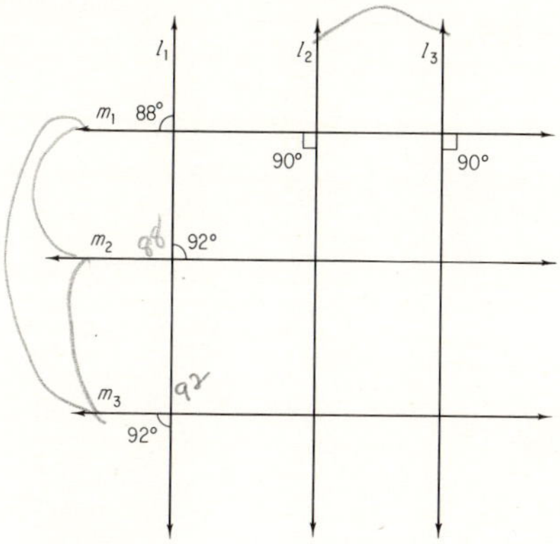

PROB. 15.

16. *Given:* $\overline{AB}$ and $\overline{CD}$ bisect each other at M.
Prove: $\overline{AC} \parallel \overline{BD}$.

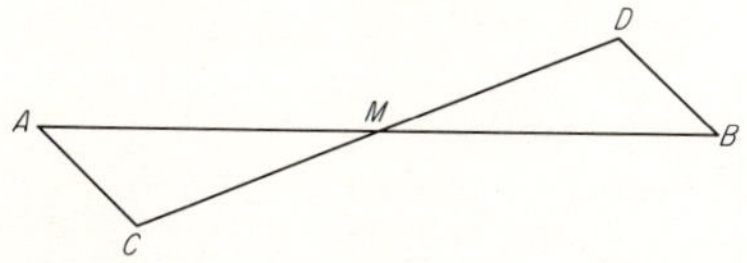

PROB. 16.

17. *Given:* $\overline{RP} \parallel \overline{TQ}$; $RP = TQ$.
Prove: $RS = TS$; $PS = QS$.

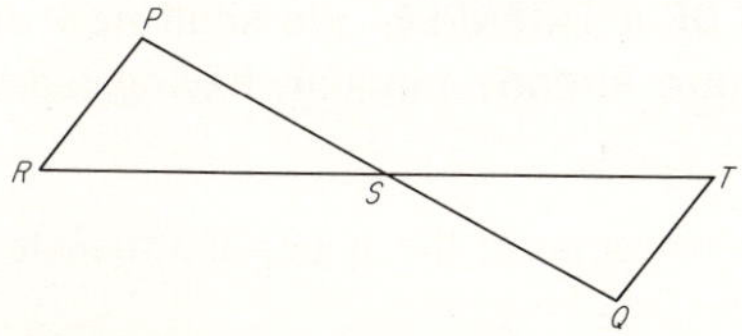

PROB. 17.

18. *Given:* $AB = CD$; $AD = CB$.
Prove: $\overline{AB} \parallel \overline{CD}$.

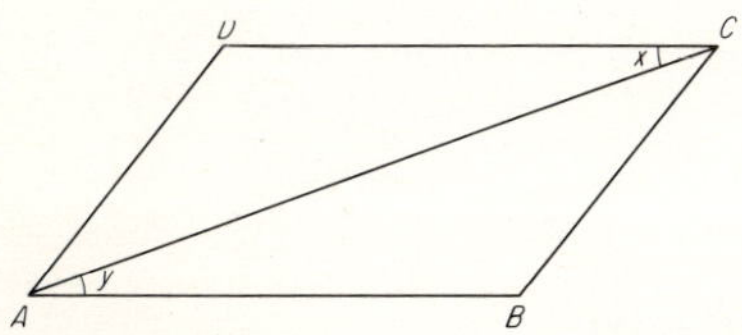

PROB. 18.

19. *Given:* $\overline{PT} \parallel \overline{RS}$; $PT = RS$.
Prove: $PR = ST$.

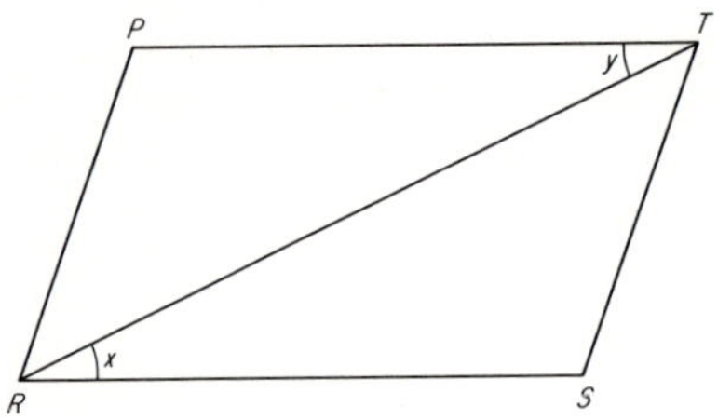

PROB. 19.

20. Determine $m\angle\beta$.

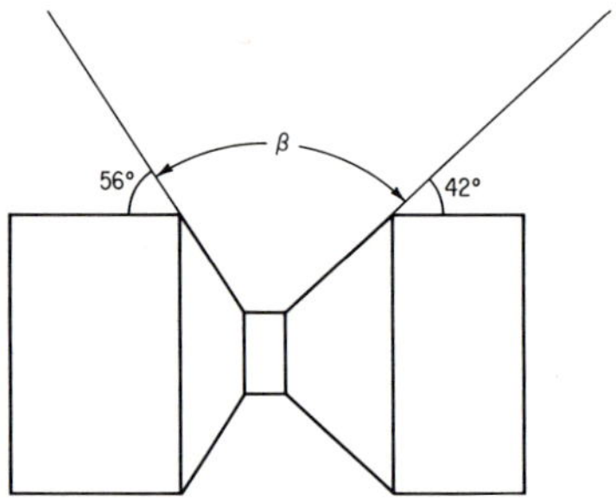

PROB. 20.

14-4 SUM OF THE ANGLES OF A TRIANGLE We shall now prove a theorem with which many students are already familiar, having learned inductively about it in other classes.

Theorem 14-6 The sum of the measures of the angles of a triangle is equal to 180°.

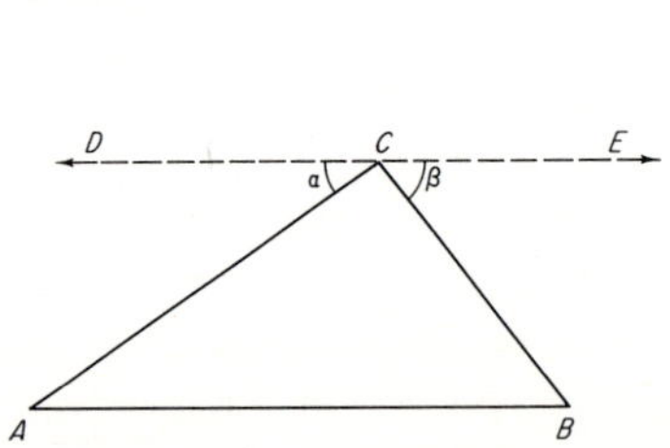

THEOREM 14-6.

Given: Any triangle *ABC*.
Prove: $m\angle A + m\angle B + m\angle C = 180°$.

PROOF

1 Through C draw $\overleftrightarrow{DE} \parallel \overline{AB}$. Postulate 8 tells us this is possible.
2 $m\angle\alpha = m\angle A$ by Theorem 14-3.
3 $m\angle\beta = m\angle B$ by Theorem 14-3.
4 $m\angle ACB = m\angle ACB$ by Axiom 1 (reflexive property).
5 $m\angle\alpha + m\angle\beta + m\angle ACB = m\angle A + m\angle B + m\angle ACB$ because, if equals are added to equals, the sums are equal.
6 But $m\angle\alpha + m\angle\beta + m\angle ACB = 180°$ because the measure of a straight angle is 180°.
7 The $m\angle A + m\angle B + m\angle ACB = 180°$, by Axiom 8 (substitution property).

14-5 EXTERIOR ANGLE OF A TRIANGLE Theorem 14-6 can now be used to prove the following important theorem.

Theorem 14-7 The measure of an exterior angle of a triangle is equal to the sum of the measures of the two nonadjacent interior angles.

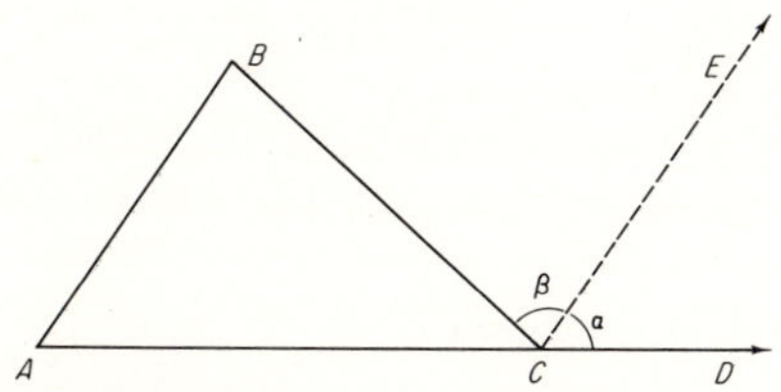

THEOREM 14-7.

Given: $\angle BCD$ is an exterior angle of $\triangle ABC$.
Prove: $m\angle BCD = m\angle A + m\angle B$.

PROOF

1 Draw $\overrightarrow{CE} \parallel \overline{AB}$. Postulate 8 allows this.
2 $m\angle\alpha = m\angle A$ by Theorem 14-5.
3 $m\angle\beta = m\angle B$ by Theorem 14-3.
4 $m\angle\alpha + m\angle\beta = m\angle A + m\angle B$ from statements 2 and 3.
5 $m\angle BCD = m\angle\alpha + m\angle\beta$ by Postulate 4.
6 $m\angle BCD = m\angle A + m\angle B$ by Axiom 8 (substitution property).

Exercise 14-2

1. *Given:* $AC = BC$; $m\angle A = 60°$.
   Find: $m\angle DCB$.

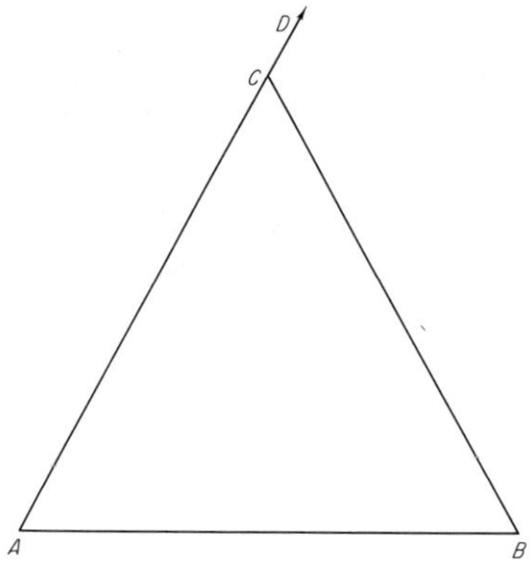

PROB. 1.

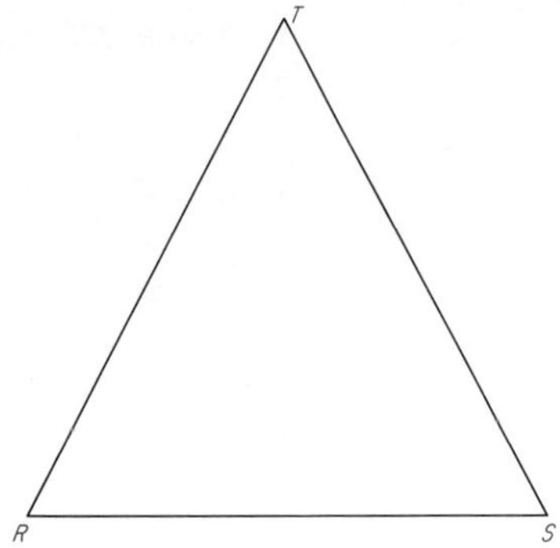

PROB. 2.

2. *Given:* $RT = ST = RS$.
   Find: $m\angle T$.

3. *Given:* $GI = HI$; $m\angle I = 40°$.
   Find: $m\angle H$.

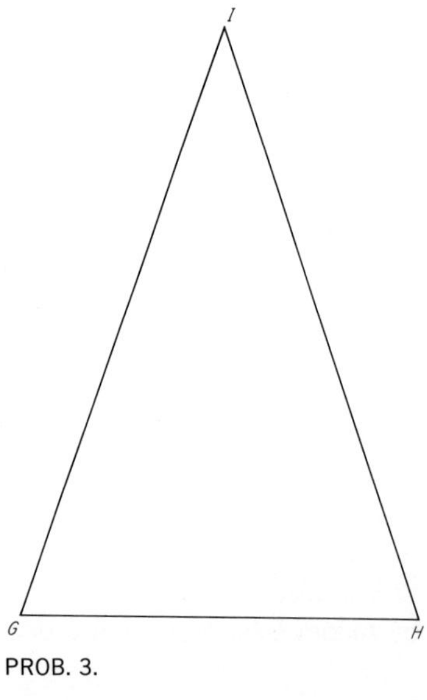

PROB. 3.

4. *Given:* $\overline{FE} \perp \overline{DE}$; $FE = DE$.
   Find: $m\angle D$.

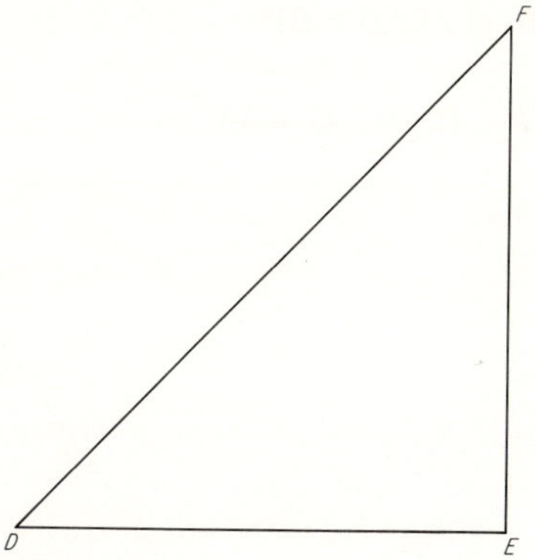

PROB. 4.

5. *Given:* $m\angle D = 80°$; $m\angle C = 60°$.
   Find: $m\angle ABD$.
6. *Given:* $m\angle P = 65°$; $m\angle SQR = 135°$.
   Find: $m\angle S$.

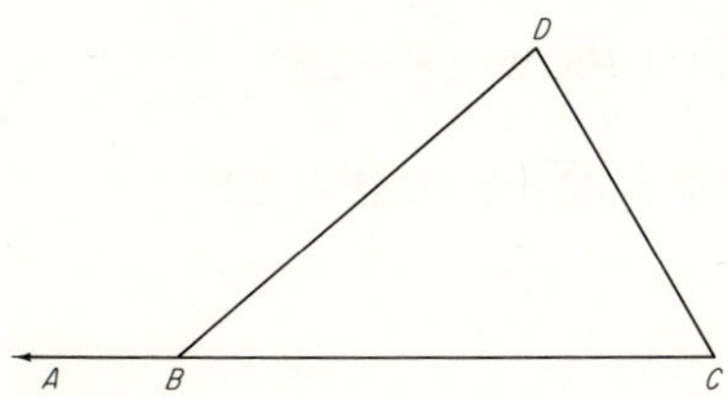

PROB. 5.

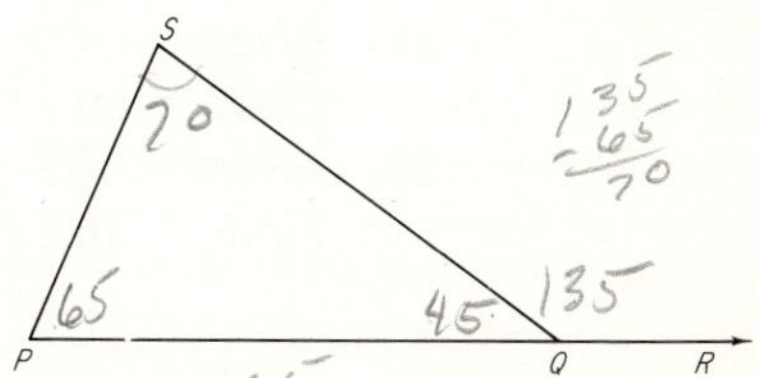

PROB. 6.

7. *Given:* $\overrightarrow{FJ}$ bisects $\angle EFH$; $m\angle H = 82°$; $m\angle G = 44°$.
   Find: $m\angle JFH$.
8. *Given:* $KN = KL$; $\overline{KN} \perp \overline{KL}$.
   Find: $m\angle MLN$.

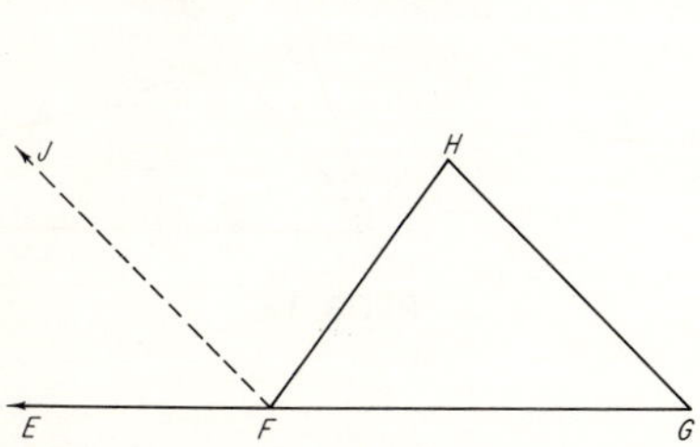

PROB. 7.

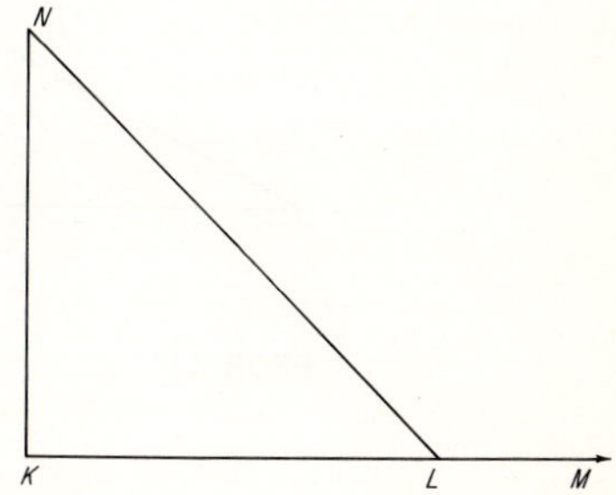

PROB. 8.

9. *Given:* $\angle CDE \cong \angle B$; $m\angle CED = 82°$.
   Find: $m\angle A$.
10. *Given:* $\overline{QS} \perp \overline{RT}$; $RL = TL$; $m\angle Q = 48°$.
   Find: $m\angle LPQ$.

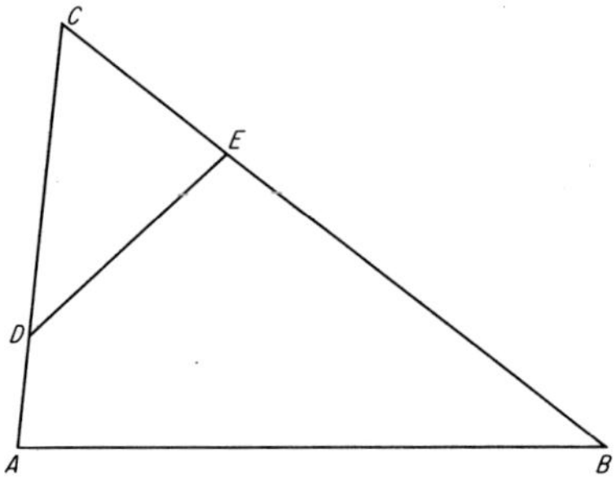

PROB. 9.

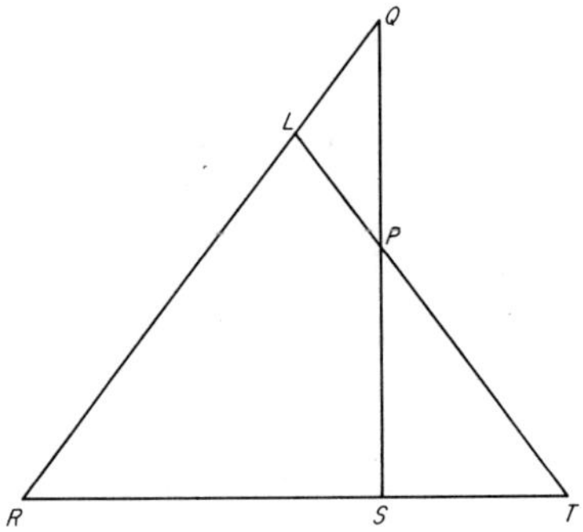

PROB. 10.

11. *Given:* $\overline{CD} \perp \overline{PT}$; $\overline{CA} \perp \overline{PR}$; $m\angle P = 28°$.
   Find: $m\angle C$.
12. *Given:* $\overline{AE} \perp \overline{BC}$; $\overline{BD} \perp \overline{AC}$; $m\angle CAE = 20°$.
   Find: $m\angle DBC$.

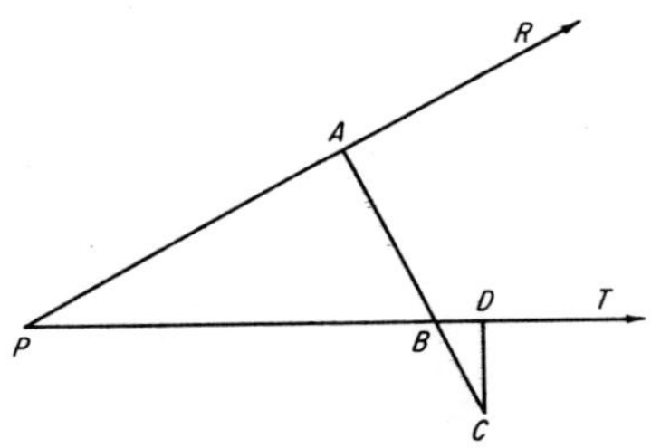

PROB. 11.

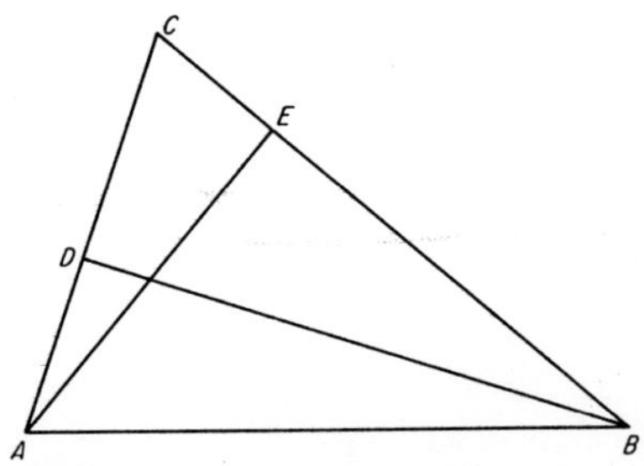

PROB. 12.

13. *Given:* $\overline{CD} \perp \overline{AB}$; $\overline{BC} \perp \overline{AC}$; $m\angle A = 40°$.
   Find: $m\angle BCD$.

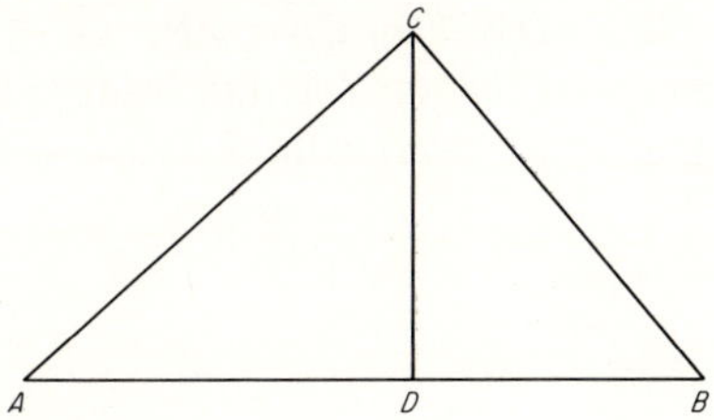

PROB. 13.

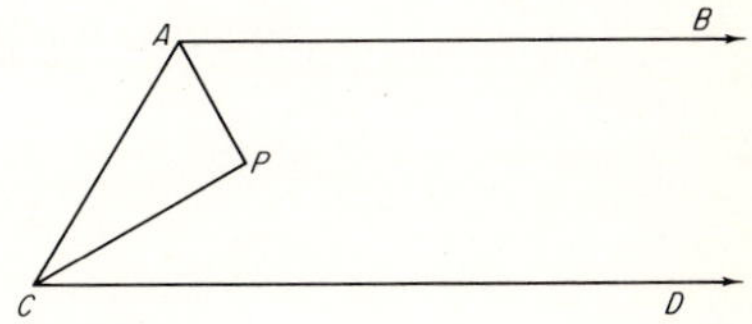

PROB. 14.

14. *Given:* $\overrightarrow{AB} \parallel \overrightarrow{CD}$; $\overline{AP}$ bisects $\angle CAB$; $\overline{CP}$ bisects $\angle ACD$.
Find: $m\angle P$.

15. *Given:* $m\angle P = 130°$; $\overline{RP}$ bisects $\angle TRS$; $\overline{SP}$ bisects $\angle RST$.
Find: $m\angle T$.

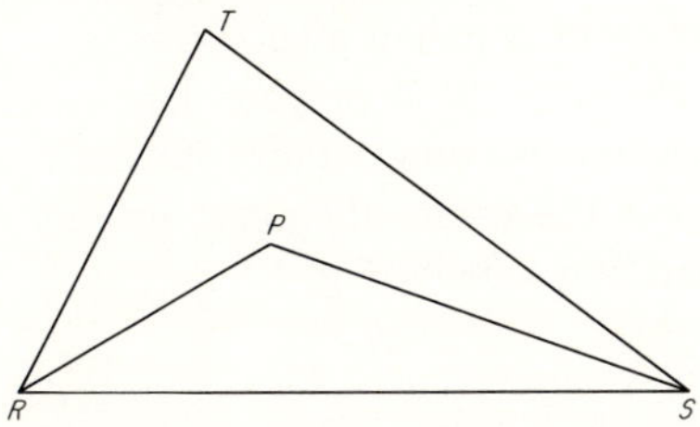

PROB. 15.

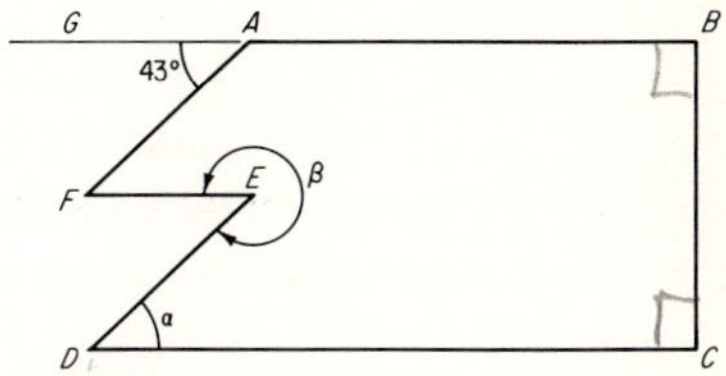

PROB. 16.

16. *Given:* $\overline{AB} \perp \overline{BC}$; $\overline{DC} \perp \overline{BC}$; $\overline{FE} \parallel \overline{AB}$; $\overline{DE} \parallel \overline{FA}$; $m\angle GAF = 43°$.
Find: (a) $m\angle \alpha$; (b) $m\angle \beta$.

17. In the hoisting crane diagram, $AB = MB = 15$ ft 8 in., $m\angle MBL = 143°$. Find (a) $m\angle KAB$; (b) $m\angle KMB$.

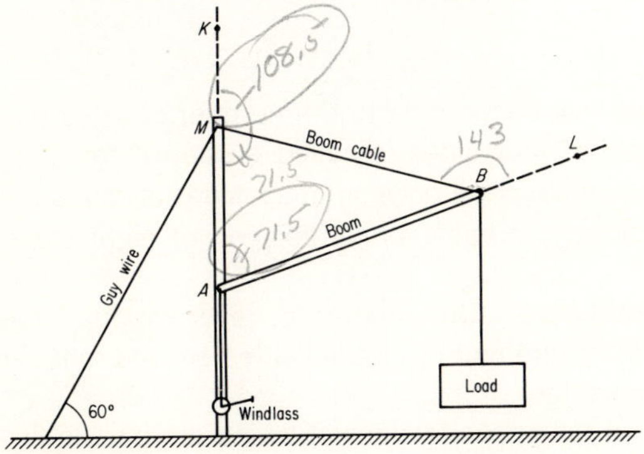

PROB. 17.

18. In the figure, $\overline{HF} \perp \overline{GE}$; $\overline{HD} \perp \overline{EC}$; $\overline{GA} \perp \overline{AB}$; $\overline{BC} \perp \overline{EC}$; $\overline{BC} \perp \overline{AB}$; $m\angle FHD = 43°$. Find $m\angle AGE$. (*Note:* You may want to draw some lines to solve the problem.)

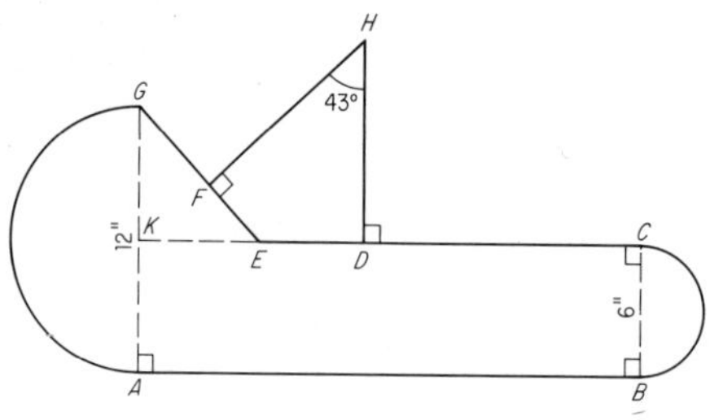

PROB. 18.

14-6 POLYGONS A closed figure bounded by straight lines lying in the same plane is called a *polygon.* The segments forming the boundary are the *sides* of the polygon. The sum of the measures of the sides is the *perimeter.* Each point where two sides meet is called a *vertex.*

The simplest polygon is the triangle. A polygon has as many angles as it has sides. A *regular polygon* is one which is both equilateral and equiangular. The meanings of the terms *diagonal, interior angle,* and *exterior angle* can be recognized by studying Fig. 14-7.

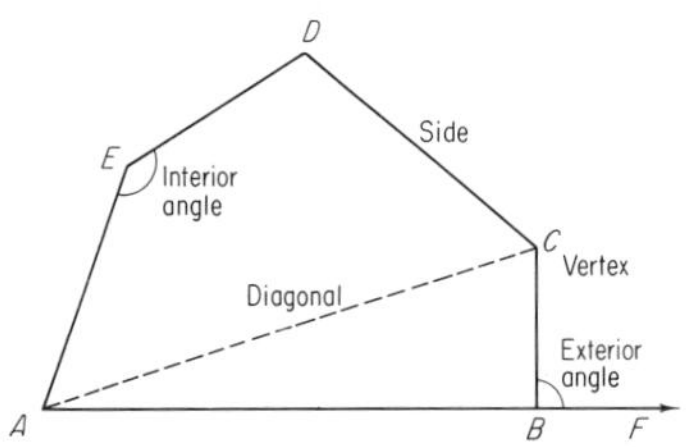

FIG. 14-7.

Polygons are named according to the number of sides they have (see Fig. 14-8). A polygon of four sides is called a *quadrilateral;* a polygon of five sides is called a *pentagon;* a polygon of six sides is called a *hexagon;* a polygon of seven sides is a *heptagon;* a polygon of eight sides is an *octagon.*

14-7 QUADRILATERALS Quadrilaterals have various shapes and sizes. Those with at least one pair of parallel sides are the most important. These we shall now consider.

A quadrilateral the opposite sides of which are parallel is a *parallelogram*

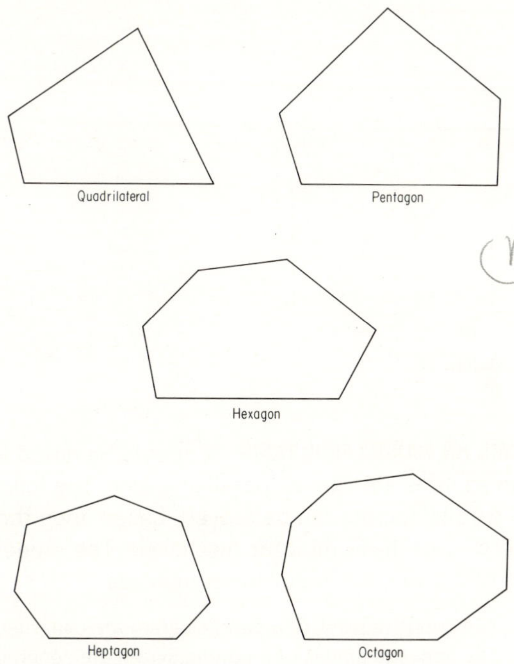

FIG. 14-8. Polygons.

(symbol ▱). Any side of the parallelogram may be called the base, as $\overline{AB}$ of ▱*ABCD* (Fig. 14-9). The *altitude* of a parallelogram is the perpendicular distance between a pair of bases, as $\overline{DE}$. Each parallelogram has two altitudes.

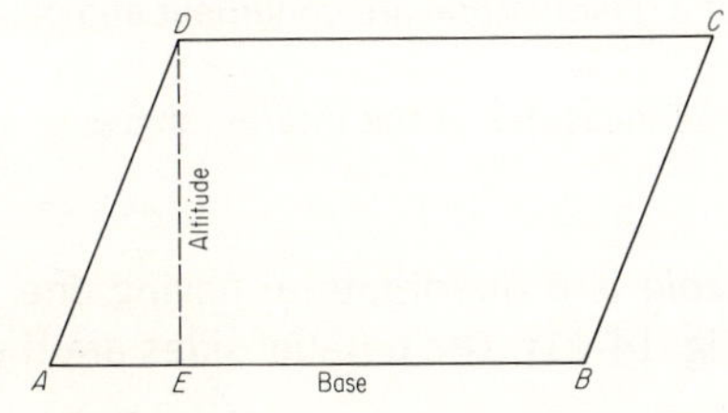

FIG. 14-9.

An equilateral parallelogram is called a *rhombus* (see Fig. 14-10).

A parallelogram whose angles are right angles is called a *rectangle* (see Fig. 14-10).

A rectangle whose sides are congruent is called a *square* (see Fig. 14-10).

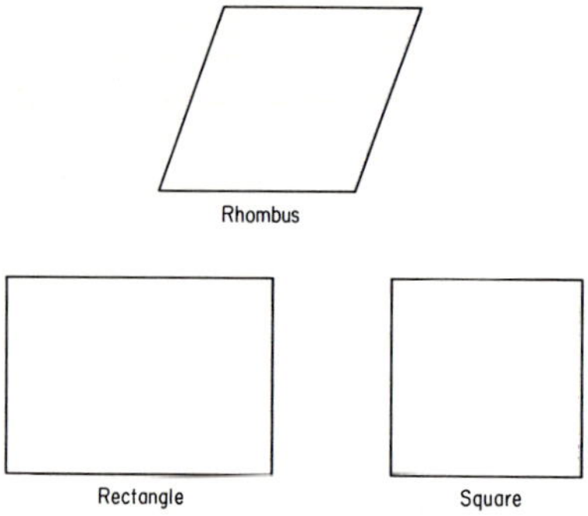

Fig. 14-10. Quadrilaterals.

14-8 THEOREMS ON PARALLELOGRAMS It should be noted that rhombuses, rectangles, and squares are special parallelograms. The following properties common to all parallelograms can be proved, based upon the axioms, postulates, and theorems we have thus far discussed. The student is advised to make a sketch which will check out the properties.

Theorem 14-8 The opposite sides of a parallelogram are congruent.
Theorem 14-9 The opposite angles of a parallelogram are congruent.
Theorem 14-10 Either diagonal divides a parallelogram into two congruent triangles.
Theorem 14-11 Any two adjacent angles of a parallelogram are supplementary.
Theorem 14-12 The diagonals of a parallelogram bisect each other.
Theorem 14-13 The diagonals of a rectangle have the same measure.
Theorem 14-14 The diagonals of a rhombus are perpendicular to each other.
Theorem 14-15 If the opposite sides of a quadrilateral are congruent, the quadrilateral is a parallelogram.
Theorem 14-16 If two sides of a quadrilateral are congruent and parallel, the quadrilateral is a parallelogram.
Theorem 14-17 The sum of the measures of the interior angles of a quadrilateral is 360°.

14-9 TRAPEZOID A *trapezoid* is a quadrilateral having one, and only one, pair of sides parallel (see Fig. 14-11). The parallel sides are the *bases* (upper

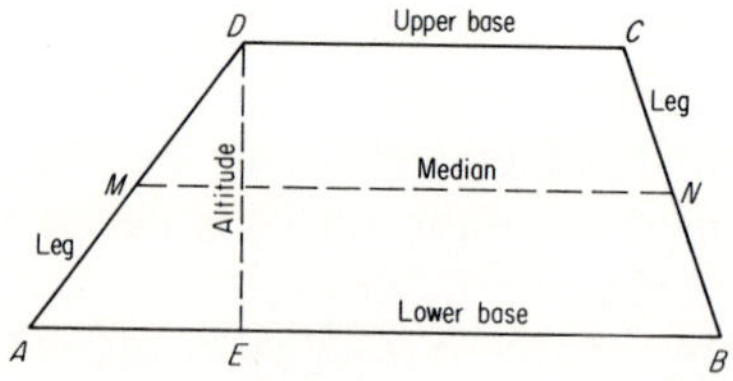

FIG. 14-11. Trapezoid.

and lower) of the trapezoid. The nonparallel sides are called *legs.* The pair of angles adjacent to a base are called *base angles.* The *altitude* is the perpendicular segment, as $\overline{DE}$, between the bases. An *isosceles trapezoid* is one the legs of which are congruent (see Fig. 14-12).

FIG. 14-12. Isosceles trapezoid.

The *median* of a trapezoid is the line segment connecting the midpoints of the legs. It can be shown that the measure of the median of a trapezoid is equal to one-half the sum of the measures of the bases.

Exercise 14-3

In Probs. 1 to 25, indicate whether the statement is *always true* (mark T) or *not always true* (mark F).

1. A square is a rectangle.
2. A square is a rhombus.
3. A square is a parallelogram.
4. A rectangle is a square.
5. A rectangle is a rhombus.
6. A rectangle is a parallelogram.
7. A quadrilateral is a polygon.
8. A quadrilateral is a trapezoid.
9. A quadrilateral is a rectangle.
10. A polygon is a quadrilateral.
11. A quadrilateral is a parallelogram.
12. A polygon is a trapezoid.
13. The diagonals of a parallelogram bisect each other.
14. The diagonals of a parallelogram are congruent.
15. The diagonals of a rhombus are perpendicular to each other.
16. The lines through the vertices of a parallelogram parallel to the diagonals form another parallelogram.
17. If the diagonals of a parallelogram are perpendicular, the parallelogram is a square.
18. If the diagonals of a quadrilateral are perpendicular to each other, the quadrilateral is a parallelogram.

19. The nonparallel sides of an isosceles trapezoid make congruent angles with either base.
20. If two sides of a quadrilateral are congruent, it is a parallelogram.
21. A pentagon has six sides.
22. The sum of the measures of the interior angles of a parallelogram is 180°.
23. The opposite angles of a parallelogram are supplementary.
24. A pentagon has three diagonals.
25. Two consecutive angles of a parallelogram are complementary.

In Probs. 26 to 36 indicate the word(s) or number that will make the statement true.

26. A trapezoid the nonparallel sides of which are congruent is called a(n) ________ trapezoid.
27. A quadrilateral that has two and only two sides parallel is called a(n) ________.
28. The bases of a trapezoid are the ________ sides of the figure.
29. An equilateral parallelogram is a(n) ________.
30. An equilateral rectangle is a(n) ________.
31. Either diagonal of a parallelogram divides it into two ________ triangles.
32. The opposite sides of a parallelogram are ________ and ________.
33. The diagonals of a rhombus are ________ to each other.
34. The measure of each interior angle of a regular hexagon is ________ degrees.
35. The measure of the median of a trapezoid whose bases measure 12 in. and 18 in. is ________ in.
36. The number of diagonals that can be drawn from a vertex in a polygon of 100 sides is ________.

CHAPTER 15

THE CIRCLE

15-1 USES OF THE CIRCLE The circle is one of the most common and useful of all curved lines. Man's perfection of the wheel as an application of the circle has been classified by many as one of the most important inventions in the history of civilization. Without the wheel most of the world's work would cease. Industry would be completely crippled without wheels, gears, and axles with circular cross sections. Transportation would revert to conditions of prehistoric times without the wheel.

Every year many millions of feet of circular pipe and wire are manufactured. Most canned goods come in circular cylinders. Circular shapes are found in many ornamental and landscape designs. Traffic circles are used to direct and divert the flow of vehicles of transportation.

15-2 DEFINITIONS A *circle* is defined as the set of points lying in one plane each of which is the same distance from a fixed point, called the *center*. The symbol for circle is ⊙ (plural ⓢ). Sometimes the term *circle* is used to refer to the region enclosed by the curve. A line segment one of whose end points is the center of the circle and the other a point on the circle is called a *radius* (plural radii) of the circle. In Fig. 15-1, O is the

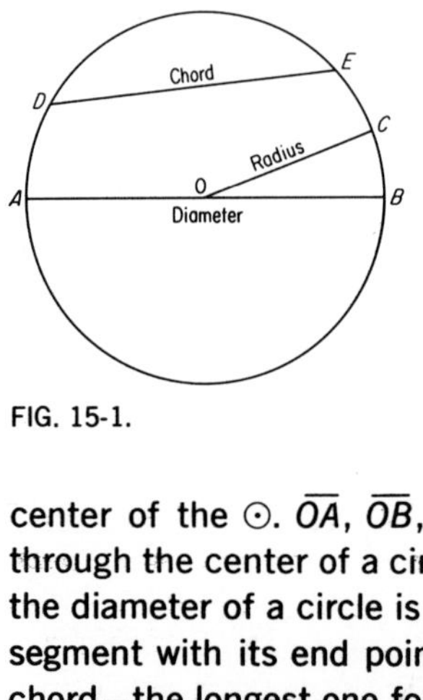

FIG. 15-1.

center of the ⊙. $\overline{OA}$, $\overline{OB}$, and $\overline{OC}$ are radii. A *diameter* is a line segment through the center of a circle with end points on the circle. The measure of the diameter of a circle is twice the measure of its radius. A *chord* is a line segment with its end points on the circle. Thus, the diameter is a special chord—the longest one for a given circle.

It will be noted that we defined "radius" and "diameter" each as a segment, that is, a set of points. Common usage, however, often lets the words denote their measures. Thus, we speak of a circle with a radius of, say, 10 in. In general, no confusion should arise in this dual meaning of the word.

A *secant*, as $\overleftrightarrow{MN}$ in Fig. 15-2, is a line which intersects a circle at two points.

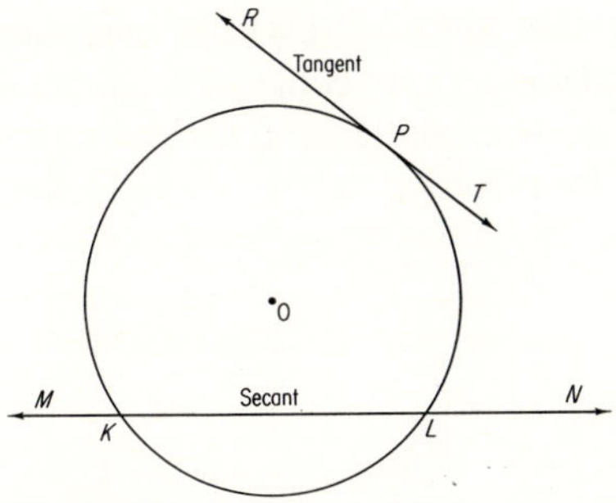

FIG. 15-2.

A *tangent* is a line which touches a circle at one, and only one, point. This point is called the *point of tangency.* In Fig. 15-2, $\overleftrightarrow{RT}$ is tangent to ⊙ *O* at *P*.

15-3 ARCS AND ANGLES The length (measure) of a circle is called its *circumference.* Any part of a circle is called an *arc* of the circle, such as arc *AE* (written $\widehat{AE}$) of Fig. 15-3. Since any two points on a circle divide it into

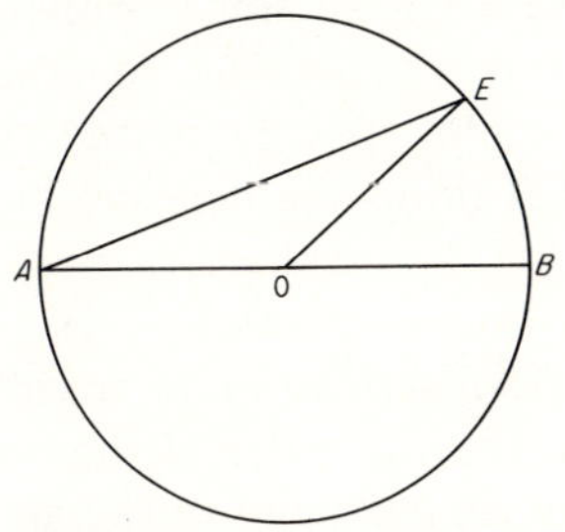

FIG. 15-3.

two arcs, we shall call the shorter arc a *minor arc* and the longer arc a *major arc.* We shall agree in this text to let $\widehat{AE}$ stand for the minor arc. We shall write $\widehat{ABE}$ for the major arc.

The diameter of a circle divides it into two congruent *semicircles.*

A *central angle,* as $\angle AOE$ of Fig. 15-3, is an angle whose vertex is the center of the circle. Angle *AOE* is said to *intercept* arc *AE* and chord *AE.* Arc *AE* and chord *AE* are said to *subtend* the central angle *AOE.*

An arc of a circle is measured by its central angle. Thus, if $m\angle BOE = 46°$, then $m\widehat{BE} = 46°$ and $m\widehat{BAE} = 314°$. Conversely, the measure of a central angle of a circle is equal to the measure of its intercepted arc. If $m\widehat{AE} = 134°$, then $m\angle AOE = 134°$.

15-4 CONGRUENT CIRCLES AND ARCS Circles are *congruent* if they have congruent radii. Two arcs of the same, or congruent, circles are *congruent* if they have the same measure. The relationship between arc length and arc degrees can be illustrated by referring to Fig. 15-4. In the figure are two

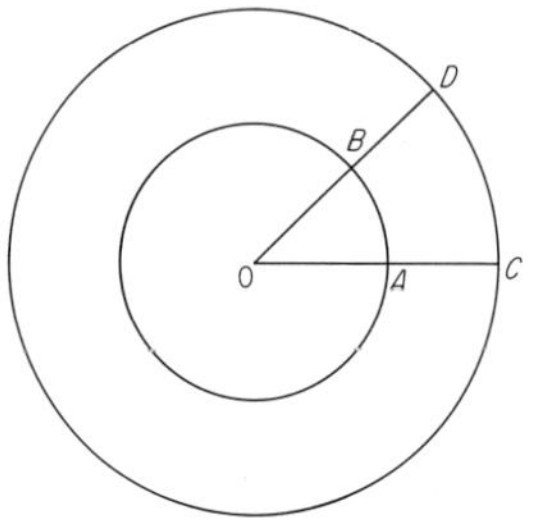

FIG. 15-4.

circles, called *concentric circles,* with the same center. Since both $\widehat{AB}$ and $\widehat{CD}$ subtend the same central angle, they must have the same measure. Thus, $m\widehat{AB} = m\widehat{CD}$. However, it will be noted that the lengths of the arcs are not equal; that is, $\widehat{AB} \neq \widehat{CD}$. Here we have two arcs of different circles with the same number of arc degrees but with unequal lengths. The student should recognize that $m\widehat{AB} = m\widehat{CD}$ states that the measures *in degrees* of the two arcs are equal while $\widehat{AB} = \widehat{CD}$ states the measure in *units of length* are equal.

15-5 INSCRIBED ANGLES An angle is *inscribed* in an arc of a circle if the end points of the arc are points on the two sides of the angle and if the vertex of the angle is a point but not an end point of the arc. In Fig. 15-5*a* $\angle RST$ is inscribed in minor arc RST. In Fig. 15-5*b* $\angle ABC$ is inscribed in major arc ABC. Angle RST is said to intercept $\widehat{RT}$; $\angle ABC$ intercepts $\widehat{AC}$.

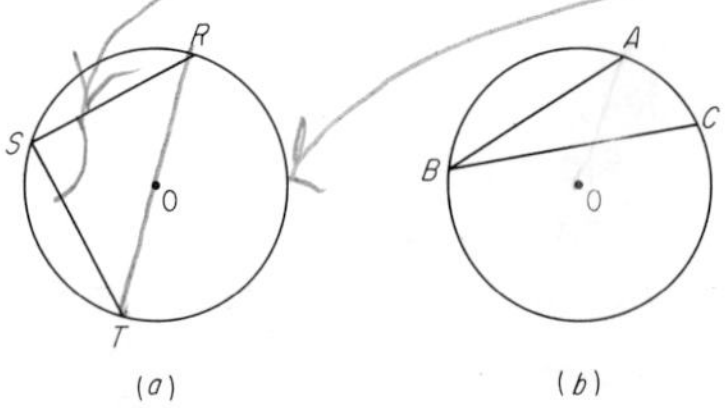

FIG. 15-5.

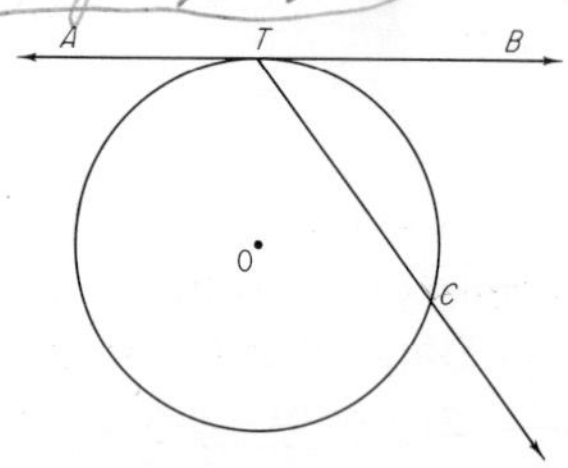

FIG. 15-6.

In Fig. 15-6 $\angle BTC$ is formed by a tangent and a secant with the common point on the circle. $\angle BTC$ is said to intercept $\widehat{TC}$.

The following statements can be proved.

Theorem 15-1 The measure of an inscribed angle is equal to half the measure of its intercepted arc. Thus if, in Fig. 15-5*b*, $m\widehat{AC} = 40°$, then $m\angle ABC = 20°$.

Theorem 15-2 The measure of the angle formed by a tangent and a secant drawn from the point of tangency is half the measure of its intercepted arc. Thus if, in Fig. 15-6, $m\widehat{TC} = 80°$, then $m\angle BTC = 40°$.

Theorem 15-3 If a line is tangent to a circle, it is perpendicular to the radius drawn to the point of tangency. If, in Fig. 15-7, $\overleftrightarrow{AB}$ is tangent to $\odot O$ at T, then $\overleftrightarrow{AB}$ is perpendicular to radius OT.

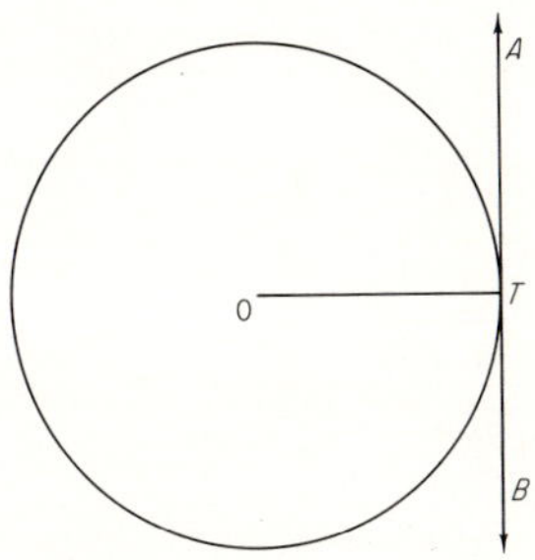

FIG. 15-7.

Theorem 15-4 If a line is perpendicular to a radius at its point on the circle, it is tangent to the circle. Theorem 15-4 is the converse of Theorem 15-3. It states that, if in Fig. 15-7, $\overleftrightarrow{AB} \perp \overline{OT}$, then $\overleftrightarrow{AB}$ is tangent to $\odot O$ at T.

Exercise 15-1

Find the number of degrees in the measures of $\angle\alpha$, $\angle\beta$, and arc m. O is the center of the circle. T is the point of tangency.

1. 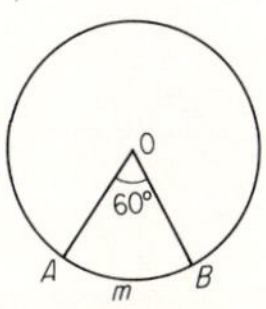

3. 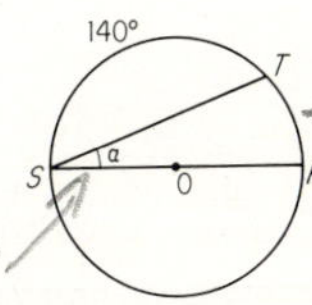

2. 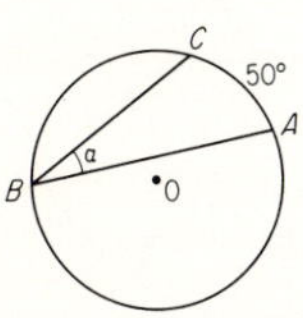

4. 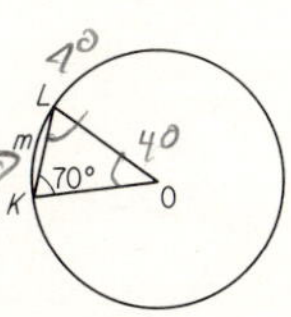

5.

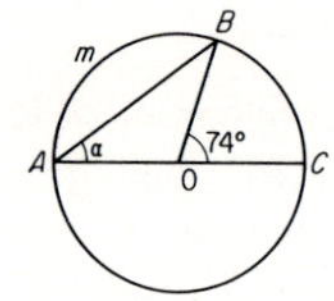

6.

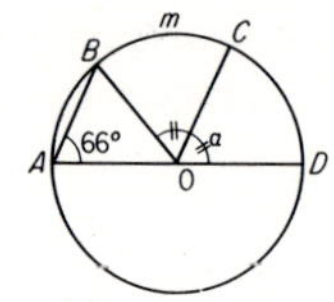

$\overline{OC}$ bisects $\angle BOD$

7.

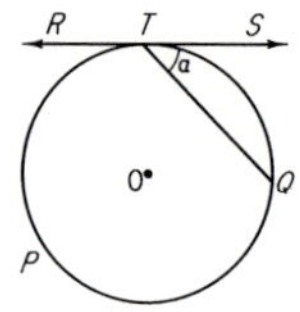

$m\widehat{TPQ} = 270°$

8.

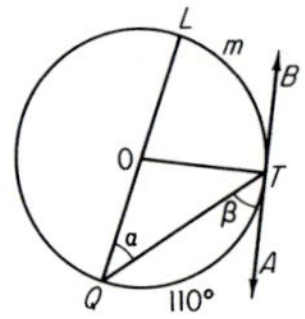

9.

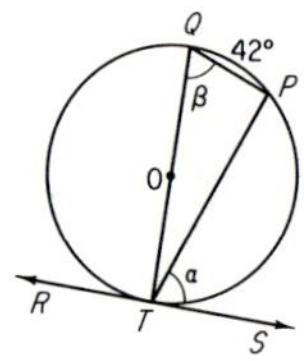

10.

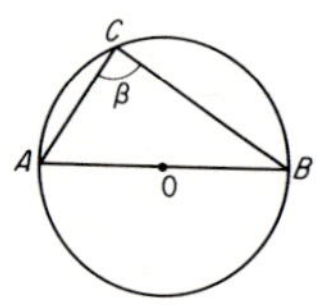

11.

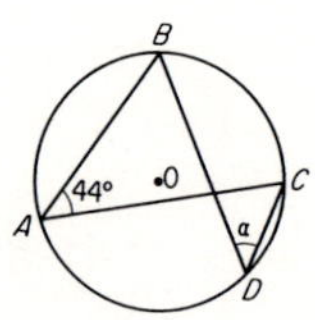

12.

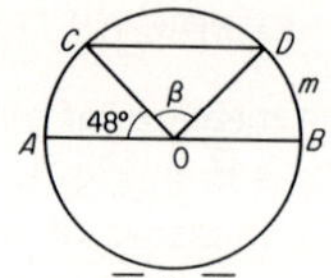

$\overline{CD} \parallel \overline{AB}$

13.

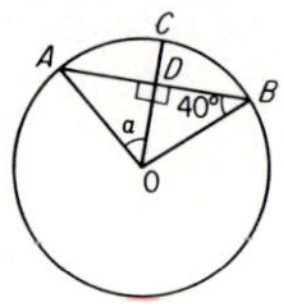

$\overline{OC} \perp \overline{AB}$

14.

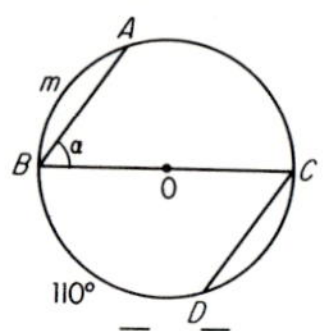

$\overline{AB} \parallel \overline{CD}$

15.

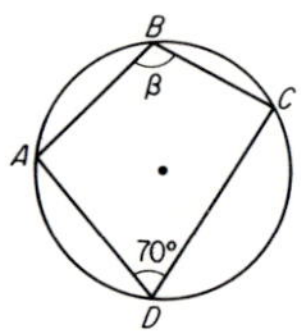

16.

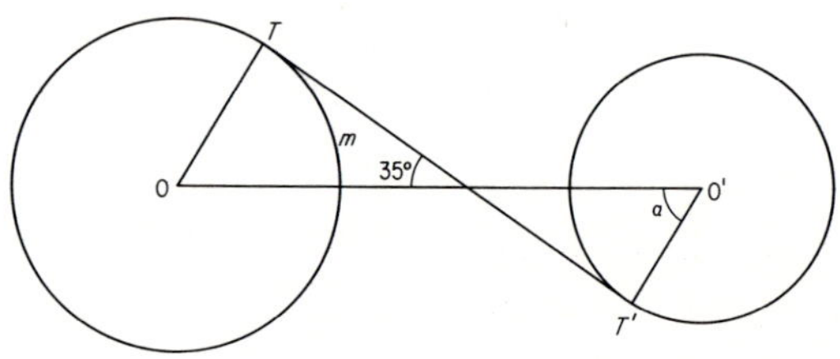

17.

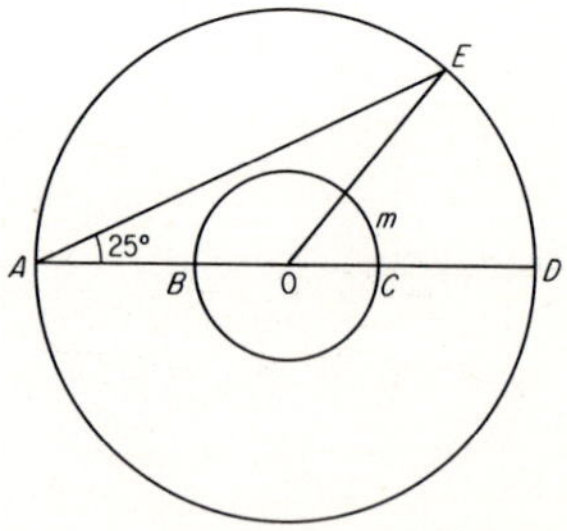

18.

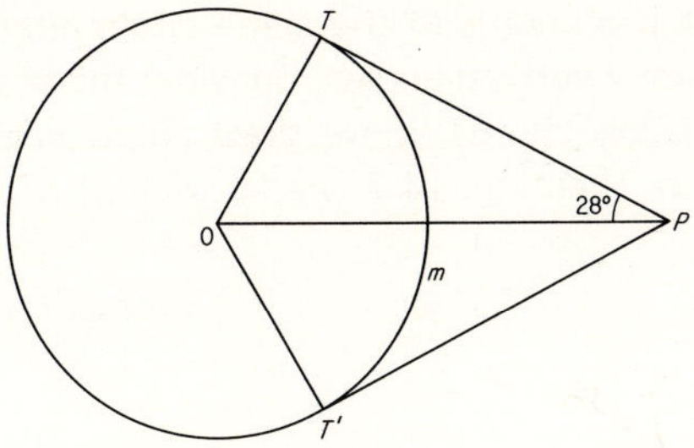

19.

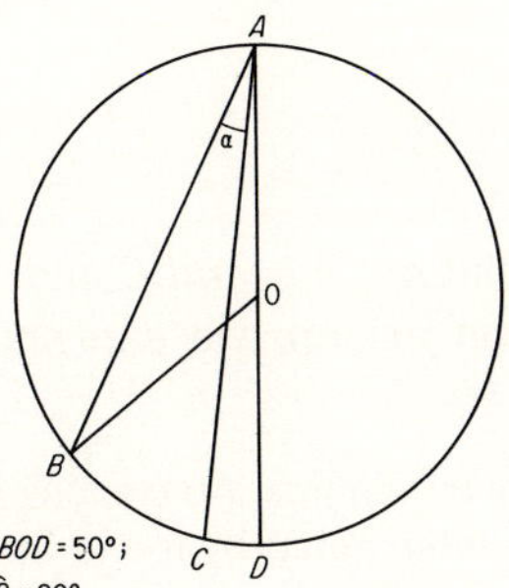

$m\angle BOD = 50°$; $m\widehat{CD} = 20°$

20.

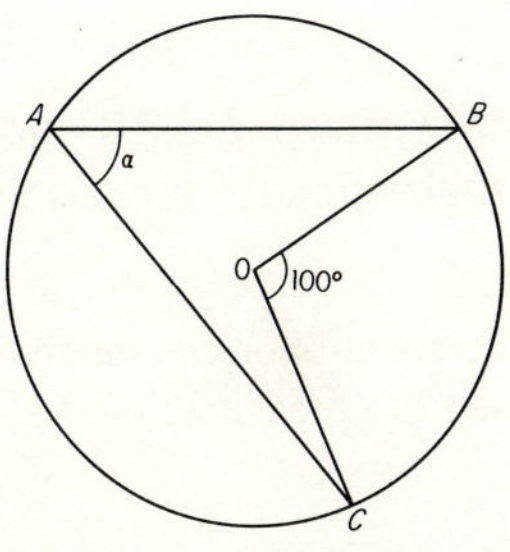

21.

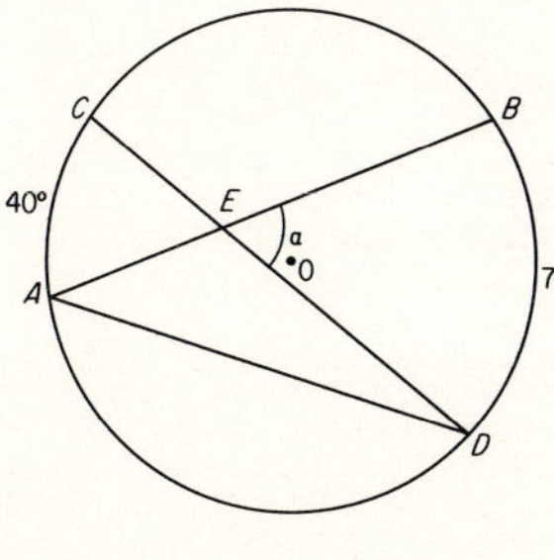

22.

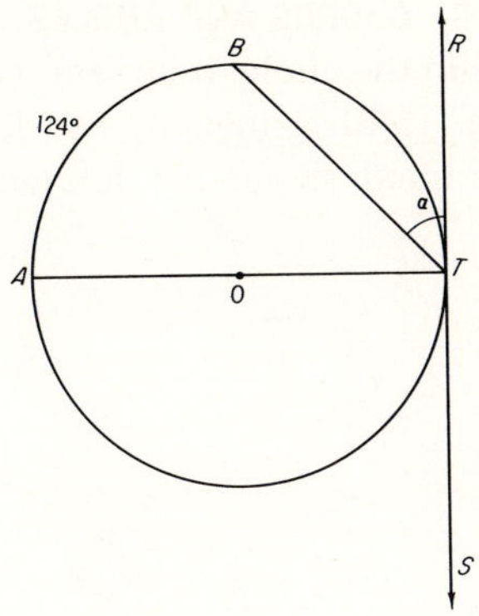

23.

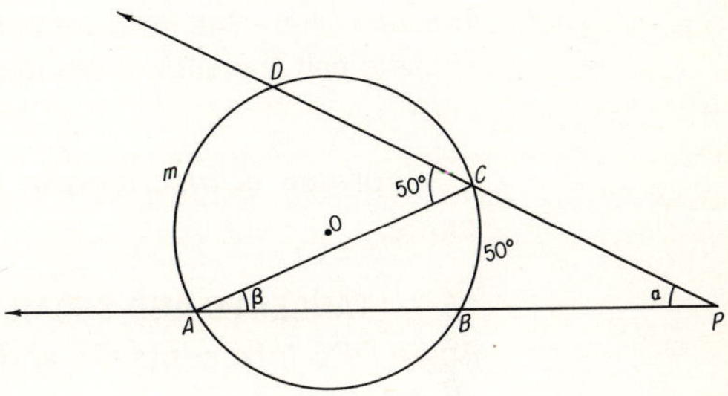

24.

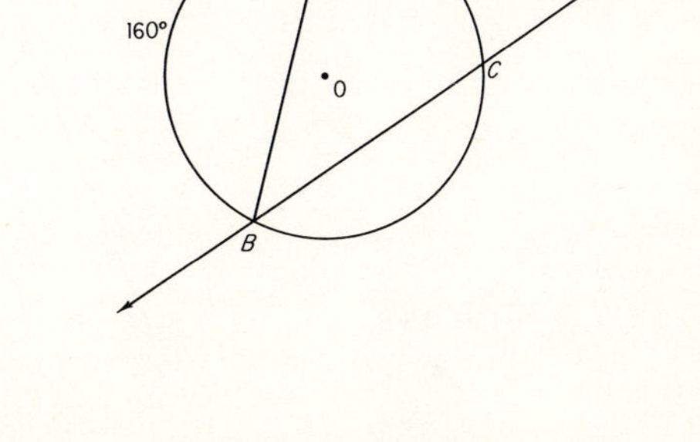

15-6 CHORDS AND ANGLES When two chords of the same circle intersect within the circle, they form two pairs of vertical angles. Consider the angle α (Fig. 15-8) formed by two intersecting chords. If we draw chord AB, two inscribed angles are formed. From Theorem 14-7 we know that $m\angle\alpha =$

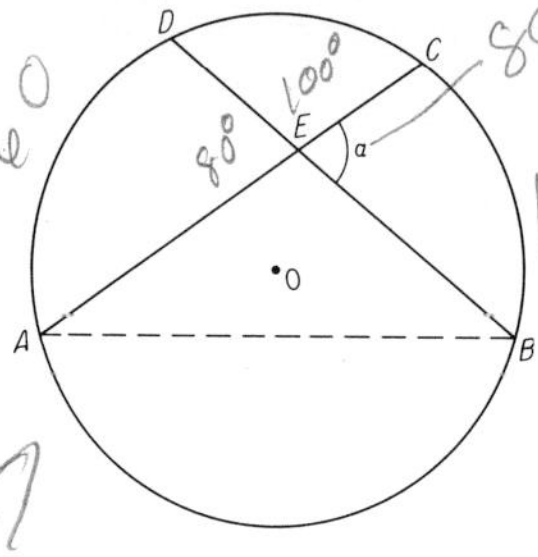

FIG. 15-8.

$m\angle A + m\angle B$. From Theorem 15-1 we know that $m\angle A = \frac{1}{2}m\widehat{BC}$ and $m\angle B = \frac{1}{2}m\widehat{AD}$. Then, using Axiom 8 (the substitution axiom), we can show that $m\angle\alpha = \frac{1}{2}(m\widehat{BC} + m\widehat{AD})$.

Theorem 15-5 The measure of an angle formed by two chords intersecting within a circle is half the sum of the measures of the arcs intercepted by it and its vertical angle.

A polygon is *inscribed in a circle* if each of its sides is a chord of the circle.

15-7 TANGENTS AND SECANTS In Fig. 15-9, $\angle APD$ intercepts $\widehat{AD}$ and $\widehat{AB}$. Angle DPE intercepts $\widehat{DE}$ and $\widehat{BC}$. The following theorems are proved in the study of geometry.

Theorem 15-6 The measure of the angle formed by two secants or two tangents intersecting outside a circle is half the difference between the measures of the inter-

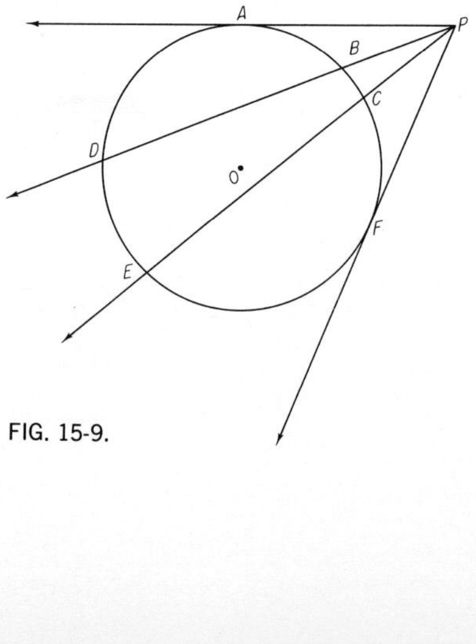

FIG. 15-9.

cepted arcs. In Fig. 15-9, $m\angle DPE = \frac{1}{2}(m\widehat{DE} - m\widehat{BC})$ and $m\angle APF = \frac{1}{2}(m\widehat{ADF} - m\widehat{ABF})$.

Theorem 15-7 The measure of the angle formed by a secant and a tangent intersecting outside a circle is half the difference between the measures of the intercepted arcs. In Fig. 15-9, $m\angle APD = \frac{1}{2}(m\widehat{AD} - m\widehat{AB})$.

A polygon is *circumscribed about a circle* if each of its sides is tangent to the circle.

Exercise 15-2

Find the number of degrees in the measures of $\angle\alpha$, $\angle\beta$, and arc m. O is the center of the circle. T is a point of tangency.

1.

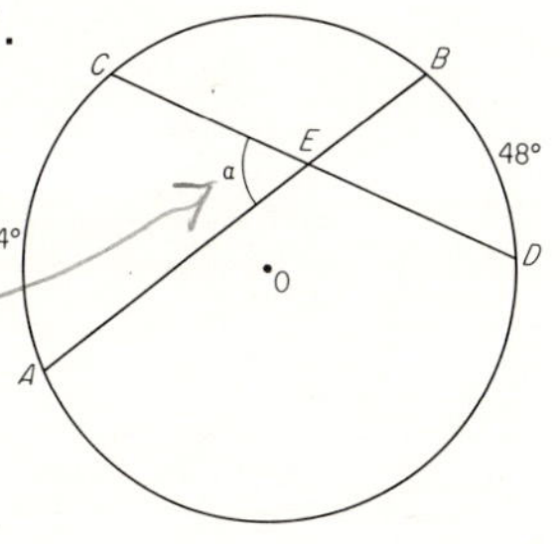

4.

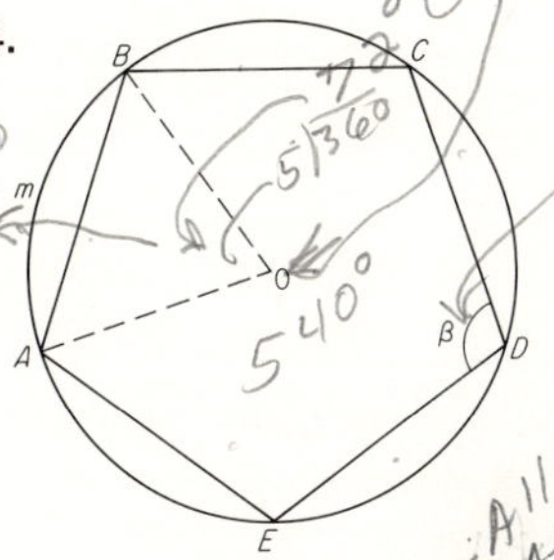

ABCDE is a regular pentagon

2.

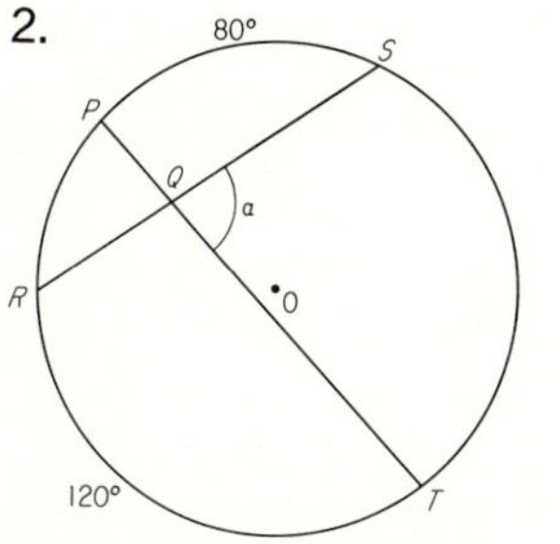

5.

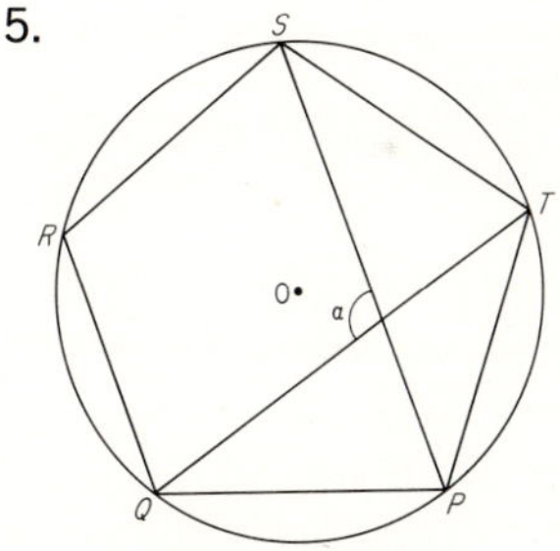

RSTPQ is a regular pentagon

3.

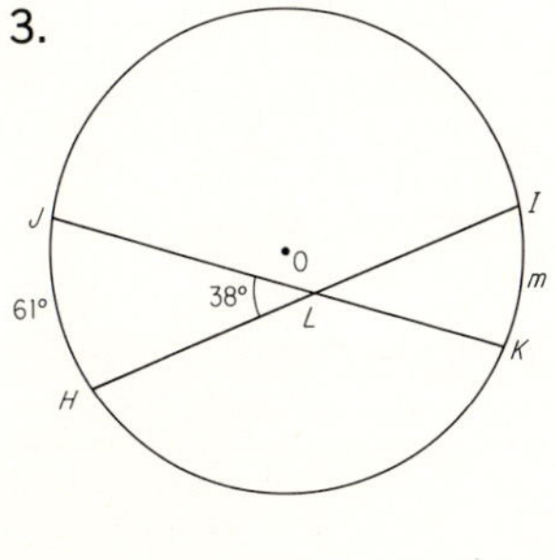

6.

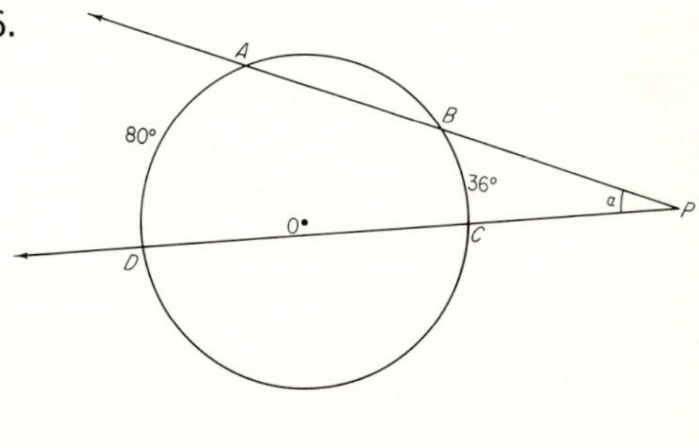

7.

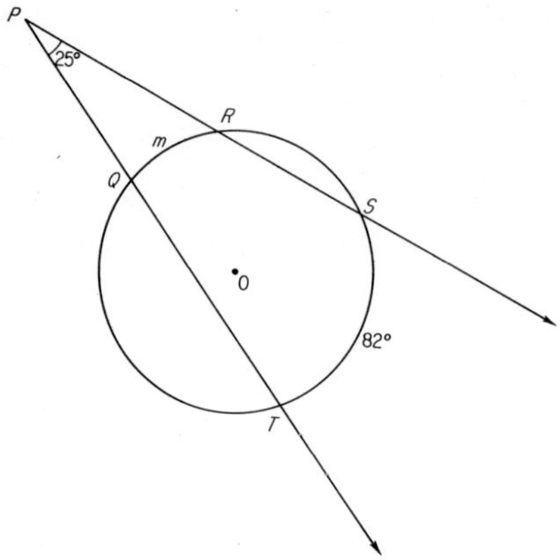

8.

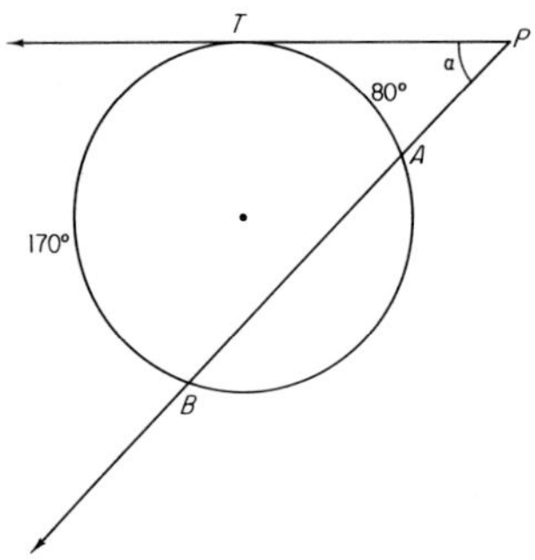

9.

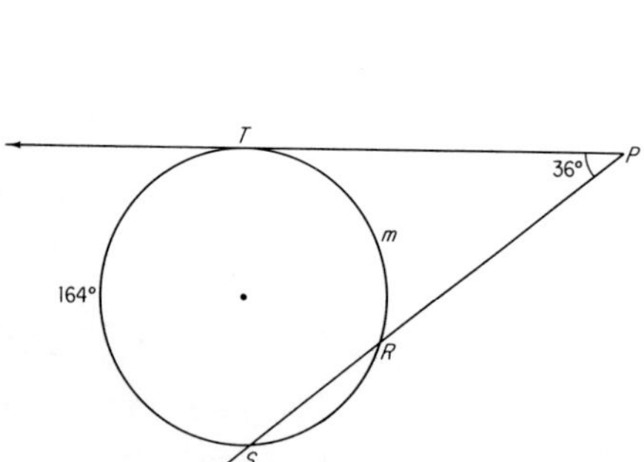

10.

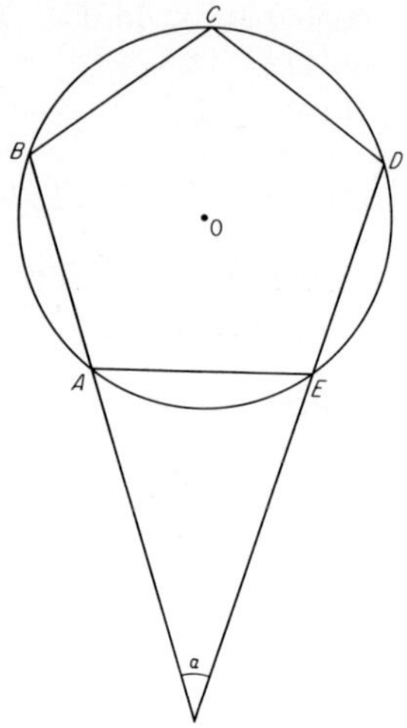

ABCDE is a regular pentagon

11.

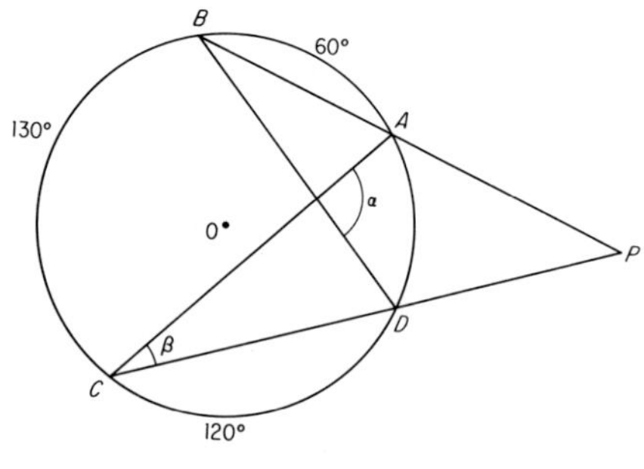

12.

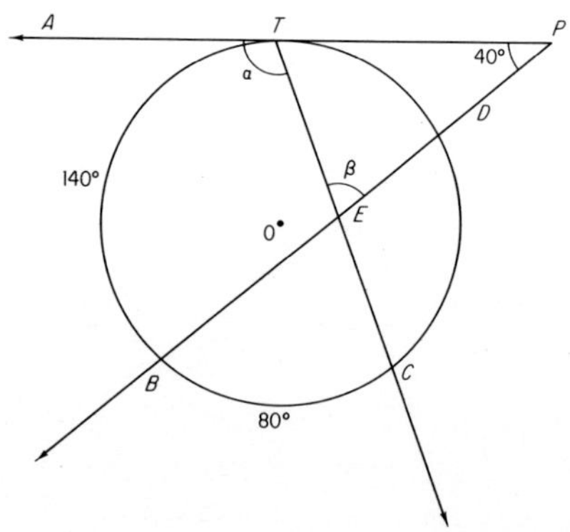

13.

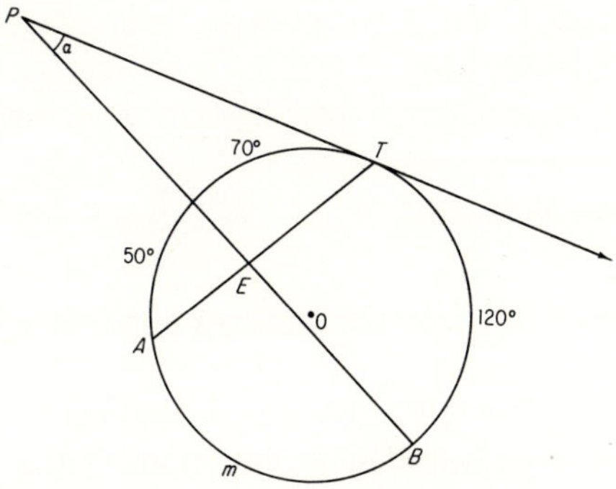

15.

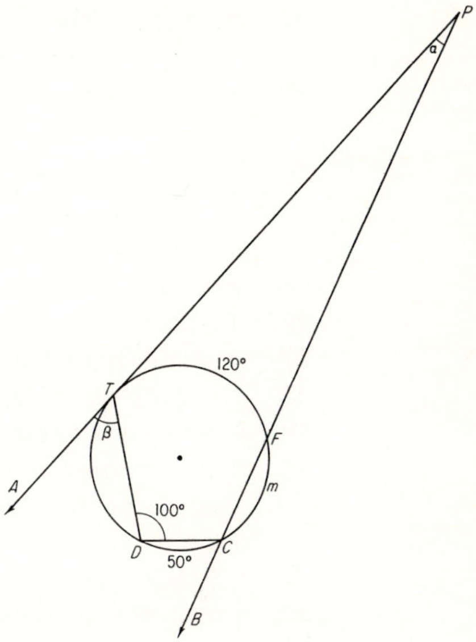

14.

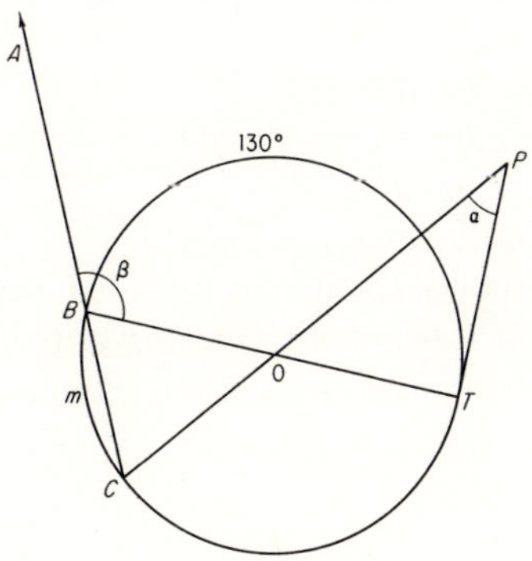

16.

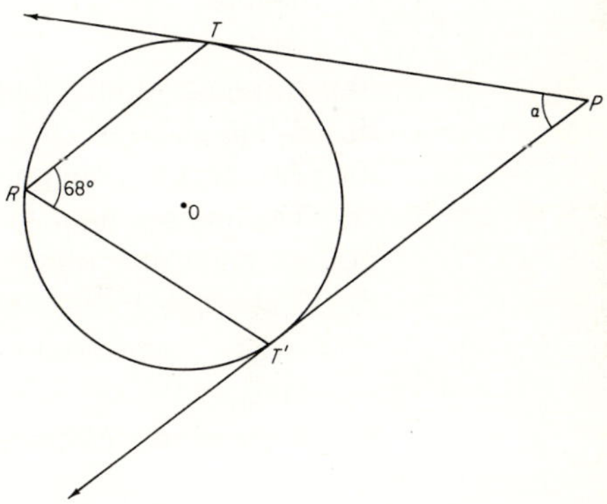

In Probs. 17 to 25, indicate the word which will make the statement true.

17. A central angle of a circle is formed by two ______.
18. An inscribed angle of a circle is formed by two ______.
19. An angle inscribed in a semicircle is a(n) ______.
20. The greatest number of obtuse angles a triangle inscribed in a circle can have is ______.
21. Tangent segments drawn to a circle from an outside point are ______.
22. The largest chord of a circle is the ______ of the circle.

23. An angle is inscribed in an arc. If the intercepted arc is increased by 10°, the inscribed angle is increased by ______.
24. The opposite angles of a quadrilateral inscribed in a circle are ______.
25. If a line is ______ to a radius at its point on the circle, it is tangent to the circle.

In Probs. 26 to 40, indicate whether the statement is *always true* (mark T) or *not always true* (mark F).

26. If a parallelogram is inscribed in a circle, it must be a rectangle.
27. Doubling the minor arc of a circle will double the chord of the arc.
28. An equilateral polygon inscribed in a circle must be equiangular.
29. A radius of a circle is a chord of the circle.
30. If an inscribed angle and a central angle subtend the same arc, the measure of the inscribed angle is twice the measure of the central angle.
31. A straight line can intersect a circle in three points.
32. A rectangle circumscribed about a circle must be a square.
33. The angle formed by two chords intersecting in a circle is equal in degrees to half the difference of the measures of the intercepted arcs.
34. A trapezoid inscribed in a circle must be isosceles.
35. A line perpendicular to a radius is tangent to the circle.
36. Two arcs are congruent if they have equal lengths.
37. The line segment joining two points on a circle is a secant.
38. An angle inscribed in an arc less than a semicircle must be acute.
39. The angle formed by a secant and a tangent intersecting outside a circle is measured by half the sum of the measures of the intercepted arcs.
40. A chord of a circle is a diameter.

CHAPTER 16

RATIO, PROPORTION, AND SIMILAR POLYGONS

16-1 RATIO Much communication today is concerned with comparing quantities. When one person weighs 150 lb and another 200 lb, we say that the first weighs $\frac{150}{200}$ or $\frac{3}{4}$ as much as the second or the second weighs $\frac{4}{3}$ as much as the first. If one person pays \$1,000 in Federal taxes and \$100 in state taxes, we compare his taxes by saying his Federal and state taxes are in the ratio of 10 to 1.

DEFINITION *The* ratio *of one quantity to another like quantity is the quotient of the first divided by the second.*

A ratio is a fraction, and all the rules governing a fraction apply to ratios. A ratio can be written with a fraction bar, a solidus, a division sign, or the symbol : (which is read "is to"). Thus the ratio of 3 to 4 is $\frac{3}{4}$, 3/4, $3 \div 4$, or $3:4$. The 3 and 4 are called *terms* of the ratio.

It is important to note that only measures of *like* quantities can be compared. We cannot compare, for instance, 200 lb and 1,000 dollars. Pounds and dollars do not measure the same kind of quantity. However, we can compare 2 yd and 5 ft by first converting 2 yd to 6 ft. The ratio, then, becomes 6/5. The ratio of the measures of two straight angles and three right angles is 4/3.

It will be noted that a ratio is always an *abstract* number; i.e., it has no units of measure. Unless there is an important reason to the contrary, a ratio should be expressed in its simplest form. Therefore, we should express, for example, a ratio as $\frac{3}{4}$ rather than $\frac{150}{200}$.

16-2 PROPORTION A *proportion* is a statement of equality of two or more ratios. For example, since 6/8 and 3/4 have the same value, the ratios can be equated to form a proportion: $\frac{6}{8} = \frac{3}{4}$ or $6:8 = 3:4$. If the ratios $a:b$ and $c:d$ are equal, the expression $a:b = c:d$ is a proportion. This is read "a is to b as c is to d." In the proportion, a is referred to as the *first term*, b as the *second term*, c as the *third term*, and d as the *fourth term.*

The first and fourth terms are called the *extremes*, and the second and third terms are called the *means*. It should be noted that four terms are necessary to form a proportion.

The *fourth proportional* to three quantities is the fourth term of the proportion, the first three terms of which are taken in order. Thus, 3 is the fourth proportional of 8, 6, and 4.

When the two means of a proportion are equal, either is said to be the *mean proportional* between the first and fourth terms of the proportion. If $a:b = b:c$, then b is the mean proportional between a and c. Since $3:6 = 6:12$, 6 is the mean proportional between 3 and 12.

16-3 THEOREMS ABOUT PROPORTIONS Since a proportion is a fractional equation, all the axioms which deal with equalities can be applied to a proportion. Applying these axioms to propositions leads to the following important theorems. The student may wish to develop his own proofs for each of the theorems.

Theorem 16-1 In a proportion, the product of the extremes is equal to the product of the means.

Example 1

If $2/3 = 4/6$

then $2(6) = 3(4)$

Example 2

If $a:b = c:d$

then $ad = bc$

Theorem 16-2 In a proportion, the means may be interchanged.

Example 1

If $12/8 = 3/2$

then $12/3 = 8/2$

Example 2

If $a:b = c:d$

then $a:c = b:d$

Theorem 16-3 In a proportion the ratios may be inverted.

Example 1

If $\frac{5}{6} = \frac{10}{12}$

then $\frac{6}{5} = \frac{12}{10}$

Example 2

If $\frac{a}{b} = \frac{c}{d}$

then $\frac{b}{a} = \frac{d}{c}$

Theorem 16-4 If the product of two quantities is equal to the product of two other quantities, either pair of quantities can be used as the means and the other pair as the extremes of a proportion.

Example 1

If $4(5) = 2(10)$

then $\frac{4}{2} = \frac{10}{5}$

Example 2

If $xy = rs$

then $x:r = s:y$

Theorem 16-5 If four quantities are in proportion, the terms are in proportion by addition or subtraction; that is, the sum (or difference) of the first and second terms is to the second term as the sum (or difference) of the third and fourth terms is to the fourth term.

Example 1

If $\frac{3}{4} = \frac{9}{12}$

then $\frac{3+4}{4} = \frac{9+12}{12}$

Example 2

If $\frac{a}{b} = \frac{c}{d}$

then $\frac{a+b}{b} = \frac{c+d}{d}$

Exercise 16-1

1. Express the following ratios in lowest terms.
   (a) 10 to 15
   (b) 64:36
   (c) $\frac{2}{3}$ to $\frac{3}{4}$
   (d) 4 right ∠s to 3 straight ∠s
   (e) 2 in. to 1 ft
   (f) 5 hr to 15 min
   (g) 2 yd to 1 ft 3 in.
   (h) 3 in.:5 cm
2. Prove Example 2 of Theorem 16-1 by using Axiom 6.
3. Prove Example 2 of Theorem 16-3 by using Axiom 7.
4. Prove Example 2 of Theorem 16-5 by adding 1 to both sides of the original proportion.
5. Find the values of x which make each proportion a true statement.
   (a) $3:x = 4:24$
   (b) $5:20 = x:100$
   (c) $9:6 = 15:x$
   (d) $x:5 = 3:8$
   (e) $1\frac{1}{4}$ ft:3 in. = 2 yd:x in.
   (f) 28 in.:x ft = 4 yd:2 ft
6. Change each of the following proportions to another whose first term is a and whose second term is b.

(a) $\dfrac{b}{a}=\dfrac{5}{6}$

(b) $\dfrac{a}{4}=\dfrac{b}{3}$

(c) $\dfrac{6}{a}=\dfrac{7}{b}$

(d) $\dfrac{4}{b}=\dfrac{12}{a}$

(e) $b:4=a:7$

(f) $a:2=b:5$

7. Find the ratio of r to s in each of the following:
   (a) $3r=4s$
   (b) $8r=6s$
   (c) $\frac{2}{3}r=1\frac{2}{3}s$
   (d) $r=\frac{1}{4}s$
   (e) $ar=bs$
   (f) $xs=yr$
8. Find the fourth proportional to 4, 5, and 8.
9. Find the mean proportional between
   (a) 2 and 18
   (b) 4 and 16
   (c) 20 and 5
   (d) 2 and 32
   (e) $\frac{3}{4}$ and $\frac{4}{27}$
   (f) 5 and 4
10. In a draftsman's scale drawing, $\frac{1}{8}$ in. represents 1 ft. What length will be represented by a segment $2\frac{5}{8}$ in. long in the drawing?
11. If a machine can produce 60 bolts in 2 min, how many bolts can it produce in 8 hr?
12. An inclined plane rises 124 ft in 620 ft. How far does it rise in 285 ft?
13. Two pulleys are connected by a belt. One pulley has a diameter of 8 in. and is rotating 160 rpm. The diameter of the second pulley is 5 in. What is the speed of rotation of the second pulley?
14. The circumference of a circle is 87.6 in. The diameter of the circle is 27.9 in. Use your slide rule to find the ratio of the circumference and the diameter for this circle.

16-4 SIMILAR POLYGONS In Chap. 13 we studied a relationship between figures called congruence. Congruent figures are alike in every respect. They have the same shape and size. We shall now define polygons which have the same shape but not necessarily the same size.

DEFINITION *Two polygons are said to be similar to each other if their corresponding angles are congruent and their corresponding sides are proportional. These polygons have the same shape.*

Design engineers and architects are continually dealing with similar figures. A newly designed structure is first drawn to scale on paper. The design is much smaller than the structure itself, but all parts have the shape of the finished product. Blueprints of these drawings are made. The blueprint can be read by the manufacturer. By using a ruler and a scale, he can determine the true dimensions of any part of the structure represented in the blueprint.

The surveyor continually uses the properties of similarity of triangles in his work.

The symbol for "similar to" or "is similar to" is $\sim$. In Fig. 16-1, $ABCDE \sim A'B'C'D'E'$ if:

1 $\angle A \cong \angle A'$, $\angle B \cong \angle B'$, $\angle C \cong \angle C'$, $\angle D \cong \angle D'$, $\angle E \cong \angle E'$.

2 $\dfrac{AB}{A'B'} = \dfrac{BC}{B'C'} = \dfrac{CD}{C'D'} = \dfrac{DE}{D'E'} = \dfrac{EA}{E'A'}$.

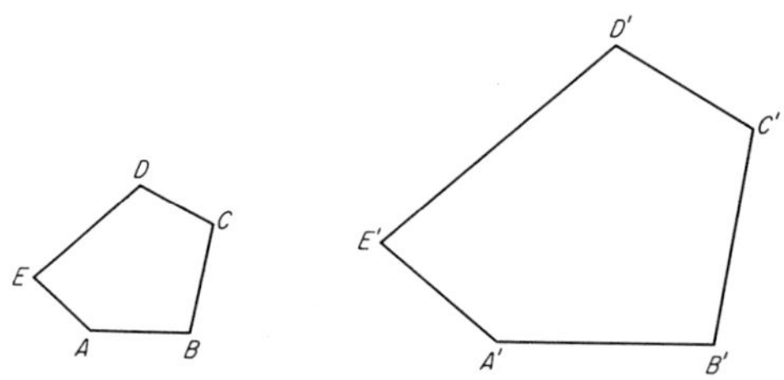

FIG. 16-1.

Conversely, if two polygons are similar, their corresponding angles are congruent and their corresponding sides are proportional.

It is important to note that the definition of similar polygons has two parts. In order for polygons to be similar, (1) corresponding angles must be congruent and (2) corresponding sides must be proportional.

16-5 THEOREMS ON SIMILAR TRIANGLES The following theorems can be proved and have wide application in mathematics, science, and engineering.

Theorem 16-6 If two triangles have two angles of one congruent to two angles of the other, they are similar.

Theorem 16-7 A line parallel to one side of a triangle and intersecting the other two sides divides these sides proportionally. This theorem states that if, in Fig. 16-2, it is given that $\overleftrightarrow{DE} \parallel \overleftrightarrow{AB}$, then we can conclude that $CD/DA = CE/EB$.

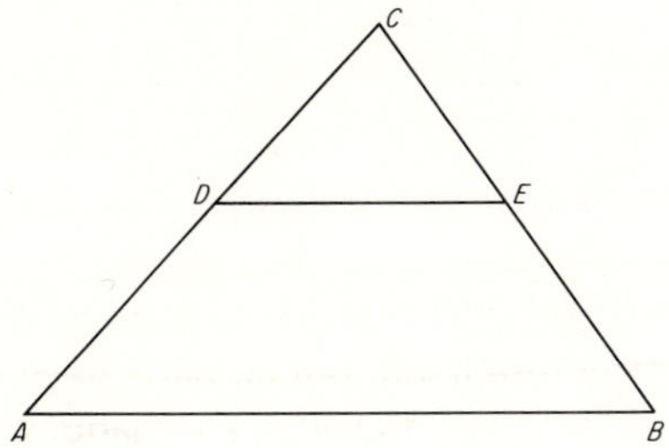

FIG. 16-2.

It is then a simple matter to apply Theorem 16-5 to show that

$$\frac{CD + DA}{DA} = \frac{CE + EB}{EB}$$

or $$\frac{CA}{DA} = \frac{CB}{EB}$$

Theorem 16-8 Parallel lines cut off proportional segments on two transversals. This theorem states that if, in Fig. 16-3, it is given that $m \parallel n \parallel p$, then $AB/BC = DE/EF$.

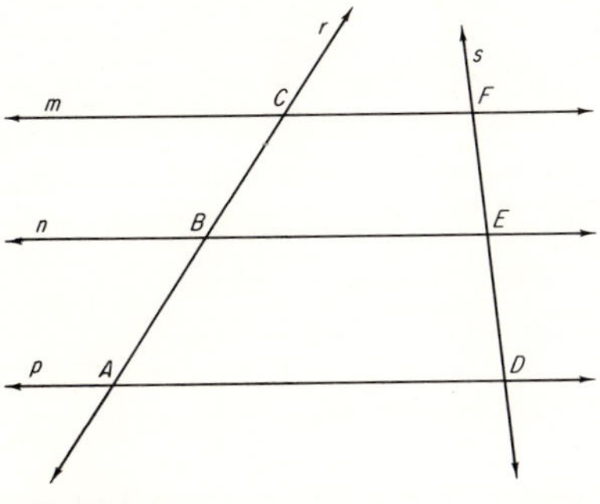

FIG. 16-3.

If we apply Theorem 16-5, we can show that

$$\frac{AB + BC}{BC} = \frac{DE + EF}{EF}$$

or $$\frac{AC}{BC} = \frac{DF}{EF}$$

Exercise 16-2

In Probs. 1 to 10, it is given that $\overleftrightarrow{DE} \parallel \overleftrightarrow{AB}$. In each problem the lengths of three segments are given. Find the value of x in each exercise:

	1	2	3	4	5	6	7	8	9	10
AD	4	5	*x*	6	8	8	3		8	
DC	6	15	9	*x*			4	10		
AC					*x*	32		*x*	12	
BE	8	*x*	12	9	10	10	6	9		6
EC	*x*	21	18	15	14			15		15
BC						*x*	*x*			
AB									*x*	21
DE									16	*x*

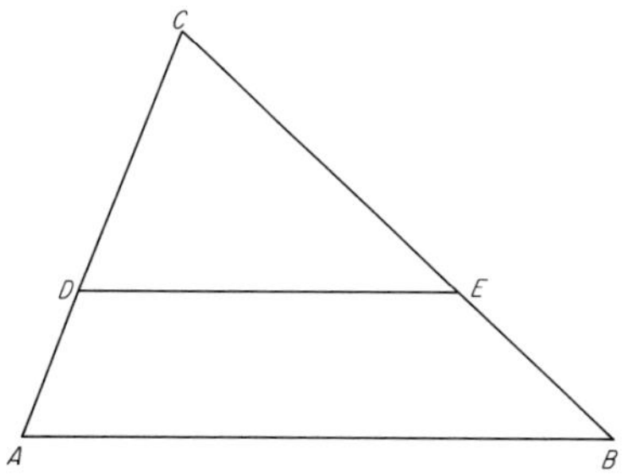

PROBS. 1–10.

In Probs. 11 to 16 it is given that $m \| n \| p$. Find the value of x in each problem.

	11	12	13	14	15	16
AB	12	*x*	3		12	
BC	*x*	25		15	18	8
AC			*x*	24		12
DE	16	24	9	*x*		
EF	12	15	15	10	*x*	10
DF					40	*x*

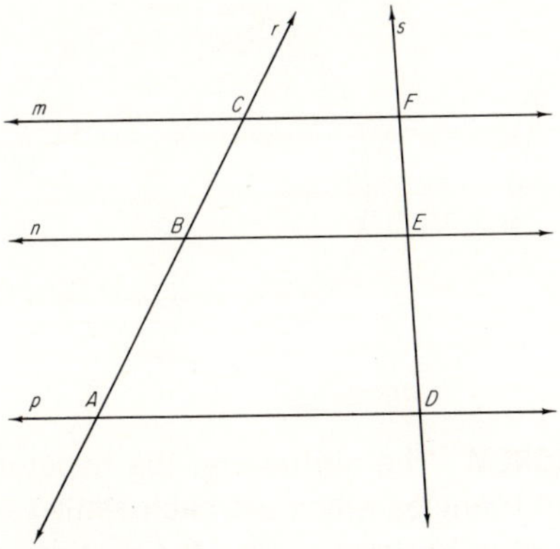

PROBS. 11–16.

In Probs. 17 to 22 determine the pairs of corresponding sides of similar triangles; then find the length of x. Right angles are marked ⊾; congruent angles are marked with identical marks.

17.

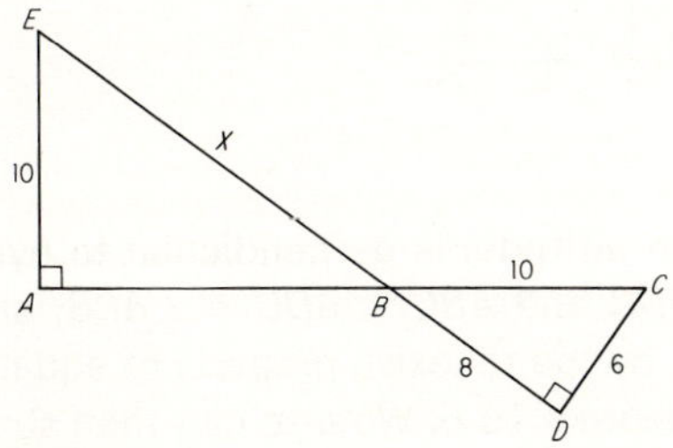

PROB. 17.

18.

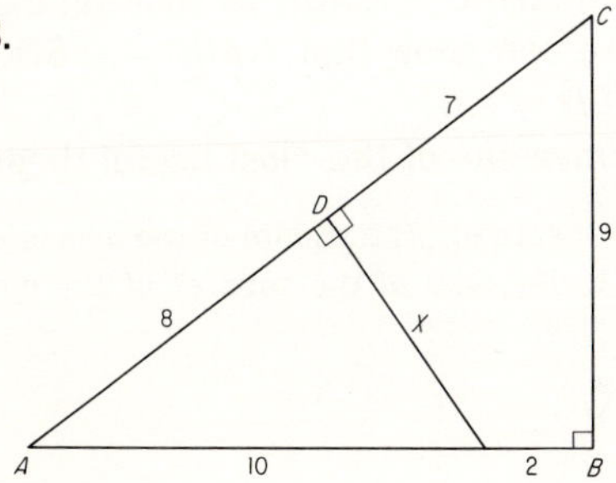

PROB. 18.

19.

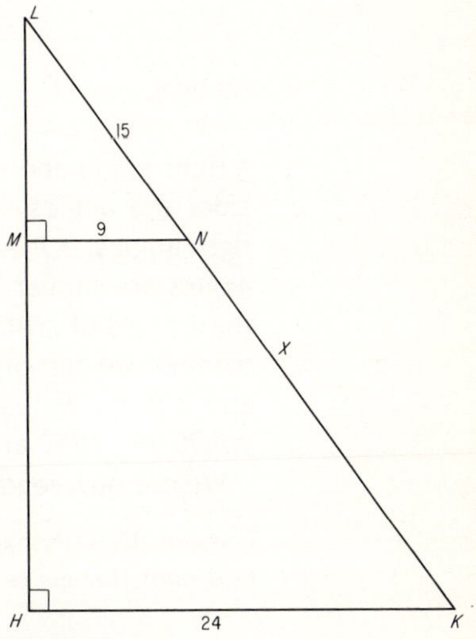

PROB. 19.

20.

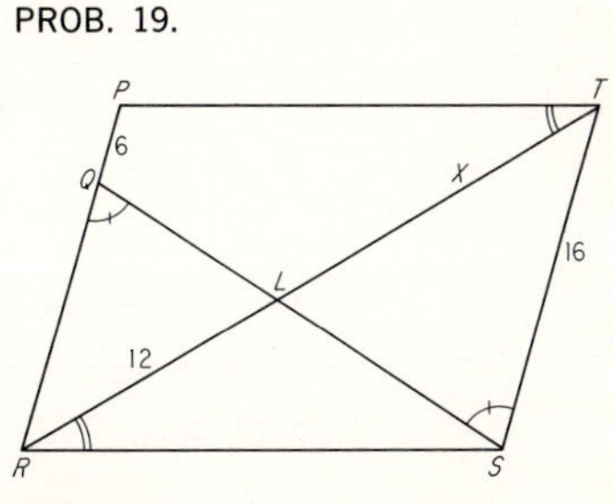

PROB. 20.

21.

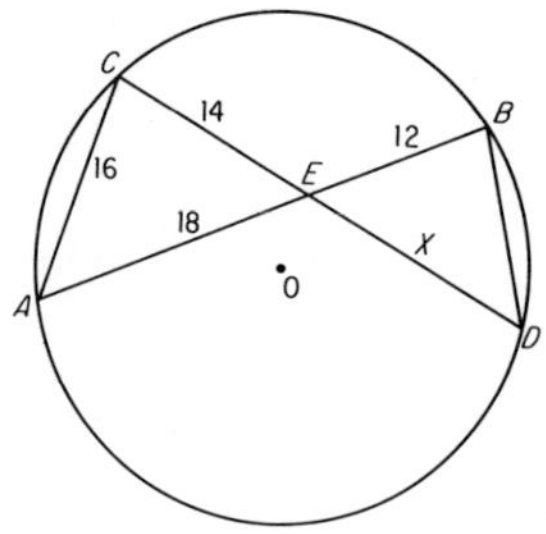

PROB. 21.

22.

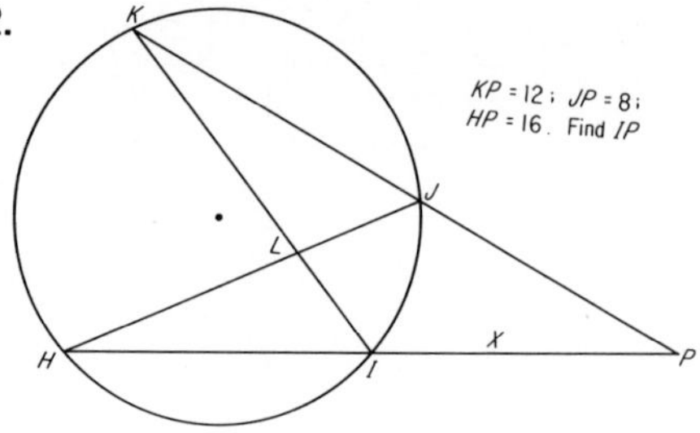

PROB. 22.

16-6 PYTHAGOREAN THEOREM The altitude on the hypotenuse of a right triangle divides it into two triangles which are each similar to the given triangle and to each other. Let us assume in Fig. 16-4 that, in $\triangle ABC$, $\angle ACB$ is

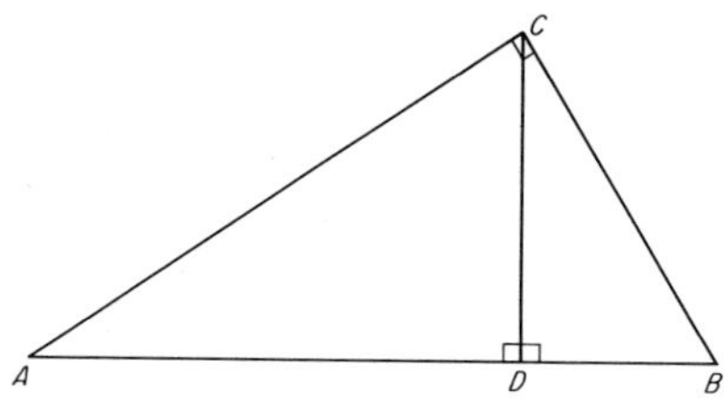

FIG. 16-4.

a right angle and $\overline{CD}$ (an altitude) is perpendicular to hypotenuse AB. Consider the angles of ⊿s ADC and ABC. $\angle ADC \cong \angle ACB$, since they are both right angles. $\angle A \cong \angle A$, by the reflexive property of equality. Hence, the triangles are similar by Theorem 16-6. We also can then show that $\angle ACD$, the third angle of $\triangle ACD$, is congruent to the third angle ($\angle B$) of $\triangle ABC$. In like manner, we can prove that $\triangle BDC \sim \triangle ABC$ by showing that $\angle BDC \cong \angle ACB$ and $\angle B \cong \angle B$. Lastly, we can show that $\triangle ADC \sim \triangle BDC$ by showing that $\angle ADC \cong \angle BDC$ and $\angle ACD \cong \angle B$.

We are now ready to prove one of the most useful theorems in geometry.

Theorem 16-9 (Pythagorean theorem) The square of the measure of the hypotenuse of a right triangle is equal to the sum of the squares of the measures of the legs.

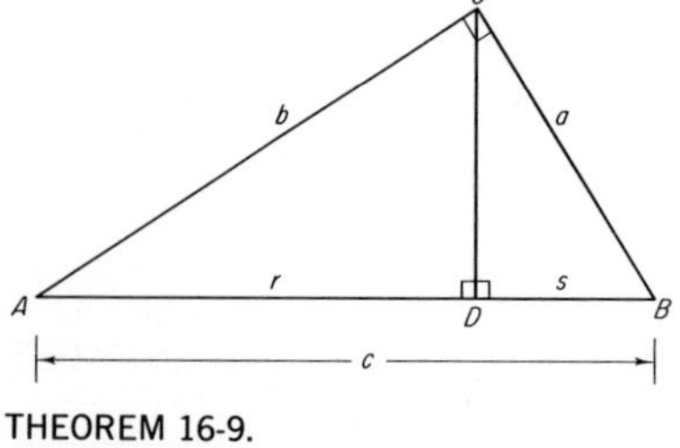

THEOREM 16-9.

1 Assume that $\triangle ABC$ is a right $\triangle$ with $\angle ACB$ a right $\angle$.
2 Draw $\overline{CD} \perp \overline{AB}$.
3 $\triangle ABC \sim \triangle BDC$, since $\angle B \cong \angle B$ and right $\angle ACB \cong$ right $\angle BDC$.
4 $\triangle ABC \sim \triangle ADC$, since $\angle A \cong \angle A$ and $rt\ \angle ACB \cong rt\ \angle ADC$.
5 Then $c:a = a:s$ and $c:b = b:r$, since corresponding sides of similar polygons are proportional.
6 $a^2 = cs$ and $b^2 = cr$ by Theorem 16-1.
7 $a^2 + b^2 = cs + cr = c(s + r)$ by adding equals and using the distributive law.
8 But $s + r = c$ by Postulate 3.
9 Then $a^2 + b^2 = c^2$ by Axiom 8.

Example 16-1

Given: $\overline{CB} \perp \overline{AB}$; $AB = 8$ in.; $BC = 15$ in.
Find: AC.

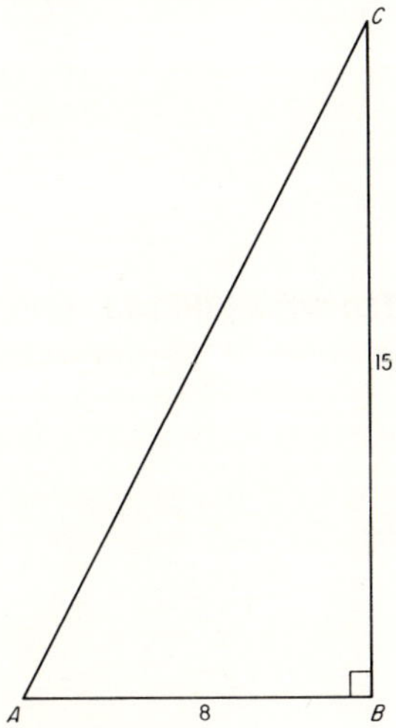

EXAMPLE 16-1.

SOLUTION

By the Pythagorean theorem $(AC)^2 = (AB)^2 + (BC)^2$

Substituting known values, $(AC)^2 = (8)^2 + (15)^2$

$= 64 + 225$

$= 289$

Then $AC = \sqrt{289} = 17$ in.

Example 16-2

Given: $\overline{TR} \perp \overline{RS}$; $TS = 45$ ft; $RT = 27$ ft.
Find: RS.

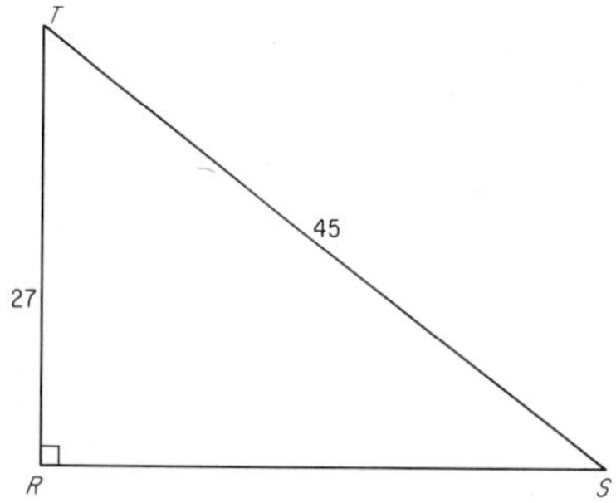

EXAMPLE 16-2.

SOLUTION

$$(TS)^2 = (RT)^2 + (RS)^2$$

or
$$\begin{aligned}(RS)^2 &= (TS)^2 - (RT)^2\\ &= (45)^2 - (27)^2\\ &= 2{,}025 - 729\\ &= 1{,}296\end{aligned}$$

$$RS = \sqrt{1{,}296} = 36 \text{ ft}$$

16-7 SIMILAR TRIANGLES ASSOCIATED WITH CIRCLES Consider Fig. 16-5 in which two chords intersect within a circle. If we draw chords AC and BD,

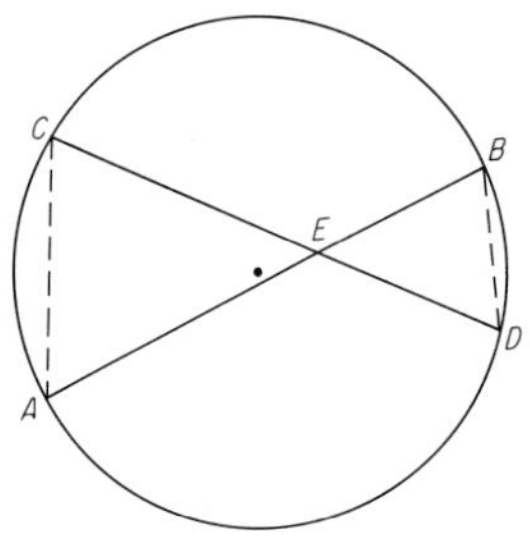

FIG. 16-5.

two triangles are formed. $\angle A \cong \angle D$, since they intercept the same arc. In like manner, $\angle C \cong \angle B$. $\angle AEC \cong \angle BED$, since they are vertical angles. Therefore $\triangle AEC \sim \triangle BED$ and $AE:DE = CE:BE$ or $AE \times BE = DE \times CE$. This can be stated as a theorem.

Theorem 16-10 If two chords intersect within a circle, the product of the measures of the segments of one chord is equal to the product of the measures of the segments of the other.

In Fig. 16-6 let $\overleftrightarrow{PA}$ and $\overleftrightarrow{PD}$ be secants of $\odot O$ drawn from the point P. If we draw chords AC and BD, two similar triangles are formed. It can be shown that $\angle A \cong \angle D$, $\angle ACP \cong \angle DBP$, and $\angle P \cong \angle P$. Therefore, we can form the proportion $PA:PD = PC:PB$. Using Theorem 16-1, we can then prove that $PA \times PB = PD \times PC$.

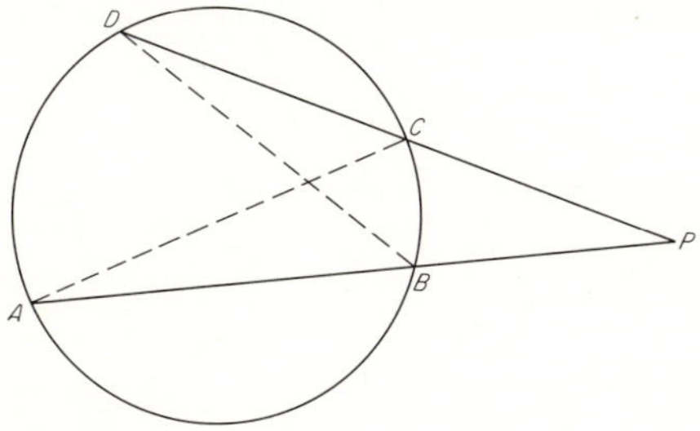

FIG. 16-6.

Theorem 16-11 If two secants are drawn from the same point outside a circle, the product of the measures of one secant and its external segment is equal to the product of the measures of the other secant and its external segment.

Exercise 16-3

In Probs. 1 to 10, $\angle ACB$ is a right $\angle$, $\overline{CD} \perp \overline{AB}$.

1. Find AB if $AC = 16$ and $BC = 12$.
2. Find BC if $BD = 9$ and $CD = 12$.
3. Find CD if $AC = 17$ and $AD = 15$.

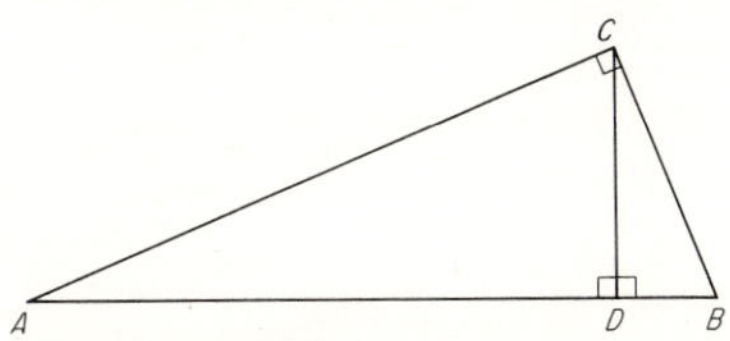

PROBS. 1–10.

4. Find AC if $AB = 40$ and $BC = 25$.
5. Find CD if $AD = 18$ and $BD = 8$.
6. Find BC if $AB = 8$ and $BD = 2$.
7. Find AC if $AD = 24$, $CD = 18$, and $BC = 22.5$.
8. Find AC if $BD = 18$, $BC = 30$, and $CD = 24$.
9. Find BD if $AD = 5$ and $CD = 10$.
10. Find DC if $AD = 16$ and $BD = 4$.

In Probs. 11 to 14 find the lengths of x. Draw a perpendicular if necessary. Right angles and congruent sides are marked.

11.

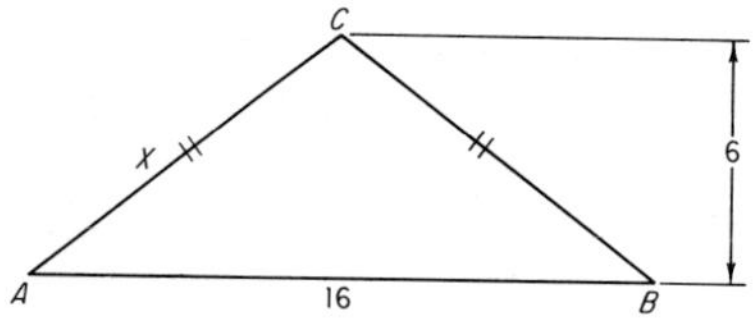

PROB. 11.

12.

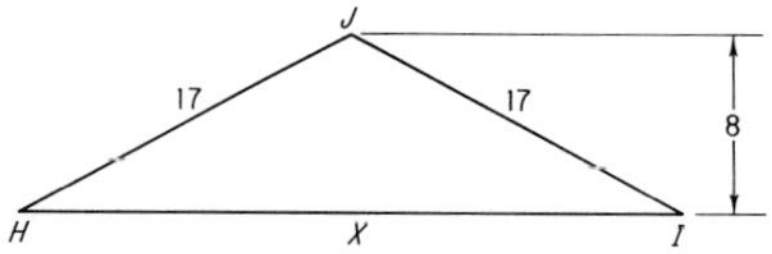

PROB. 12.

13.

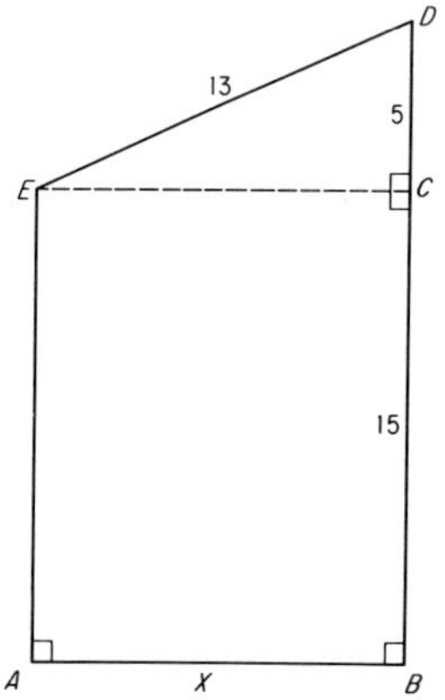

PROB. 13.

14.

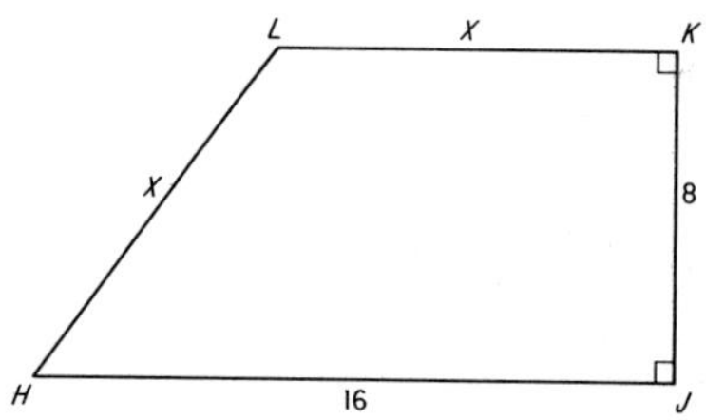

PROB. 14.

15. Find CE if $AE = 4$, $EB = 45$, $ED = 15$.
16. Find ED if $CE = 20$, $EB = 12$, $AE = 15$.
17. Find AE if $EB = 4$, $CE = 8$, $ED = 5$.
18. Find AC if $BD = 16$, $AE = 15$, $BE = 20$, $DE = 25$.

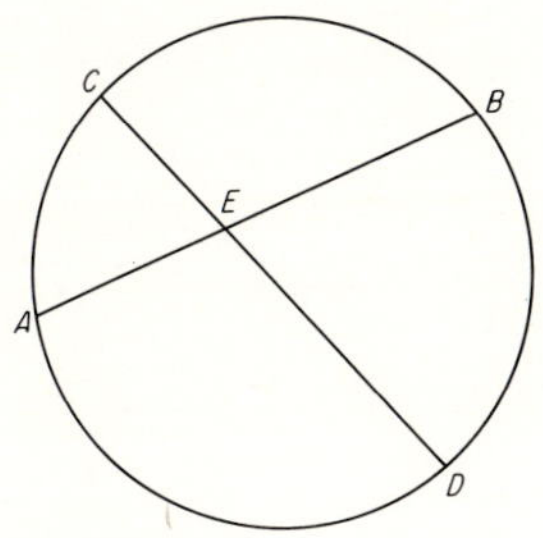

PROBS. 15–18.

19. Find RP if $PT = 24$, $PS = 10$, $PQ = 8$.
20. Find PS if $RP = 36$, $PT = 48$, $PQ = 12$.

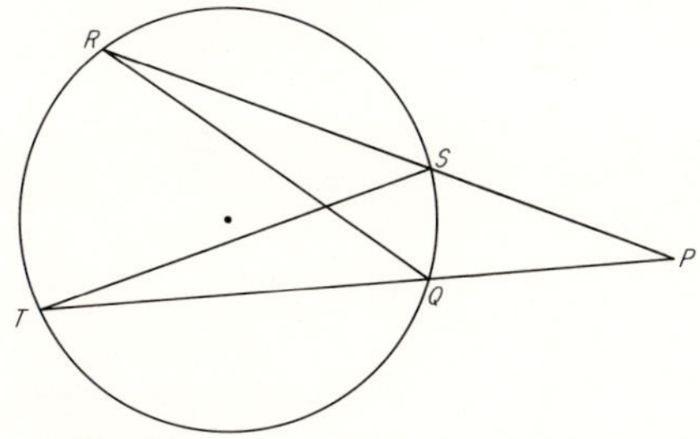

PROBS. 19–22.

21. Find PT if $PQ = 3$, $PS = 4$, $RS = 5$.
22. Find PQ if $PS = 4$, $RQ = 5$, $TS = 8$.
23. Find QS if $OR = 16$, $QT = 6$, $RQ = 10$.

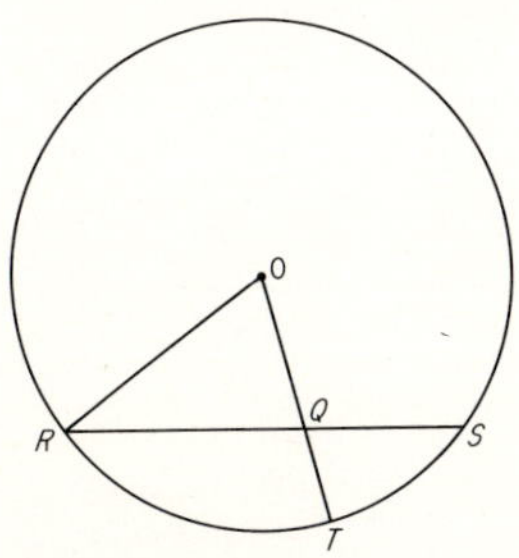

PROBS. 23, 24.

24. Find OR if $RQ = 16$, $QS = 10$, $QT = 8$.
25. A 25-ft ladder is placed against a vertical wall of a building. The foot of the ladder is 7 ft from the wall. If the top of the ladder slips 4 ft, how far will the foot of the ladder slide?

CHAPTER 17
PERIMETERS AND AREAS

17-1 THE CIRCUMFERENCE OF A CIRCLE Finding the length and area of a circle have been two of the greatest historic problems in mathematics.

DEFINITION *The* circumference *of a circle is the length of the circle* (*sometimes called the* perimeter).

The following theorem can be proved:

Theorem 17-1 The ratio of the circumference to the diameter is the same for all circles. This ratio is represented by the Greek letter π (pi). Thus, in Fig. 17-1, if we let C and C' be circumferences and d and d' diameters of Ⓢ O and O', respectively, $C/d = C'/d' = \pi$.

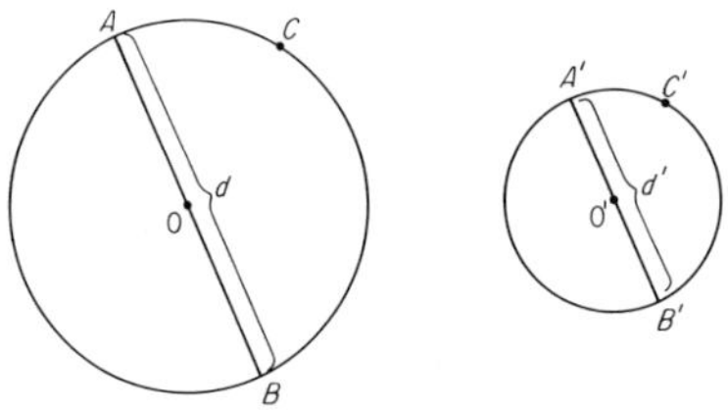

FIG. 17-1.

If we multiply each side of the equation $C/d = \pi$ by d, we obtain the formula for the circumference of the circle. Since the measure of a diameter is two times the measure of the radius, we can express the following rule.

FORMULA *The circumference of a circle is expressed by the formula* $C = \pi d$ *or* $C = 2\pi r$.

The number π is what is called an irrational number. Any number which cannot be written in the form p/q, where p and q are integers, and q is not zero, is defined as an *irrational number.* An irrational number can never be expressed as an *exact decimal fraction.* In advanced studies it can be shown that $\pi = 4(1 - \frac{1}{3} + \frac{1}{5} - \frac{1}{7} + \frac{1}{9} - \frac{1}{11} + \cdots)$. The right-hand expression is called an *infinite series.*

By the use of modern computers, the value of π has been found accurate to more than 100,000 figures. This is a degree of accuracy which has no practical value. The value of π accurate to 10 decimals is 3.1415926536.

In using the formula for circumference it is best to use π rounded off to one more digit than the given data. Then the computed values should be rounded off to the degree of accuracy of the given data.

Example 17-1 Find the circumference of a circle the length of whose diameter is 6.8 in.

SOLUTION

$$C = \pi d$$
$$= (3.14)(6.8)$$
$$= 21.352$$
$$\approx 21 \text{ in.}$$

Example 17-2 Find the radius of a circle whose circumference is 7.65 ft.

SOLUTION

$$C = 2\pi r$$
$$r = \frac{C}{2\pi}$$
$$= \frac{7.65}{2(3.142)}$$
$$= 1.218$$
$$\approx 1.22 \text{ ft}$$

17-2 MEANING OF AREA A *unit of area* is the measure of the surface within a square whose sides have unit lengths. Thus, if *ABCD* in Fig. 17-2 is a

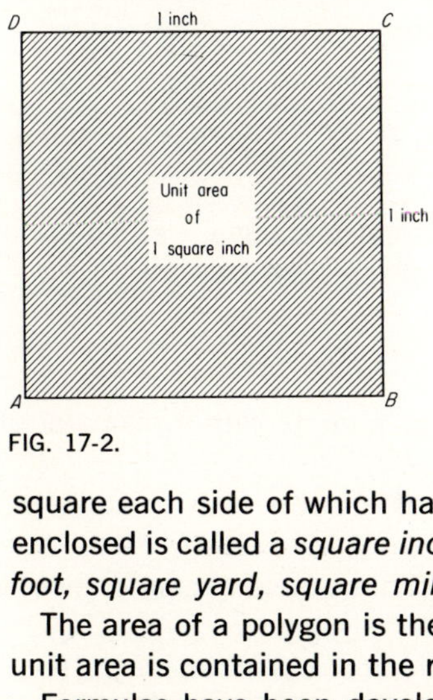

FIG. 17-2.

square each side of which has a length of 1 in., the measure of the region enclosed is called a *square inch.* Other common units of area are the *square foot, square yard, square mile, square centimeter,* and *square kilometer.*

The area of a polygon is the number which tells how many times a given unit area is contained in the region bounded by the polygon.

Formulas have been developed by which areas can be computed when certain linear measurements are known. It should be noted that the length of a line segment can be *measured* directly by using a ruler or tapemeasure but the area of a region is *computed by formula.* Formulas have been developed for areas of the rectangle, triangle, parallelogram, trapezoid, and circle. The areas of other figures can often be found by partitioning them into triangles, rectangles, and trapezoids and then summing up the areas of these figures.

17-3 AREA OF A RECTANGLE If in rectangle $ABCD$ of Fig. 17-3 $AB = 5$ in. and $BC = 3\frac{1}{2}$ in., we can draw the lines shown and then count 15 square-

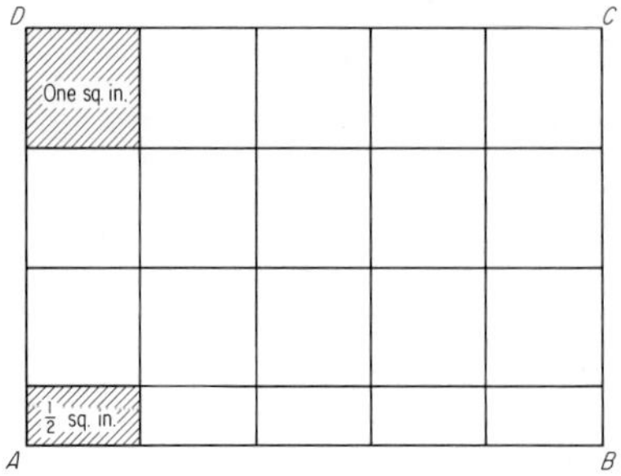

FIG. 17-3.

inch units and 5 half-square-inch units. These 5 half-square-inch units are equivalent to $2\frac{1}{2}$ square-inch units. The total area is then $17\frac{1}{2}$ in.^{2}. This number could also be obtained by multiplying $5 \times 3\frac{1}{2}$. This example suggests the following postulate.

POSTULATE 9 *The area of a rectangle is equal to the product of its base and its altitude ($A = bh$).*

Since a square is an equilateral rectangle, we can state: The area of a square is equal to the square of the sides ($A = s^2$).

Exercise 17-1

1. Find the circumference of a circle with a diameter of (a) 3.7 yd, (b) 5.07 in., (c) 1.912 mi.
2. Find the circumference of a circle with a radius of (a) 2.5 in., (b) 6.37 ft, (c) 41.08 cm.
3. Find the diameter of a circle whose circumference is (a) 4.5 ft, (b) 2.51 in., (c) 6.283 yd.
4. Find the radius of a circle whose circumference is (a) 28 in., (b) 503 ft, (c) 90.14 yd.
5. Find the area of a rectangle with base 7.3 ft and altitude 2.4 ft.
6. Find the altitude of the rectangle with area of 12.4 in.^{2} and base 2.6 in.
7. Find the area of a floor, in square feet, that is 15 ft 8 in. long and 12 ft 9 in. wide.
8. How many square inches are in the six sides of a box $7\frac{1}{2}$ by 5 by $3\frac{1}{2}$ in.?
9. What is the cost of carpeting the floor of a room with dimensions 15 ft 4 in. by 24 ft 6 in. if carpeting costs \$7.75 a square yard?

10. How long must the floor of a factory be if it is 45 ft 6 in. wide and has an area of 5,584.25 ft^2?

In Probs. 11 to 20 divide the figure into rectangles and then find the areas of each figure.

11.

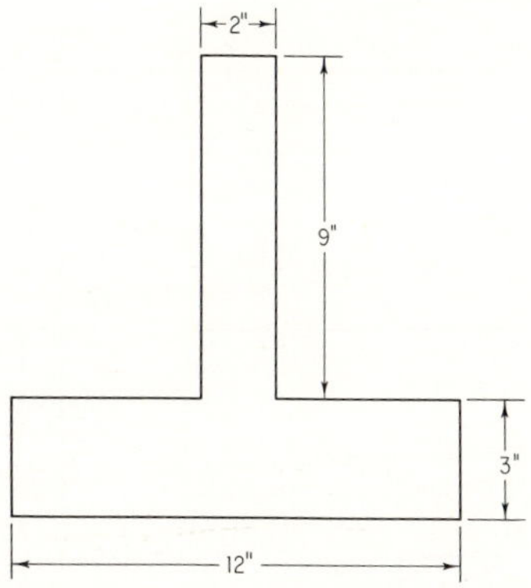

12.

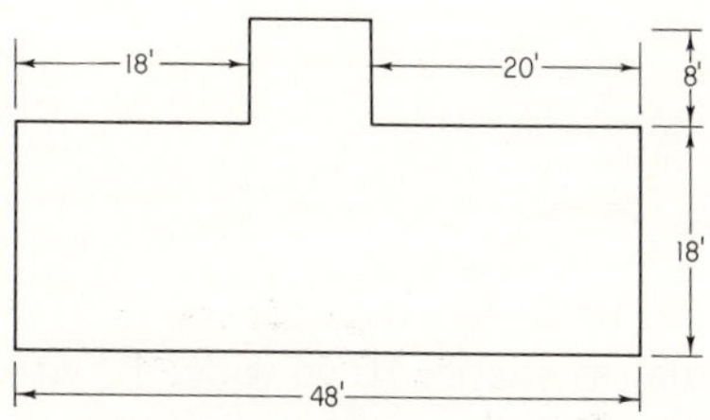

13.

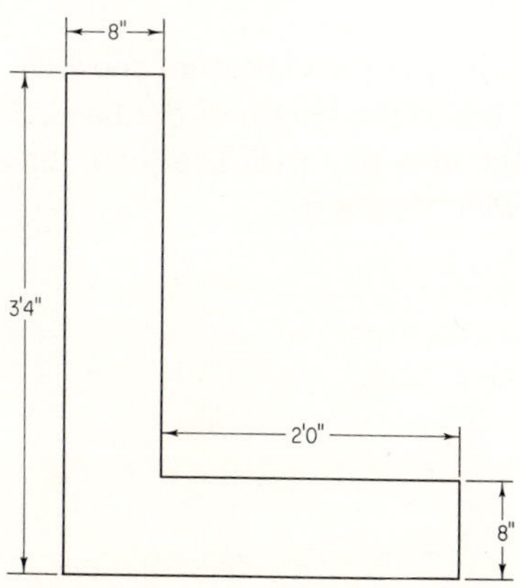

14.

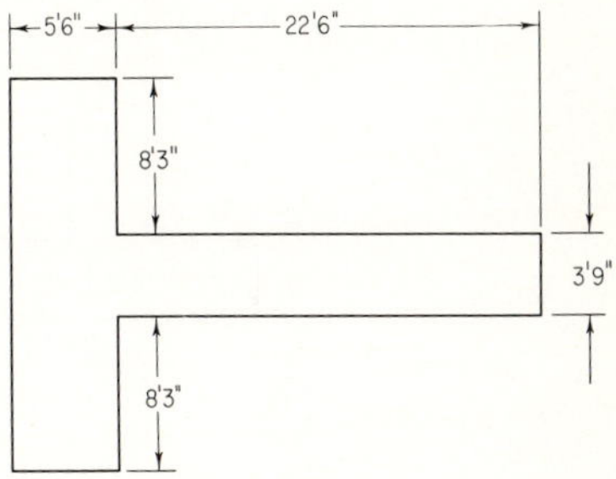

15.

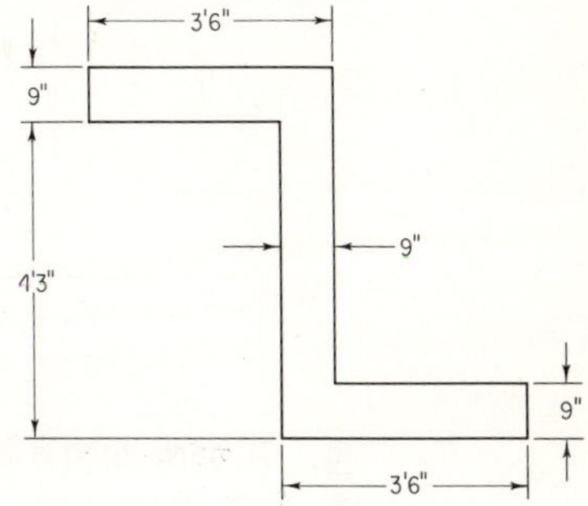

16.

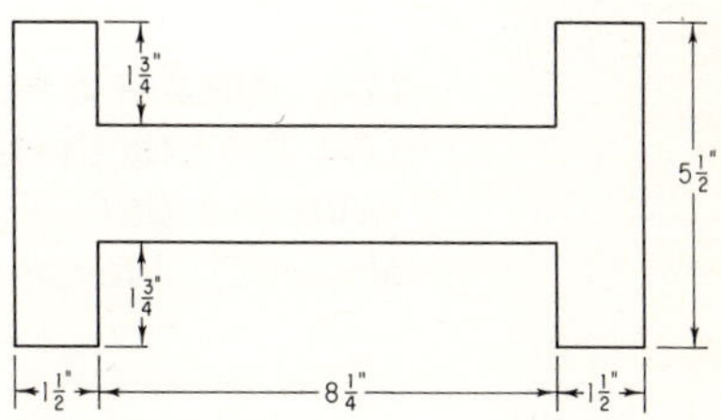

17.

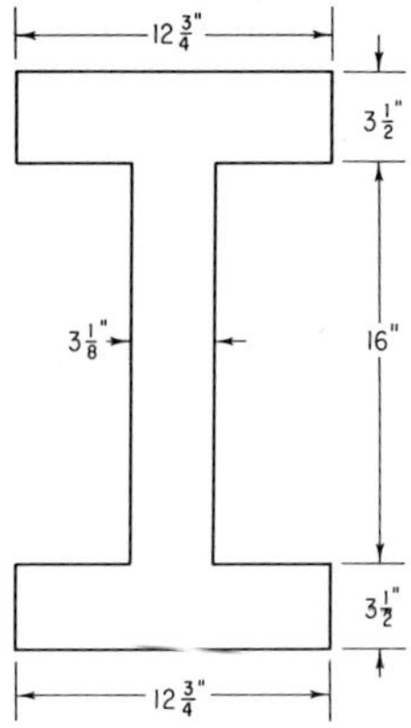

19.

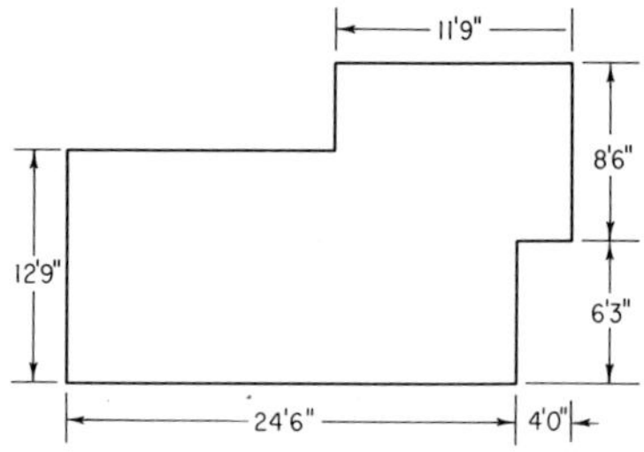

18.

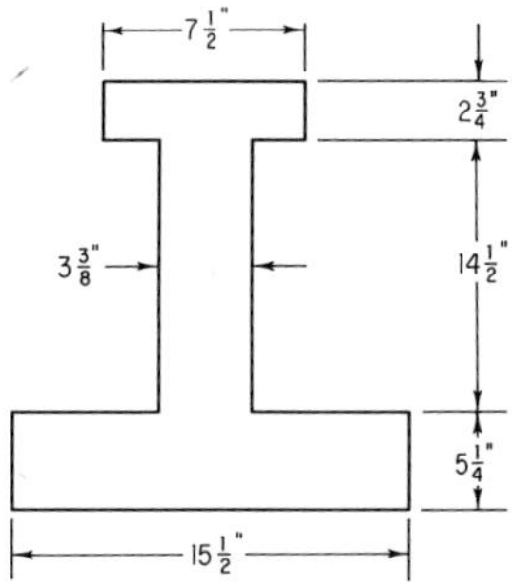

20.

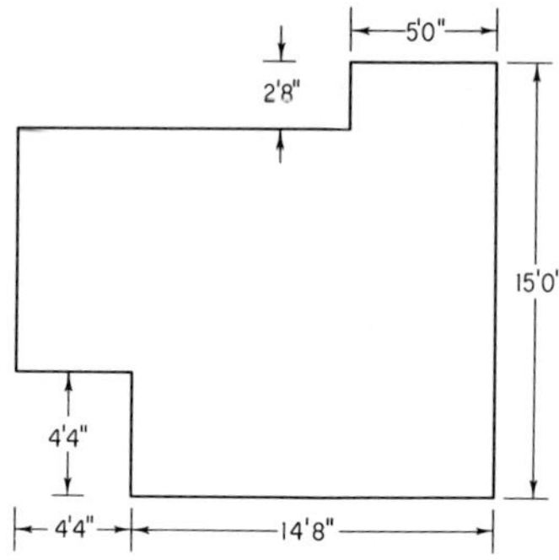

21. A barn with a flat roof is rectangular in shape, 10 yd wide, 12 yd long, and 5 yd high. It is to be painted inside and outside and on the ceiling but not on the roof or floor. Find the total number of square yards to be painted.

17-4 AREA OF A PARALLELOGRAM Consider the parallelogram shown in Fig. 17-4. Let h be the length of the altitude and b the length of the base. Since $\triangle ADE \cong \triangle BCF$, it can be shown that the area of $\square ABCD$ equals the area of $\square EFCD$. We can then state the following theorem.

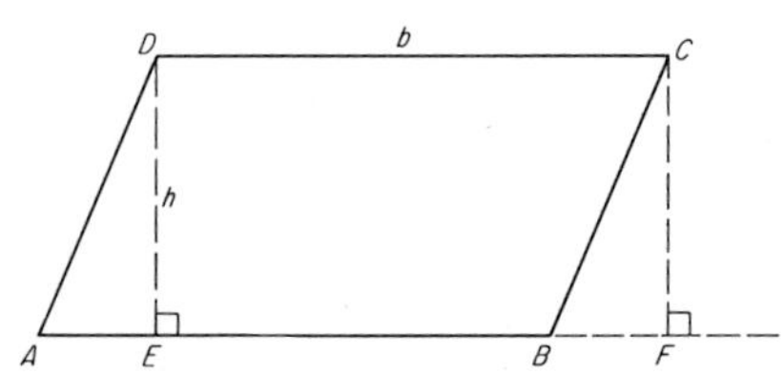

FIG. 17-4.

Theorem 17-2 The area of a parallelogram is equal to the product of its base and its altitude ($A = bh$).

17-5 AREA OF A TRIANGLE Consider $\triangle ABC$ of Fig. 17-5. If we draw $\overleftrightarrow{CD} \parallel \overleftrightarrow{AB}$ and $\overleftrightarrow{BD} \parallel \overleftrightarrow{AC}$, $\square ABDC$ will be formed. Theorem 17-2 states that the area of $\square ABDC = bh$. Diagonal BC separates the parallelogram into two triangles which can be proved congruent. Hence it can be shown that the area of $\triangle ABC = \frac{1}{2}bh$.

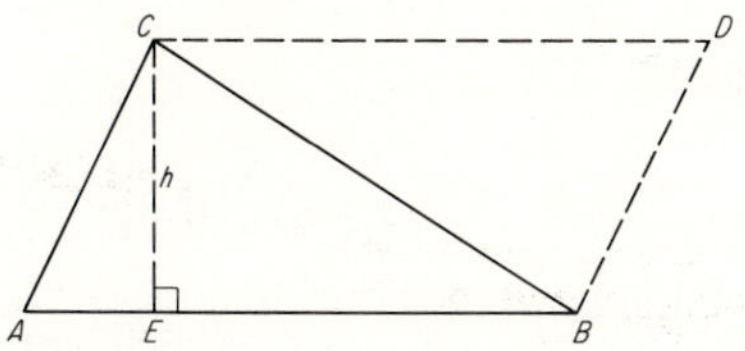

FIG. 17-5.

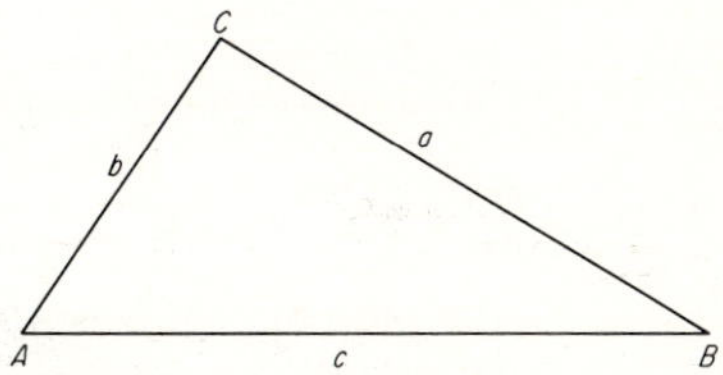

FIG. 17-6.

Theorem 17-3 The area of a triangle is equal to half the product of its base and its altitude ($A = \frac{1}{2}bh$).

Occasionally the three sides of a triangle are known but not the altitude or base. It is possible to find the area of such a triangle by using a formula attributed to Heron (first century A.D.) of Alexandria. The proof of the formula can be found in many geometry texts.

Theorem 17-4 (Heron's formula) The area of a triangle with sides a, b, c is given by the formula $A = \sqrt{s(s-a)(s-b)(s-c)}$ where $s = \frac{1}{2}(a+b+c)$.

Example 17-3 Find the area of $\triangle ABC$ if $a = 12$ ft, $b = 15$ ft, $c = 19$ ft.

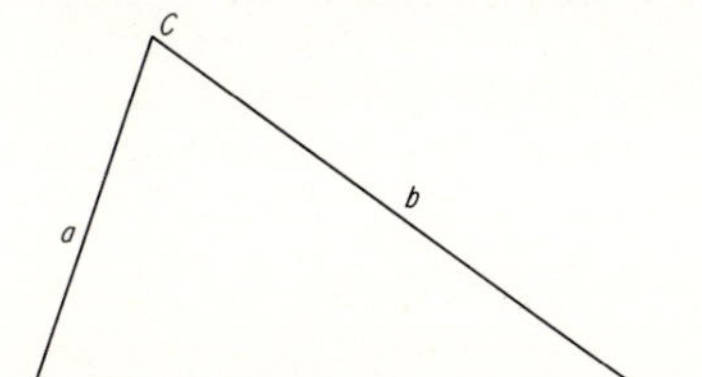

EXAMPLE 17-3.

SOLUTION

$$\begin{aligned} s &= \tfrac{1}{2}(a+b+c) \\ &= \tfrac{1}{2}(12 \text{ ft} + 15 \text{ ft} + 19 \text{ ft}) \\ &= 23 \text{ ft} \end{aligned}$$

$$\begin{aligned} A &= \sqrt{23(23-12)(23-15)(23-19)} \\ &= \sqrt{8{,}096} \\ &= 89.98 \\ &\approx 90 \text{ ft}^2 \end{aligned}$$

17-6 AREA OF A TRAPEZOID Let trapezoid *ABCD* (Fig. 17-7) have bases b_1 and b_2 and altitude h. If we draw diagonal *AC*, the trapezoid will be partitioned into $\triangle$s *ABC* and *ACD*. By Theorem 17-3, the area of $\triangle ABC = \frac{1}{2}b_1h$ and

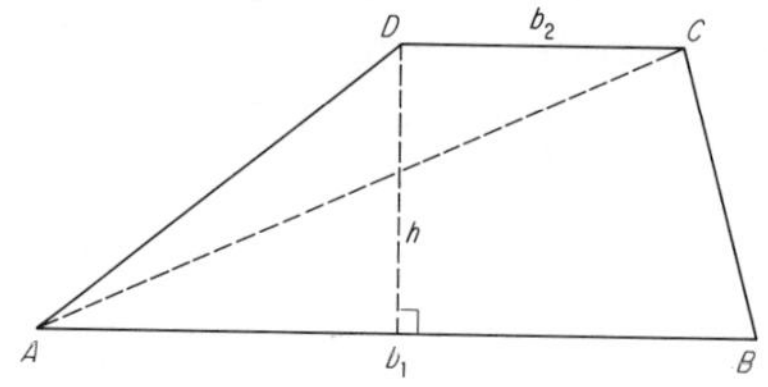

FIG. 17-7.

the area of $\triangle ACD = \frac{1}{2}b_2h$. Adding, we get the area of $ABCD = \frac{1}{2}b_1h + \frac{1}{2}b_2h$. Then, by applying the distributive law of multiplication (see Sec. 1-10), we get area $ABCD = \frac{1}{2}h(b_1 + b_2)$.

Theorem 17-5 The area of a trapezoid is equal to half the product of its altitude and the sum of its bases.

Exercise 17-2

1. Find the area of a parallelogram with base 18 in. and altitude 24 in.
2. Find the area of a parallelogram with base 15.4 in. and altitude 10.3 in.
3. Find the altitude of the parallelogram with an area of 408 in.^{2} and a base of 26 in.
4. Find the area of the square whose diagonal has a length of 32 in.
5. Find the area of the rhombus the diagonals of which are 36 and 24 in.
6. Find the area of $\triangle ABC$, in square feet, if $\overline{CD} \perp \overline{AB}$, $AB = 15$ ft 6 in., $CD = 8$ ft 3 in.

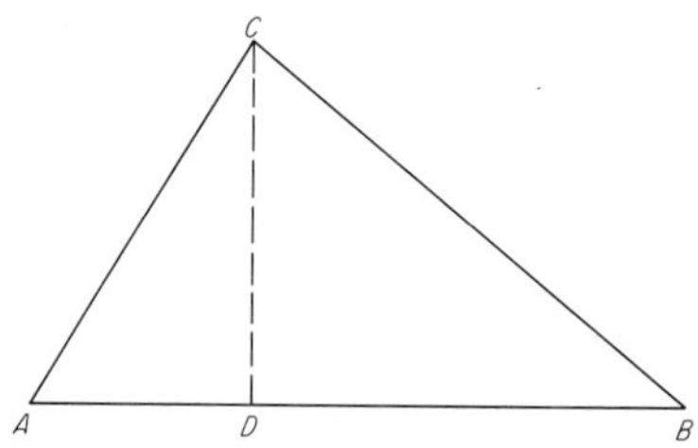

PROB. 6.

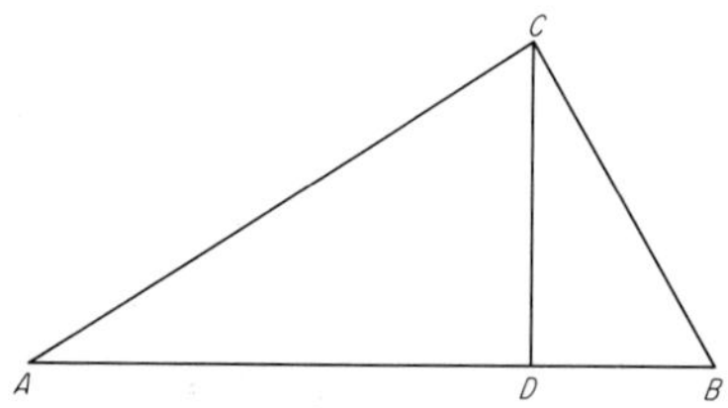

PROB. 7.

7. Find the area of $\triangle ABC$ if $\angle ACB$ is a right angle, $\overline{CD} \perp \overline{AB}$, $AD = 16$ ft, $BD = 4$ ft. (*Note:* Prove $\triangle ADC \cong \triangle BDC$ and then $AD : CD = CD : BD$.)
8. Find the length of $\overline{ST}$ if $RS = ST$. $\angle S$ is a right $\angle$, and the area of $\triangle RST$ is 128 in.^{2}

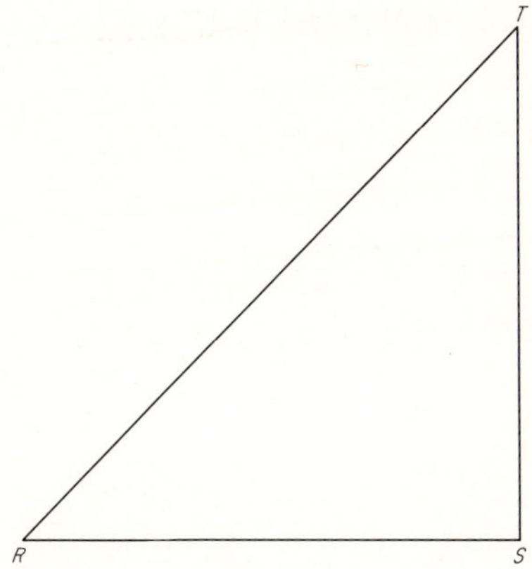

PROB. 8.

9. Find the altitude of $\triangle ABC$.
10. Find the area of $\triangle ABC$.

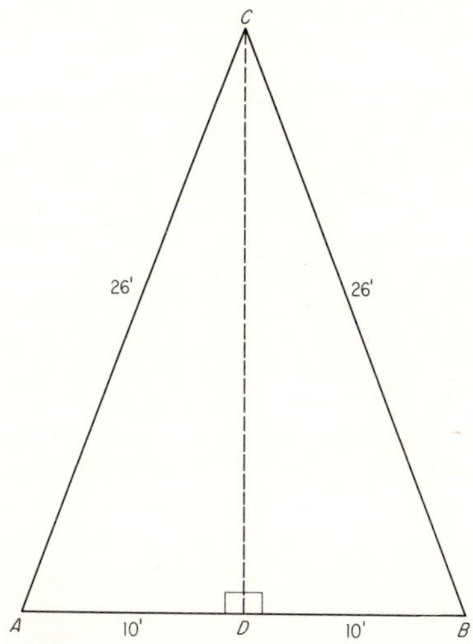

PROBS. 9 and 10.

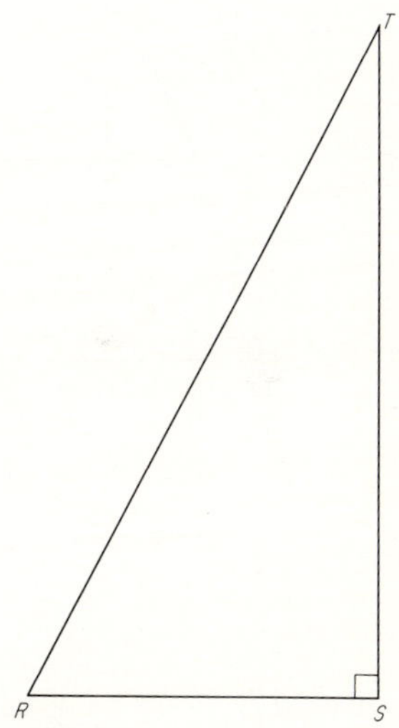

PROB. 11.

11. Find the area of $\triangle RST$ if $m\angle S = 90°$, $RT = 34$ ft, $ST = 30$ ft.
12. Find the area of a triangle the lengths of whose sides are 9, 12, and 15 ft.
13. Find the area of $\triangle RST$ (accurate to one decimal place) if $RS = 8$ in., $ST = 12$ in., $RT = 16$ in.

14. Find the altitude of equilateral $\triangle ABC$ if $AC = 9$ ft.

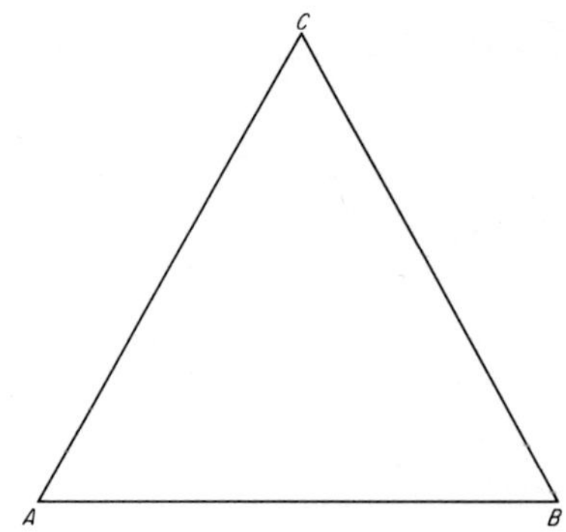

PROB. 14.

In Probs. 15 to 20 find the areas of the trapezoids.

15.

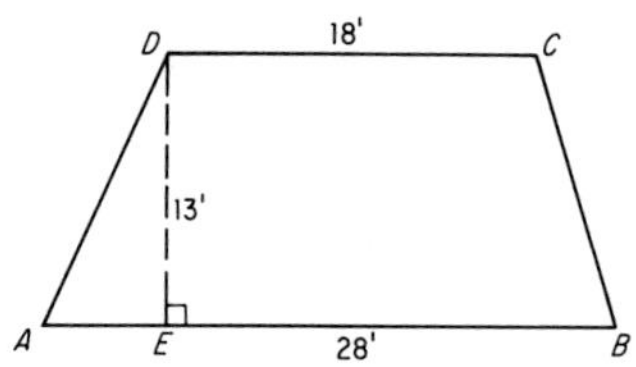

16.

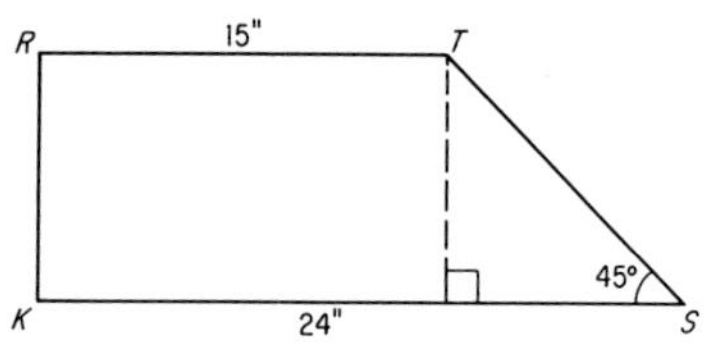

17.

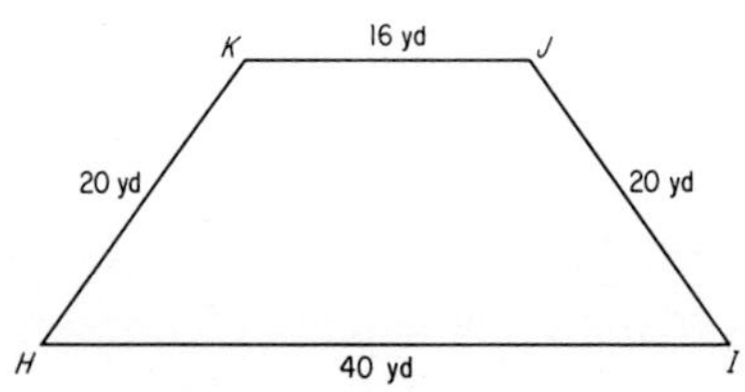

18.

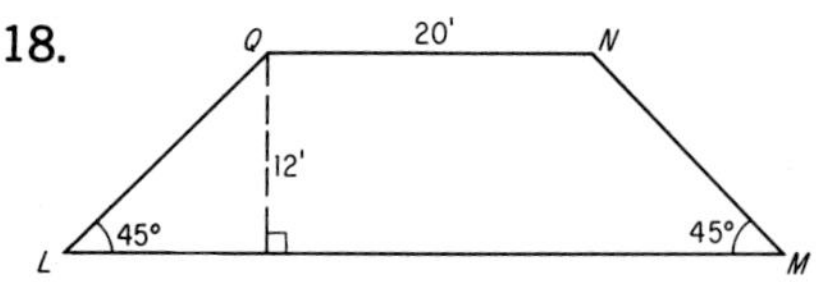

19.

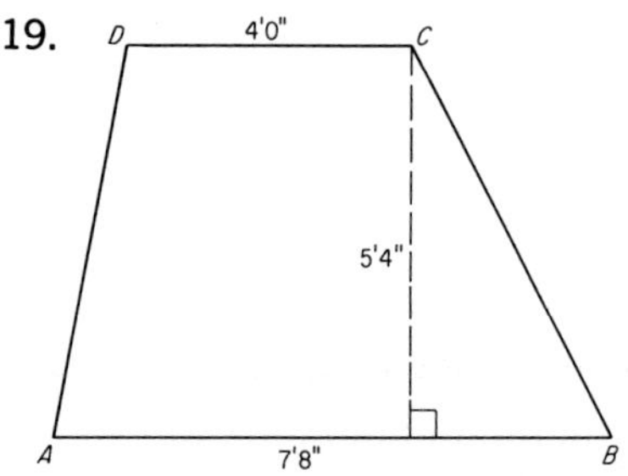

20.

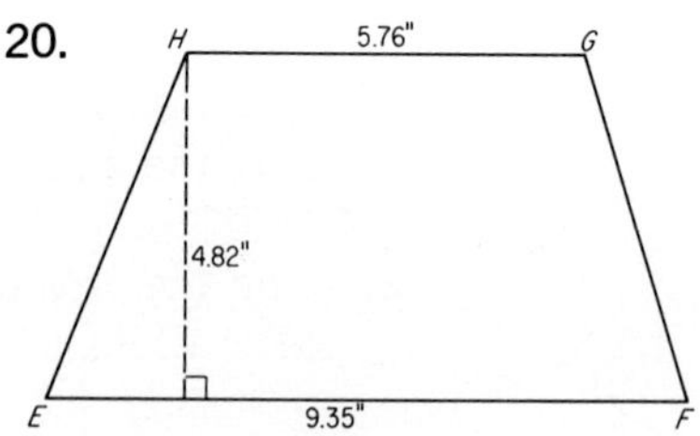

In Probs. 21 and 22 find the area of each figure.

21.

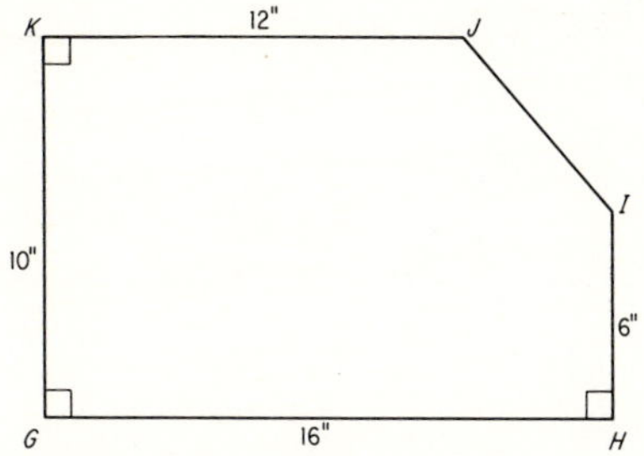

22.

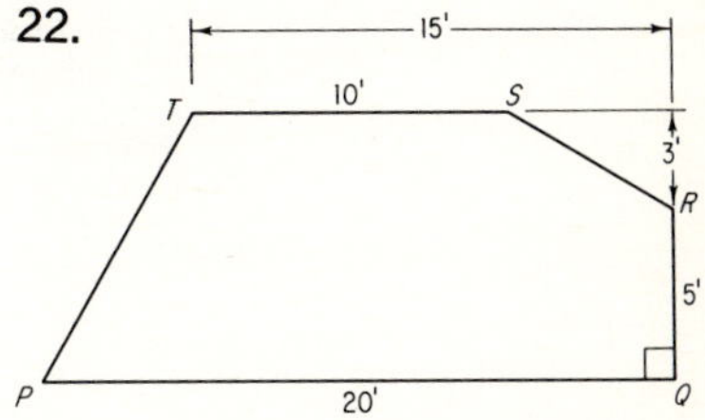

17-7 AREA OF A CIRCLE In Sec. 17-1 we learned that the ratio of the circumference to diameter of any circle is equal to the constant π. It can be proved that this constant is also equal to the ratio of the area to the square of the radius for any circle.

Theorem 17-6 The ratio of the area to the square of the radius is the same for all circles.

In Fig. 17-8

$$\frac{A}{r^2} = \frac{A'}{r'^2} = \pi$$

Multiplying each term of the equation $A/r^2 = \pi$ by r^2, we obtain the formula for the area of a circle.

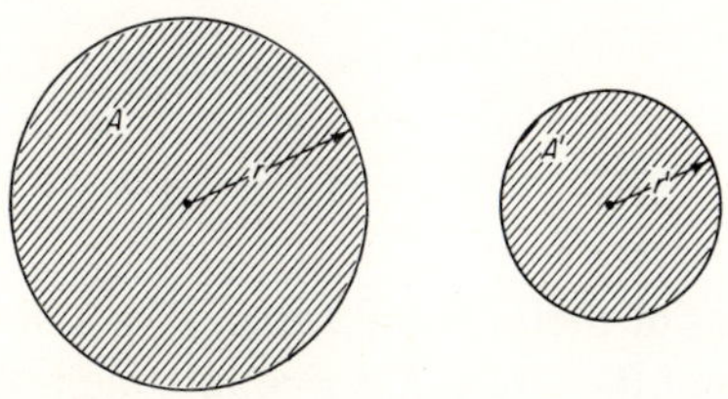

FIG. 17-8.

FORMULA *The formula for the area of a circle is $A = \pi r^2$. Since $r = d/2$, we can substitute in the formula and get $A = \pi d^2/4$.*

17-8 LENGTH OF AN ARC Since the measure of an arc is equal to the measure of its central angle, we know that the lengths of two arcs of a circle are proportional to the measures of their central angles. Thus, in Fig. 17-9, $l/s = \theta/\phi$. If we let C be the circumference of the circle, $l/C = \theta/360$ if θ is

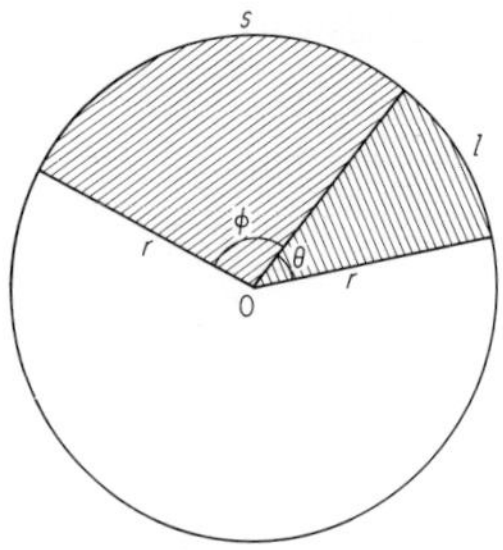

FIG. 17-9.

expressed in degrees. Multiplying each term by C and then replacing C by $2\pi r$, we get $l = \pi(\theta/360)C = \theta/360(2\pi r)$.

FORMULA *The formula for the length of an arc of a circle is $l = \pi r\theta/180$, where θ is expressed in degrees.*

17-9 AREA OF A SECTOR A *sector of a circle* is a region bounded by two radii and an arc (see Fig. 17-10). The areas of two sectors of a circle are

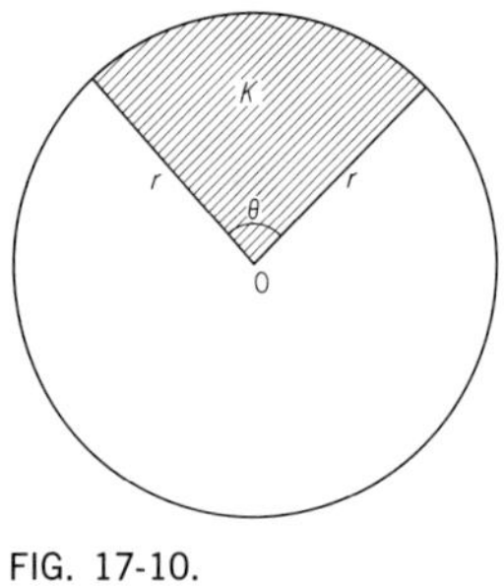

FIG. 17-10.

proportional to the measures of their central angles. Thus, if K is the area of the sector and A is the area of the circle, $K/A = \theta/360$. Multiplying each term by A and then replacing A by πr^2, we get the following.

FORMULA *The formula for the area of a sector of a circle is $K = \pi r^2\theta/360$, where θ is expressed in degrees.*

17-10 AREA OF A SEGMENT A *segment of a circle* is the region bounded by a chord and an arc of the circle. The shaded region of Fig. 17-11 represents a segment of the circle. The area of the segment can be found by subtracting the area of the triangle formed by the chord and the two radii to its end points from the area of the sector. Thus the area is found by $K = \pi r^2\theta/360 - \frac{1}{2}bh$, where θ is expressed in degrees.

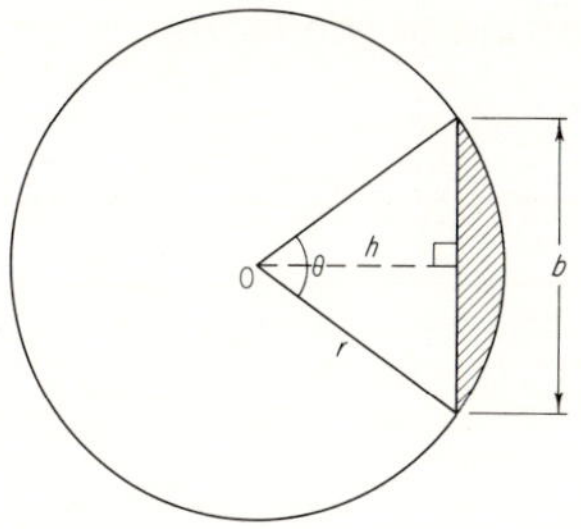

FIG. 17-11.

Exercise 17-3

1. The radius of a circle is 18 in. What is the length of an arc of (a) 30°, (b) 90°, (c) 36°, (d) 72°?
2. The radius of a circle is 15 in. What is the area of a sector with a central angle whose measure is (a) 60°, (b) 90°, (c) 36°, (d) 45°?
3. The radius of a circle is 12 in. What is the area of the segment whose arc subtends a central angle with a measure of (a) 60°, (b) 90°, (c) 120°?
4. Find the area of the shaded portion of the square.

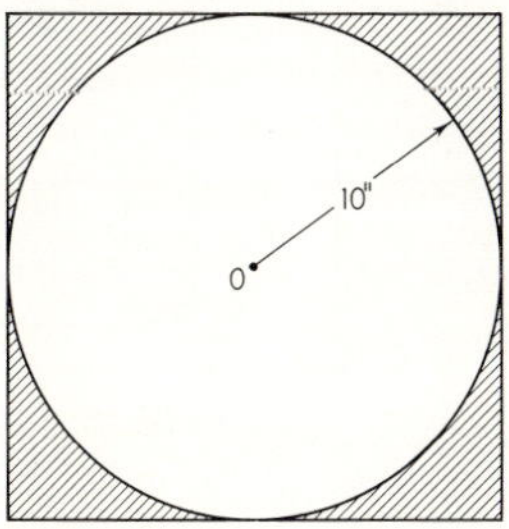

5. *ABCD* is a square. Find the area of the shaded portion of the circle.

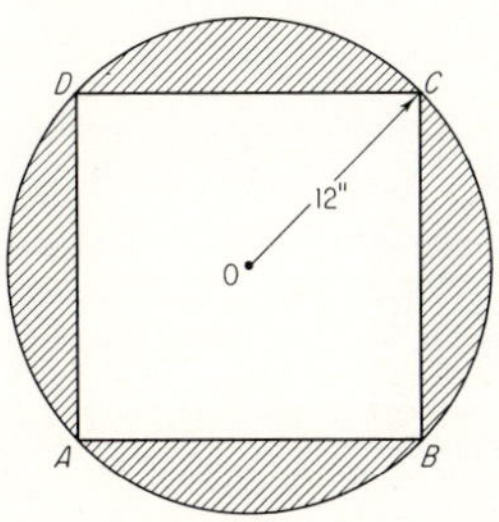

6. Find the area of the shaded portion of the figure.

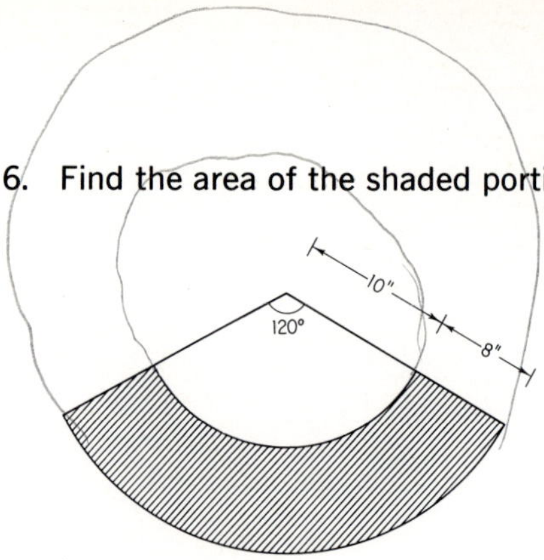

7. Find the area of the shaded portion (called a fillet) of the square.

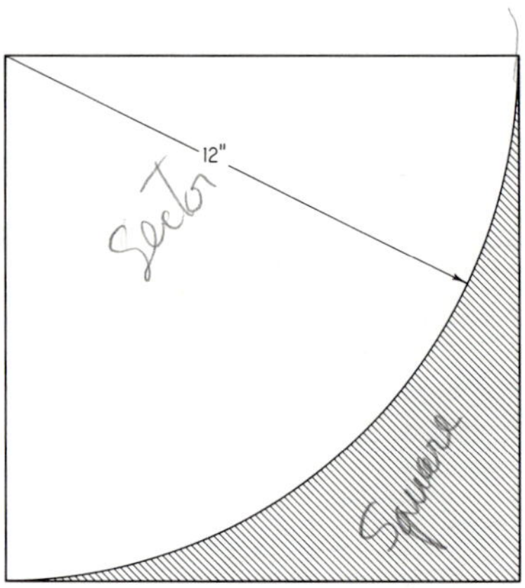

PROB. 7.

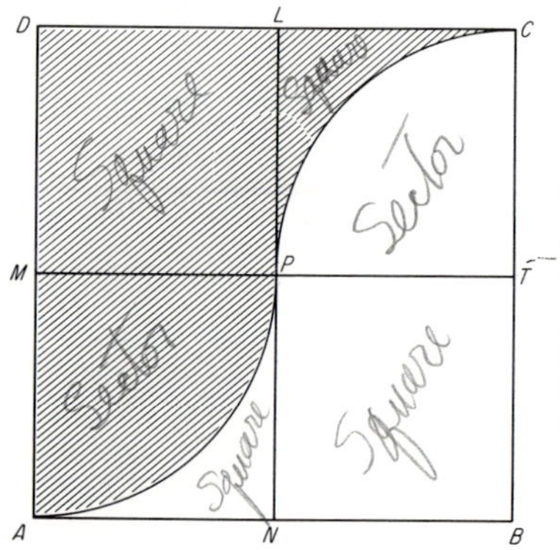

PROB. 8.

8. In the figure, L, M, N, T are midpoints of the square $ABCD$, which has sides each 4 in. long. $\widehat{PC}$ and $\widehat{AP}$ are circular arcs with centers at T and M, respectively. Find the area of the shaded portion.
9. Find the area of the shaded portion of the semicircle.

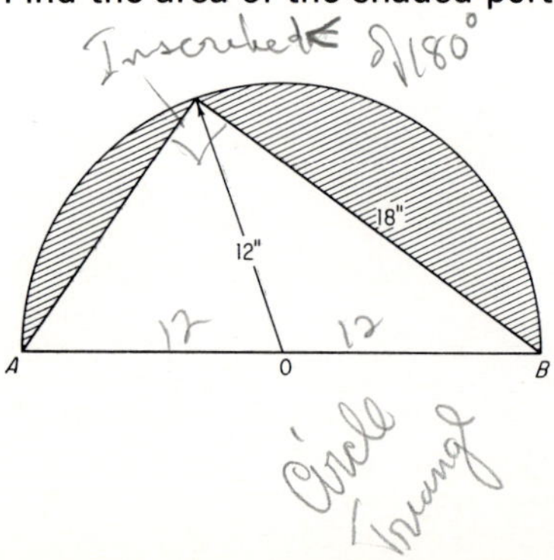

10. A continuous belt runs around two wheels as shown in the figure. Compute the length of the belt.

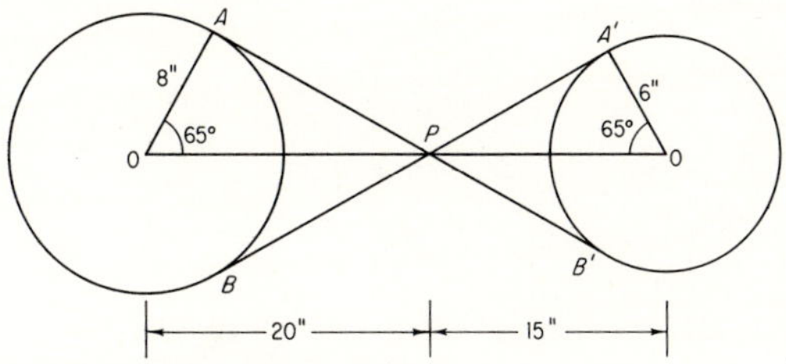

11. Find the area of the shaded portion of the larger circle.

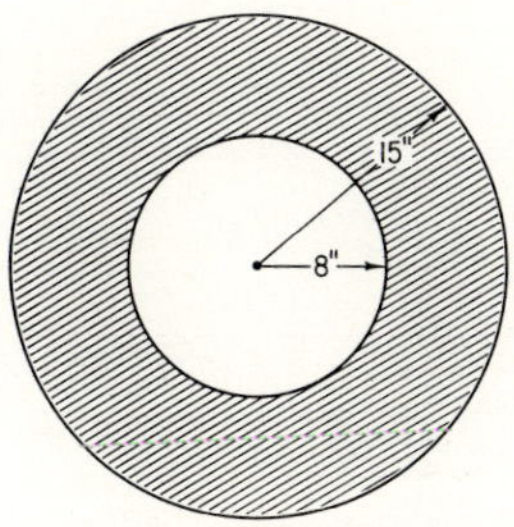

12. Find the area of the shaded portion of the trapezoid.

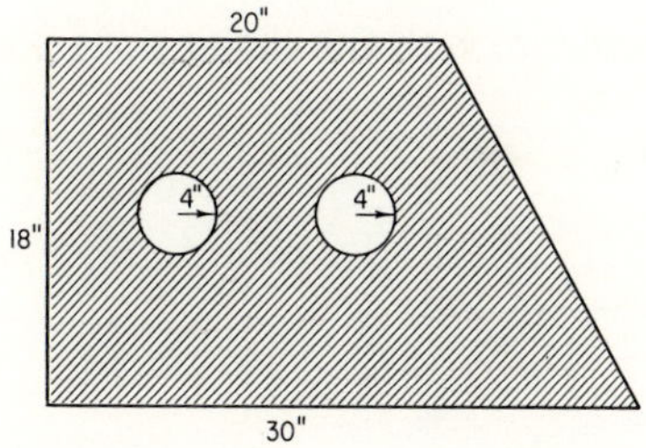

13. Find the area of the shaded region in the figure on p. 284.

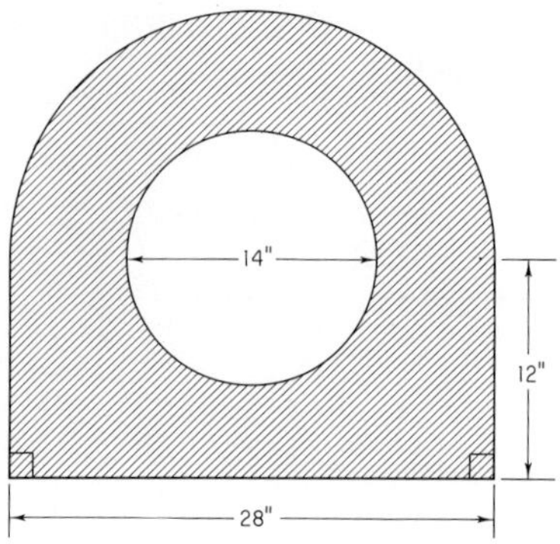

PROB. 13.

14. Find the area of the shaded region if the diameter of each of the smaller circles is 3 in. and the larger circle has a diameter of 8 in.

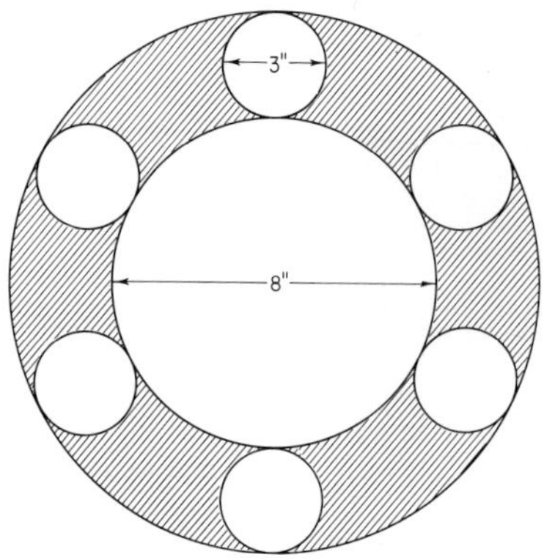

PROB. 14.

15. Find the area of the figure with the dimensions shown in the diagram.

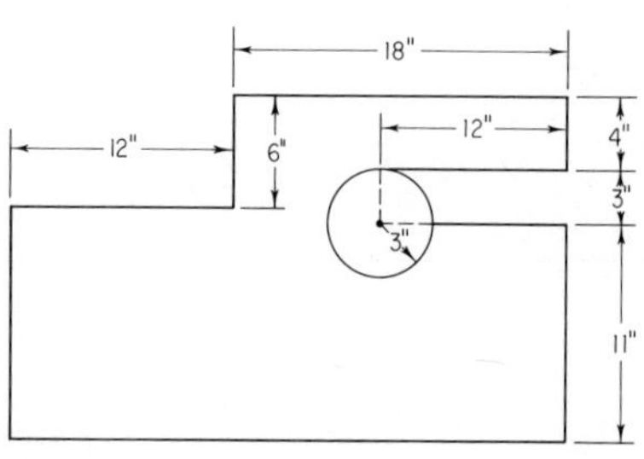

PROB. 15.

16. Find the number of circular pipes with an inside diameter of 1 in. which will carry the same amount of water as a pipe with an inside diameter of 4 in.
17. *Given:* $\triangle ABC$ is equilateral with $AB = 8$ in.
Find: The area of the equilateral arch ACB.

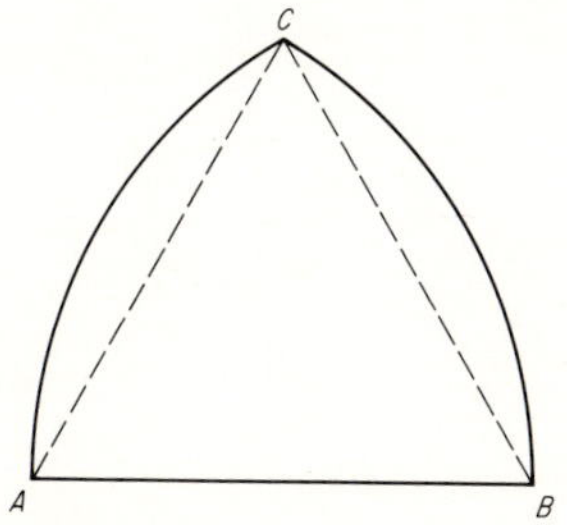

PROB. 17.

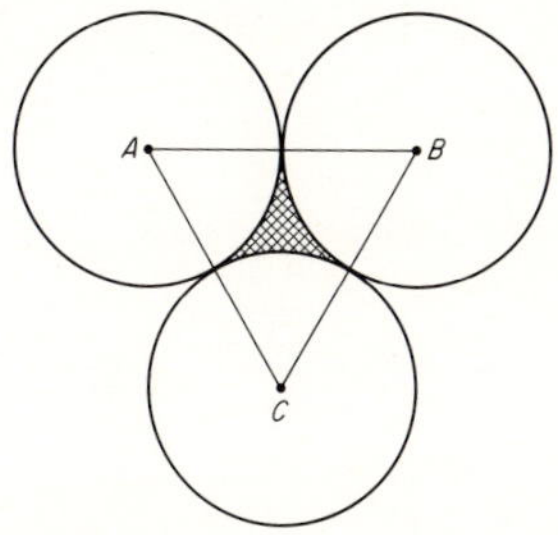

PROB. 18.

18. *Given:* The diameter of each of the three mutually tangent circles is 8 ft.
Find: The shaded area between the three circles.
19. *Given:* A regular hexagon inscribed in a circle with a radius of 12 in.
Find: The area of the hexagon.

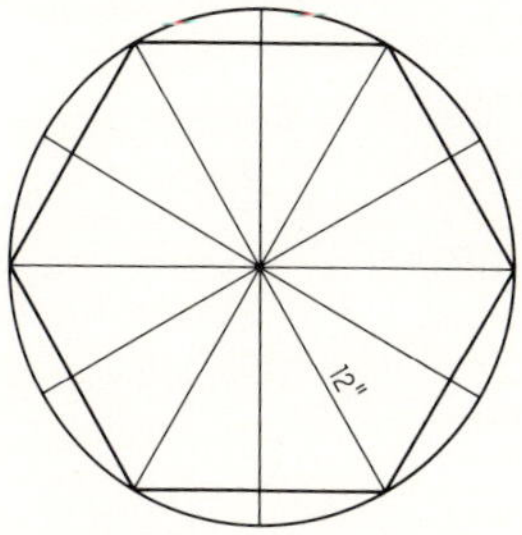

PROB. 19.

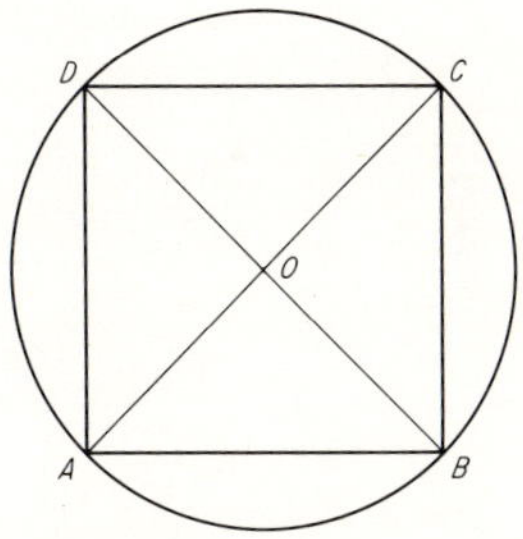

PROB. 20.

20. *Given:* The area of the circle is 16π yd^2.
Find: The area of the inscribed square.

CHAPTER 18

SURFACE AREAS AND VOLUMES

18-1 GEOMETRIC SOLIDS The figures we have studied thus far are called plane figures, since all the points and lines of each figure lie in the same plane.

Solid geometry (often called *geometry of space*) is the study of figures whose parts do not lie in the same plane. Such figures are called *geometric solids.* Examples of solid figures are the cube, sphere, cone, and pyramid.

The word *solid* in the study of geometry has a meaning different from that in everyday conversation. We often speak of a block of solid wood or solid metal. A geometric solid does not refer to the material of which it is made but refers to the configuration formed by points in space. Figure 18-1 repre-

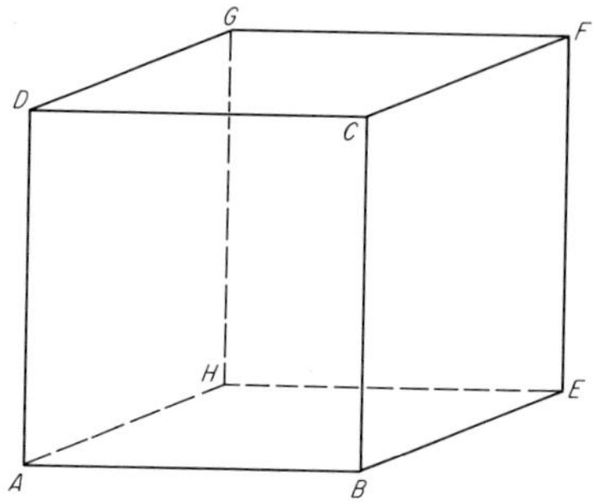

FIG. 18-1. Geometric solid.

sents a solid figure. It is a combination of points, lines, and surfaces. The 8 corners of the figure are points; the 12 edges, such as $\overline{AB}$ and $\overline{EF}$, are line segments; the 6 faces, such as *ABCD*, are portions of flat surfaces or planes.

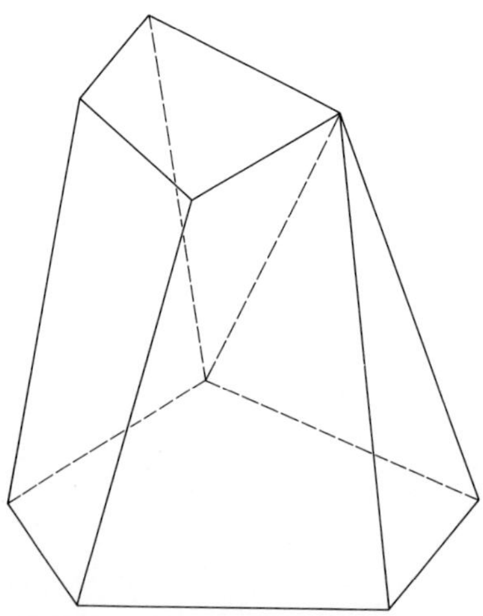
FIG. 18-2. Polyhedron.

18-2 POLYHEDRON A *polyhedron* is a solid bounded by portions of plane surfaces. The *faces* are the portions of the bounding planes; the *edges* are the lines of intersection of the faces; the *vertices* are intersections of the edges. Figure 18-2 represents a polyhedron of 8 faces, 15 edges, and 9 vertices.

18-3 PRISM A prism is a polyhedron having two parallel and congruent faces, called *bases*, and the remaining faces are parallelograms (see Fig. 18-3). The bases may be triangles, quadrilaterals, or any other kind of polygon. In the prism, base *ABCDE* is congruent to base *A′B′C′D′E′*. The faces which are parallelograms are called *lateral faces*, and their intersections (such as *DD′*) are called *lateral edges*. It should be evident that the lateral edges are congruent and parallel. The *lateral area* is the sum of the areas of the lateral faces. The *total area* is the sum of the lateral areas and the areas of the two bases. The *altitude h* of a prism is the perpendicular distance between the planes of the bases.

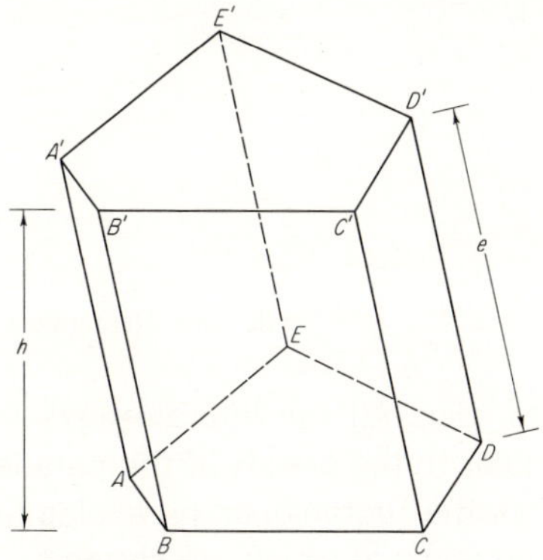

FIG. 18-3. Prism.

A *right prism* is a prism having its lateral edges perpendicular to the bases (Fig. 18-4). It can be shown that, in a right prism, the lateral faces are

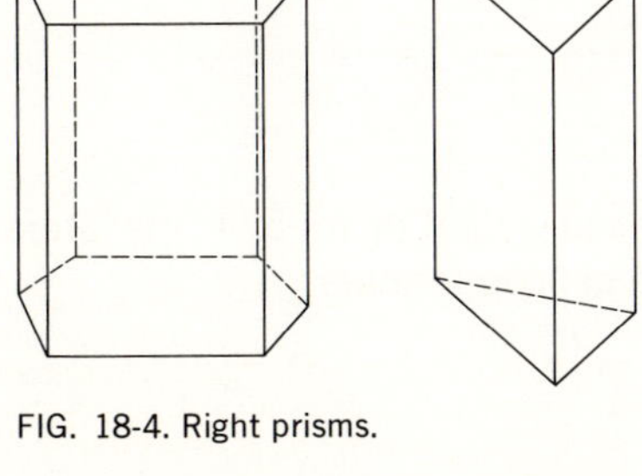
FIG. 18-4. Right prisms.

rectangles and the lateral edges equal the altitude. A *regular prism* is a right prism whose bases are regular polygons.

If the edges of a prism are not perpendicular to the bases, the prism is called an *oblique prism* (see Fig. 18-3).

18-4 PARALLELEPIPED A parallelepiped is a prism with parallelograms as faces (Fig. 18-5). A *rectangular parallelepiped* (Fig. 18-6) is a parallelepiped

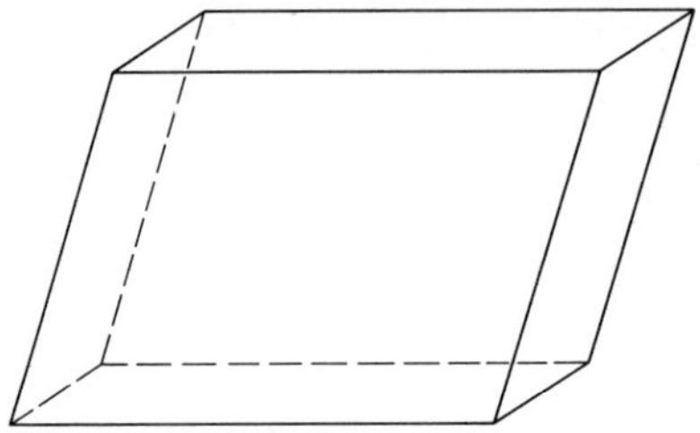

FIG. 18.5. Parallelepiped.

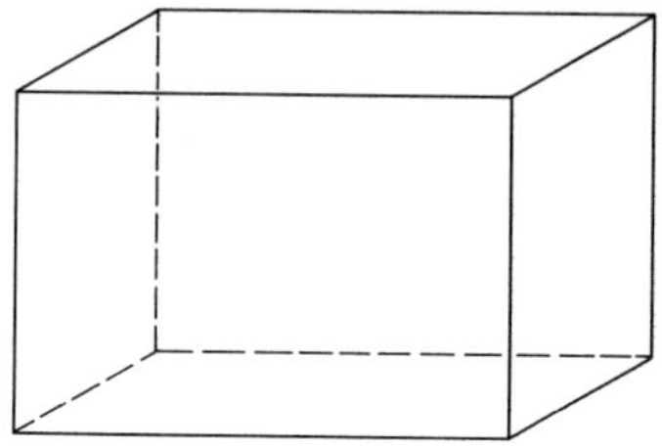

FIG. 18-6. Rectangular parallelepiped.

the faces of which are rectangles. All the lateral edges of a rectangular parallelepiped are perpendicular to the planes of the parallel faces.

The most common prism is the rectangular parallelepiped. A cube is a rectangular parallelepiped the faces of which are squares.

18-5 AREA OF A RIGHT PRISM The lateral area of a right prism is found by computing the areas of each face and then summing the areas. It can also be found by multiplying the perimeter of the base by the altitude (see Fig. 18-7).

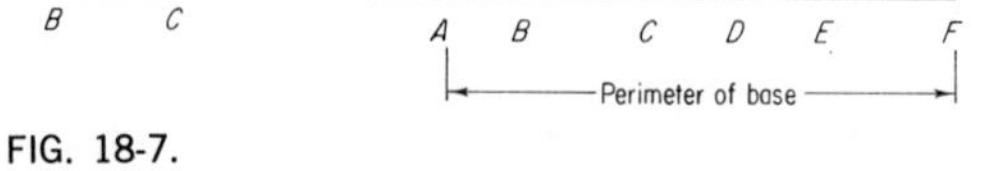

FIG. 18-7.

If we denote the total area of a right prism by T, the lateral area by L, and the area of a base by B, we get the formula

$$T = L + 2B$$

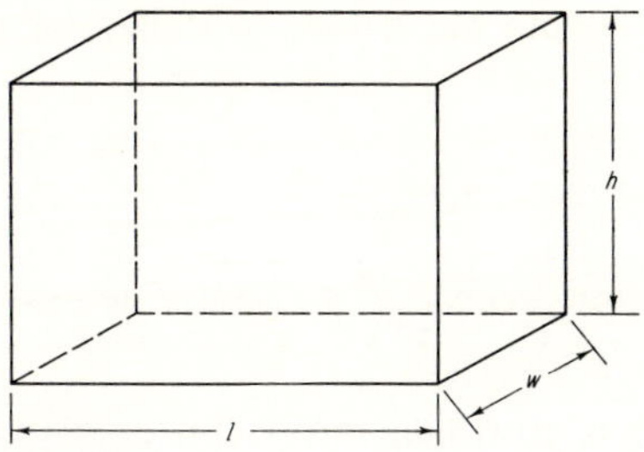

FIG. 18-8.

The total area of a rectangular parallelepiped (Fig. 18-8) with length, width, and height, l, w, and h, respectively, is equal to $2lw + 2wh + 2lh$ or

$$T = 2(lw + wh + lh)$$

For a cube with lateral edge e

$$T = 6e^2$$

18-6 VOLUME OF A PRISM The volume of a solid is defined as the number of units of space measured in the solid. This unit of space, called a cubic unit, is that of a cube the edges of which are equal to some unit of measurement for length. Consider the rectangular parallelepiped, shown in Fig. 18-9, which is 4 units long, 3 units wide, and 2 units high. If planes are

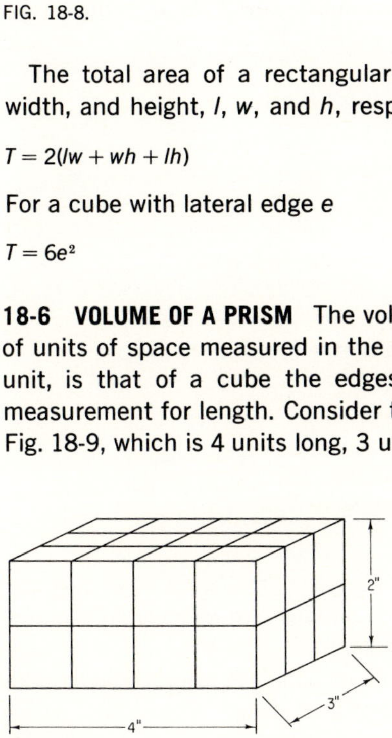

FIG. 18-9.

passed parallel to the faces of the solid as shown, the solid will consist of two layers, each layer containing 4×3, or 12, cubic units.

The two layers contain 2×12, or 24, cubic units. Thus the volume of the solid equals 24 cubic units. Common units for volume are cubic inches, cubic feet, cubic yards, and cubic centimeters. It will be noted that the numerical value for the volume above could have been obtained by multiplying together the linear dimensions or by multiplying the area of the base by the altitude.

Thus, in Fig. 18-8, if we denote the volume by V and the area of the base by A, we get the formulas

$$V = lwh \qquad \text{or} \qquad V = Ah$$

FORMULA *The volume of any prism is the product of the area of its base and its altitude.*

Example 18-1 A classroom is 40 ft long, 30 ft wide, and 12 ft high. What is the volume of the room in cubic yards? What is the lateral area in square yards?

SOLUTION

$$\begin{aligned} V &= lwh \\ &= (40 \text{ ft})(30 \text{ ft})(12 \text{ ft}) \\ &= 14{,}400 \text{ ft}^3 \\ &= 14{,}400 \text{ ft}^3 \times \frac{1 \text{ yd}^3}{27 \text{ ft}^3} \\ &= 533\tfrac{1}{3} \text{ yd}^3 \end{aligned}$$

$$\begin{aligned} L &= 2(lw + wh + lh) \\ &= 2(40 \text{ ft} \times 30 \text{ ft} + 30 \text{ ft} \times 12 \text{ ft} + 40 \text{ ft} \times 12 \text{ ft}) \\ &= 4{,}080 \text{ ft}^2 \\ &= 4{,}080 \text{ ft}^2 \times \frac{1 \text{ yd}^2}{9 \text{ ft}^2} \\ &= 453\tfrac{1}{3} \text{ yd}^2 \end{aligned}$$

Exercise 18-1

1. How many cubic yards of earth are removed in digging a pit 24 ft long, 18 ft wide, and 10 ft deep?
2. How many cubic yards of concrete are needed to pour a driveway 45 yd long, 15 ft wide, and 4 in. deep?
3. Find the volume of a cube with edge 15 in. long.
4. Find the total area of a cube with edge 15 in. long.
5. The inside dimensions of a fruit box are $22\frac{1}{2}$ by 17 by 6 in. How many cubic inches will it hold?
6. Find the volume in cubic feet of a beam 6 by 8 by 15 in.
7. What must be the dimensions of a cube that will give the same volume as a rectangular parallelepiped with dimensions 8 by 18 by 24 in.?
8. Find the volume of the right prism shown in the figure. The base is a right triangle with legs at 8 and 15 in.
9. Find the lateral area of the right prism.

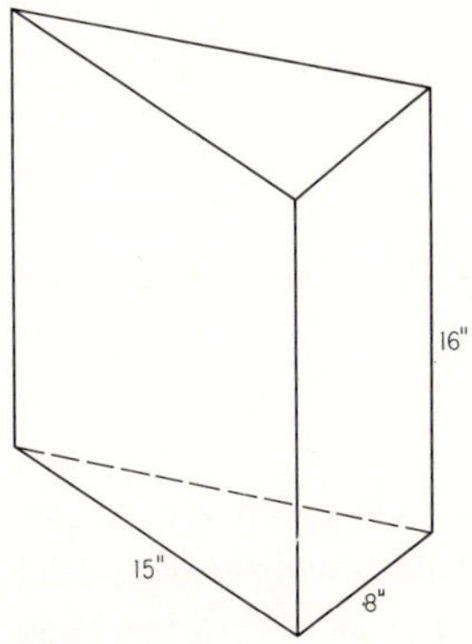

PROBS. 8–10.

10. Find the total area of the right prism.
11. Find the volume of a triangular right prism 20 in. high if the sides of the bases are 6, 8, and 12 in. (*Hint:* Use Heron's formula for the area of the base.)
12. Find the total area of the prism of Prob. 11.
13. Find the weight of a block of ice 2 ft by 4 ft by 16 in. if ice weighs 57.2 lb per cubic foot.
14. Find the volume of a right prism if its altitude is 30 in. long and the bases are rhombuses having diagonals 10 and 16 in. long.
15. Find the lateral area of the prism of Prob. 14.
16. Find the total area of the prism of Prob. 14.
17. How many gallons in a reservoir 50 ft long, 30 ft wide, and 10 ft deep? (*Note:* 1 gal is equivalent to 231 in.^{3})
18. How many gallons of paint will be needed to paint the walls and ceiling of a room 45 ft long, 30 ft wide, and 12 ft high if 1 gal will cover 400 ft^2?

18-7 CYLINDRICAL SURFACE A surface generated by a straight line that moves parallel to itself is called a *cylindrical surface.* The moving line is called the *generatrix.* A *cylinder* is a geometric solid formed by that part of a cylindrical surface bounded by two parallel planes (see Fig. 18-10).

The intersections of the cylindrical surface and the parallel planes are the *bases* of the cylinder formed. The bases of a cylinder are congruent and have equal areas. The bounding cylindrical surface is the *lateral surface* of the cylinder. The *altitude* of a cylinder is the perpendicular distance between the bases. The *radius r* of a cylinder is the radius of the bases.

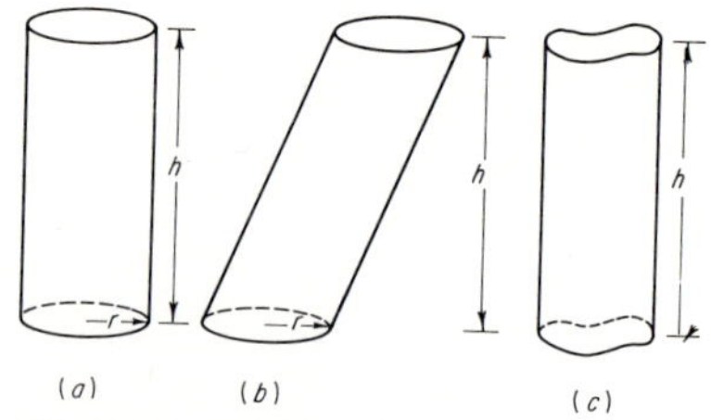

FIG. 18-10. Cylinders.

A *circular cylinder* is one the bases of which are circles. It is the most common form of cylinder. A *right circular cylinder* (Fig. 18-10*a*) is a circular cylinder in which the generatrix is perpendicular to the bases. If the generatrix is not perpendicular to the bases, the cylinder is called an *oblique cylinder* (see Fig. 18-10*b*).

18-8 AREA OF A RIGHT CYLINDER If the wrapper is removed from a can and spread along a plane (see Fig. 18-11), a rectangle will be formed with height

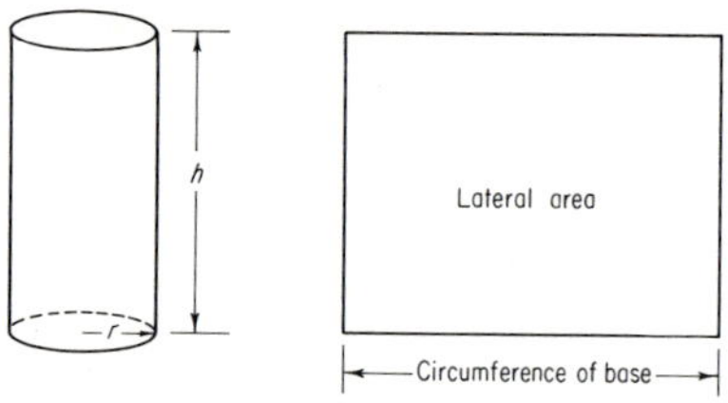

FIG. 18-11.

equal to that of the can and length equal to the circumference of the base of the can. This will point up the following formula.

FORMULA *The lateral area L of a right cylinder is equal to the circumference of the base × altitude. (For a circle,* $L = 2\pi rh$.*)*

The total area of a cylinder is equal to the sum of the lateral area and the area of the two bases.

FORMULA *The total area T of a right circular cylinder is* $T = 2\pi rh + 2\pi r^2 = 2\pi r(h + r)$.

18-9 VOLUME OF A CYLINDER The volume of a cylinder is equal to the product of the area of its base and the length of its altitude. This will be true whether the cylinder is a right or an oblique cylinder.

FORMULA *The volume of a circular cylinder is* $V = Ah = \pi r^2 h$.

A common type of problem involves finding the volume of cylindrical shells or hollow cylinders (see Fig. 18-12). If we let R be the outer radius, r

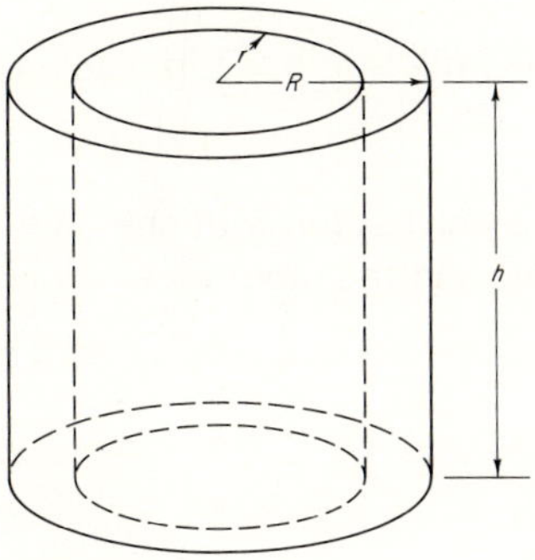

FIG. 18-12.

the inner radius, and h the height of the shell, the volume can be found by subtracting the volume of the smaller cylinder from that of the larger one.

FORMULA *The volume V of a circular cylindrical shell with outside radius R, inside radius r, and altitude h is*

$$\begin{aligned} V &= V_1 - V_2 \\ &= \pi R^2 h - \pi r^2 h \\ &= \pi h(R^2 - r^2) \end{aligned}$$

Exercise 18-2

1. A silo is 10 ft in diameter and 30 ft tall. What is the volume of the silo?
2. A cylindrical water tank is 16 ft in diameter and 10 ft tall. How many gallons will it hold? (1 gal = 231 in.^{3})
3. Find the lateral surface area of a right circular cylinder whose base has a diameter of 16 in. and whose height is 12 in.
4. Find the total area of the right circular cylinder with a radius of 10 ft and height of 8 ft.
5. A 100-gal cylindrical water tank is 5 ft high. What is its diameter?
6. A 50-gal tank has a radius of 16 in. What is its height?
7. How many gallons of gasoline will a right circular cylindrical tank hold that is 6 ft in diameter and 25 ft long? (*Note:* 1 ft^3 = 7.5 gal.)
8. How much does a mile-long copper wire weigh if the wire has a diameter of $\frac{3}{16}$ in.? Assume that copper has a density of 555 lb/ft^3.

9. An iron pipe has an inside diameter of 1 in. and is $\frac{1}{8}$ in. thick. How much will 1,000 ft of the pipe weigh? Assume that the density of the pipe is 450 lb/ft^3.
10. How many steel rods $\frac{5}{8}$ in. in diameter and 15 ft long can be made from a steel ingot with dimensions 14 in. by 16 in. by 4 ft?
11. How many square feet of asbestos are needed to wrap 500 ft of 6-in. pipe?
12. How much water, in gallons, will flow in a 1-in. pipe in a day if the flow speed is 2 ft/sec? (1 gal = 231 $in.^3$)

18-10 PYRAMID A pyramid is a polyhedron with one face, called the *base,* a polygon of any number of sides, and the other faces triangles that meet in

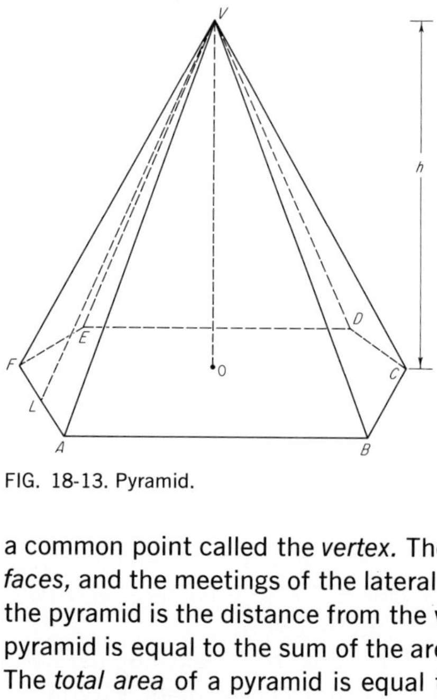

FIG. 18-13. Pyramid.

a common point called the *vertex.* The triangular faces are called the *lateral faces,* and the meetings of the lateral faces are *lateral edges.* The *altitude* of the pyramid is the distance from the vertex to the base. The *lateral area* of a pyramid is equal to the sum of the areas of the lateral faces of the pyramid. The *total area* of a pyramid is equal to the sum of the lateral area and the area of the base.

A *regular pyramid* is one whose base is a regular polygon and whose altitude from the vertex passes through the center of the base. The lateral edges of a regular pyramid are congruent. The lateral faces of a regular pyramid are congruent isosceles triangles. The *slant height* of a regular pyramid is the altitude of any of its lateral faces.

18-11 VOLUME OF A PYRAMID The volume of any pyramid is equal to one-third the product of the area of its base and the altitude.

Volume $= \frac{1}{3}$ area of base $\times$ altitude

$= \frac{1}{3}Ah$

18-12 AREA OF A REGULAR PYRAMID The *lateral area* of a regular pyramid is equal to half the product of its slant height *s* and the perimeter *p* of its base.

Lateral area $= \frac{1}{2}$ slant height $\times$ perimeter of base

$$L = \frac{1}{2}sp$$

Example 18-2 A regular pyramid has a regular hexagon (six sides) with sides equal to 6 in. and an altitude of 10 in. Find (a) the volume and (b) the lateral area of the pyramid.

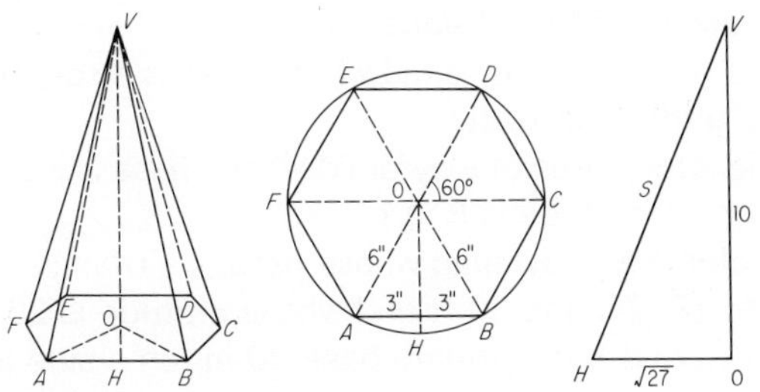

EXAMPLE 18-2.

SOLUTION (a) To solve for the volume, we must first find the area of the base. Let us imagine a circle with a 6-in. radius. If we divide the central angle into six congruent angles of 60° measure each, and then join the six chords as shown in the figure, regular polygon *ABCDEF* will be formed. It can be proved that this polygon is divided into six congruent equilateral triangles with sides of 6 in.

$$OA = 6 \text{ in.} \qquad HB = 3 \text{ in.}$$

$$(OH)^2 = (OB)^2 - (HB)^2$$
$$= (6)^2 - (3)^2$$
$$OH = \sqrt{27}$$
$$= 5.196 \text{ in.}$$
$$\text{Area of } AOB = 3(5.196)$$
$$= 15.588 \text{ in.}^2$$
$$\text{Area of } ABCDEF = 6(15.588) \text{ in.}^2$$
$$= 93.528 \text{ in.}^2$$
$$\text{Volume} = \tfrac{1}{3}Ah$$
$$= \tfrac{1}{3}(93.528)(10)$$
$$= 311.76 \text{ in.}^3$$

(b) To find the lateral area, we must next find the slant height *VH*.

$$
\begin{aligned}
(VH)^2 &= (OH)^2 + (OV)^2 \\
&= (\sqrt{27})^2 + (10)^2 \\
&= 127 \\
VH &= 11.27 \text{ in.} \\
L &= \tfrac{1}{2}sp \\
&= \tfrac{1}{2}(11.27)(6 \times 6) \\
&= 202.86 \text{ in.}^2
\end{aligned}
$$

Exercise 18-3

1. Find the volume of a pyramid whose base has an area of 54 in.^{2} and whose altitude is 8 in.
2. Find the lateral area of a pyramid whose base has a perimeter of 108 in. and whose slant height equals 15 in.
3. What is the volume of a pyramid whose base is a square 18 ft on a side and whose altitude is 10 ft?
4. What is the lateral area of a pyramid whose base is a square 12 in. on a side and whose altitude is 8 in.?
5. Find the volume of a pyramid whose base is a triangle with sides equal in length to 18, 24, and 30 ft and whose altitude is 20 ft.
6. A marble pyramid has a square base 20 in. on a side and is 6 ft high. How heavy is the pyramid if the density of marble is 170 lb/ft^3?
7. A pyramid has an equilateral triangle for its base and is 8 ft tall. What is the volume of the pyramid if the side of the triangular base is 3 ft?
8. A regular pyramid has an equilateral triangle with sides equal to 10 in. as a base and slant height of 20 in. What is its lateral area?
9. What is the volume of a regular pyramid whose base is a regular hexagon 10 in. on a side and whose altitude is 20 in.?
10. What is the lateral area of a regular pyramid with a regular hexagon with sides of 8 in. for a base and an altitude of 12 in.?

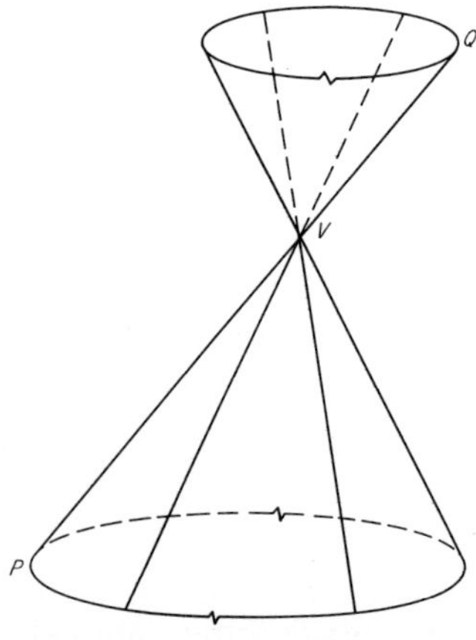

FIG. 18-14. Conical surface.

18-13 CONICAL SURFACE A *conical surface* is a surface that is generated by moving a straight line so that it turns around one of its points (see Fig. 18-14). The moving line, such as $\overleftrightarrow{PQ}$, is called the *generatrix*. *V* is the *vertex* of the surface.

18-14 CONE A *cone* is that part of a conical surface which is bounded by the vertex and a plane cutting the surface (Fig. 18-15). There are many uses

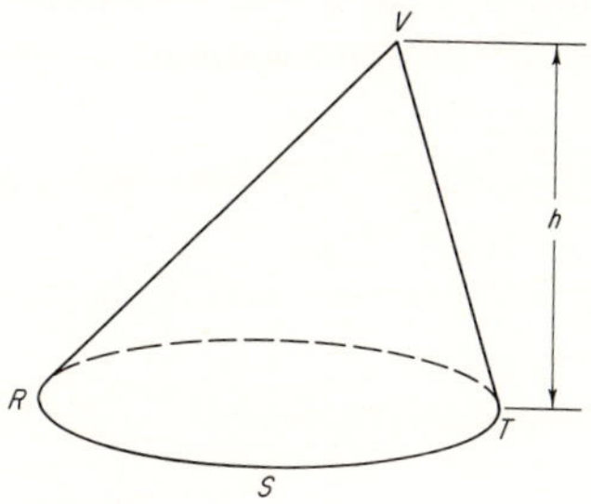

FIG. 18-15. Cone.

for cones in industry and everyday affairs. We have ice-cream cones, conical tents, conical horns, etc.

The *base RST* of the cone is the curve cut from the conical surface by the plane. The *altitude* of a cone is the perpendicular distance from the vertex to the plane of the base. The *lateral area* of a cone is the area of the lateral surface.

A cone whose base is a circle is called a *circular cone*. If the line passing through the vertex of the cone and the center of its base is perpendicular to the base, the figure is called a *right circular cone* (see Fig. 18-16). The

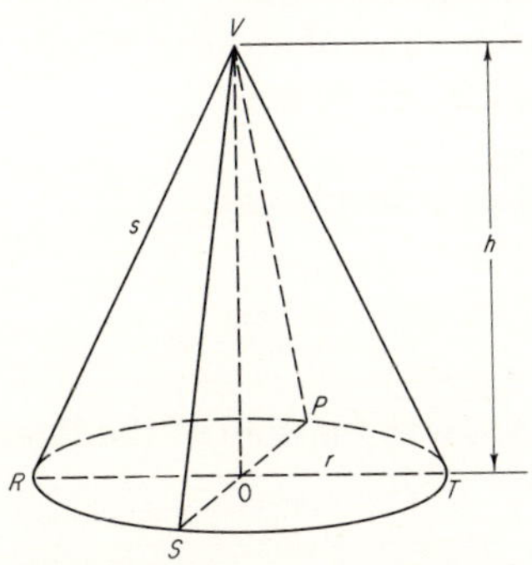

FIG. 18-16. Right circular cone.

length of $\overline{VR}$ (or $\overline{VS}$, $\overline{VP}$, etc.) is called the *slant height* of the right circular cone.

18-15 VOLUME OF A CONE It can be shown that the volume of a cone is equal to one-third the product of the area of its base and its altitude. For the circular cone with radius of the base r and altitude h we get

FORMULA $\text{Volume} = \frac{1}{3}\text{ area of base} \times \text{altitude}$

$$= \frac{1}{3}\pi r^2 h$$

The student should notice that, if a cone and a cylinder have the same base and the same altitude (Fig. 18-17), the volume of the cone is exactly

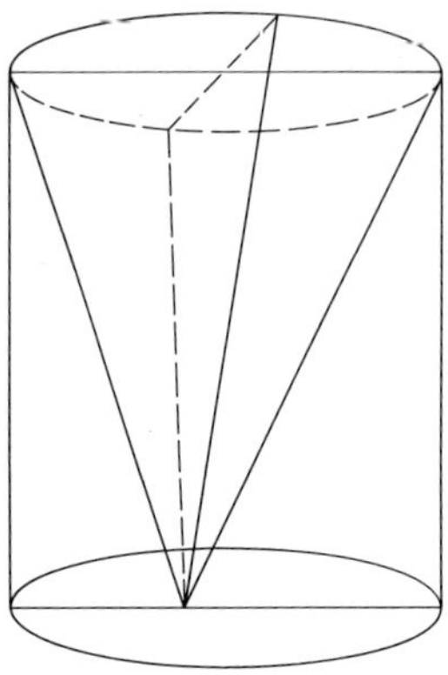

FIG. 18-17.

one-third the volume of the corresponding cylinder. This same relationship exists between a pyramid and a prism with equal bases and altitudes.

18-16 LATERAL AREA OF A RIGHT CIRCULAR CONE The lateral area of a right circular cone is equal to half the product of its slant height and the circumference of its base. Thus, in Fig. 18-16, we get

FORMULA $\text{Surface area} = \frac{1}{2}\text{ circumference of base} \times \text{slant height}$

$$L = \frac{1}{2}(2\pi r)s$$

$$L = \pi rs$$

The total area T of a cone is equal to the sum of the lateral area and the area of the base.

FORMULA $\text{Total area} = \text{lateral area} + \text{area of base}$

$$T = \pi rs + \pi r^2$$

Exercise 18-4

1. Find the volume of a circular cone 36 in. high and having a base diameter of 18 in.
2. What is the lateral area of the right circular cone having an altitude of 36 in. and a base radius of 27 in.?
3. Find the total area of a right circular cone having a slant height of 30 ft and a base radius of 18 ft.
4. Find the volume of a right circular cone with slant height of 17 ft and an altitude of 15 feet.

V-ABCD is a right circular cone.

5. Find the volume of cone *V-ABCD.*
6. Find the volume of cone *V-EFGH.*
7. Find the lateral area of cone *V-ABCD.*

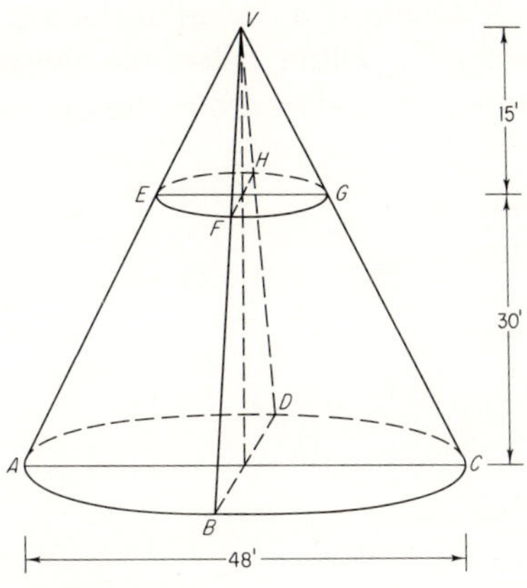

PROBS. 5–9.

8. Find the total area of cone *V-ABCD.*
9. Find the lateral area of cone *V-EFGH.*

V-RST (p. 302) is a right circular cone. Plane *PQL* ∥ plane *RST.*

10. Find the volume of *V-RST.*
11. Find the volume of the solid *PRSTLQ.*
12. Find the lateral area of *V-PQL.*
13. Find the total surface area of *V-RST.*
14. Find the total surface area of the solid *PRSTLQ.*

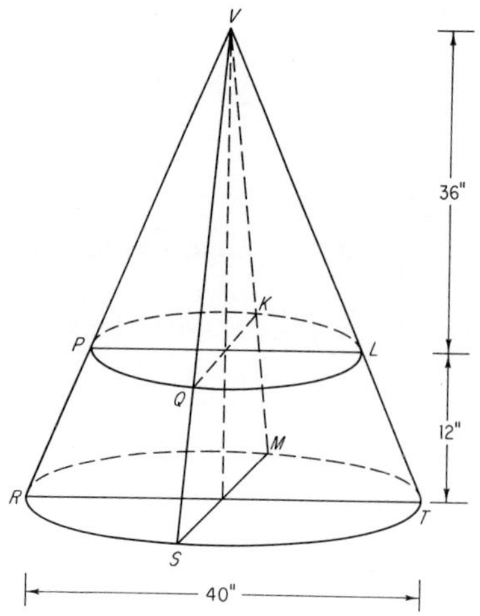

PROBS. 10–14.

18-17 THE SPHERE A *sphere* is a closed surface every point of which is equidistant from a fixed point within, called the *center* (see Fig. 18-18). If a circle is rotated about one of its diameters, the solid generated is a sphere.

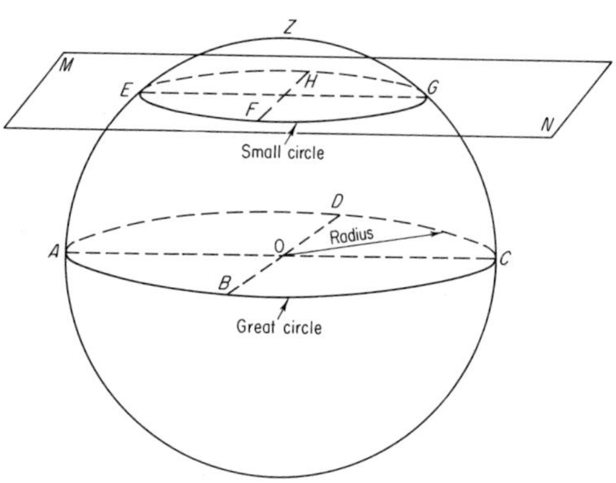

FIG. 18-18. Sphere.

A *radius* of a sphere is a line segment joining the center to any point of the sphere.

A *diameter* of a sphere is a straight-line segment through the center and having its ends on the sphere. The measure of the diameter is twice that of the radius.

If a plane intersects a sphere, the intersection is a circle. If the plane contains the diameter of the sphere, the circle is a *great circle;* otherwise the circle is a *small circle.* All great circles of a sphere are equal. Every great circle bisects the sphere into surfaces called *hemispheres.*

The portion of a sphere cut off by a plane is called a *segment* of the sphere. *Z-EFGH* of Fig. 18-18 is a segment.

18-18 THE SURFACE AREA OF A SPHERE The surface area of a sphere is equal to four times the area of the great circle of the sphere.

FORMULA *Area of sphere* = *4 times area of great circle*

$$S = 4\pi r^2$$

18-19 VOLUME OF A SPHERE The volume of a sphere is equal to $\frac{4}{3}\pi$ times the cube of the radius of the sphere.

FORMULA *Volume of sphere* = $\frac{4}{3}\pi$ *(radius)*^{3}

$$V = \frac{4}{3}\pi r^3$$

18-20 TORUS If a slender rod, such as Fig. 18-19*a*, is bent into a ring (Fig. 18-19*b*), the solid formed is called a *torus.* The surface of the torus will

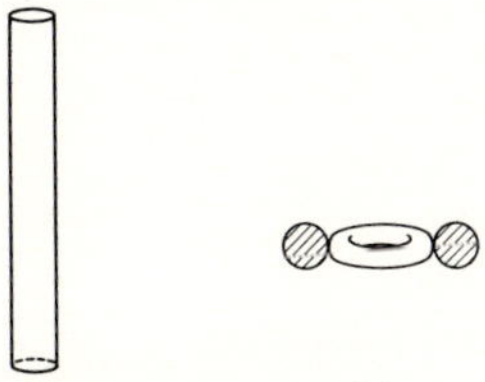

(*a*) (*b*)

FIG. 18-19. Torus.

equal the lateral surface area of the cylinder. Since $h = 2\pi R$, we can find the surface area by multiplying the circumference of the center line of the torus by the circumference of its cross section.

FORMULA *Surface area of torus* = $2\pi rh$

$$S = 2\pi r(2\pi R)$$
$$= 4\pi^2 rR$$

In like manner, the volume of the torus is equal to the volume of the cylinder. Thus, the volume of the torus is found by multiplying the circumference of the center line of the torus by the area of the cross section.

FORMULA *Volume of torus* = $2\pi R(\pi r^2)$

$$V = 2\pi^2 r^2 R$$

Exercise 18-5

1. Find the volume of a sphere having a 10-ft diameter.
2. Find the surface area of a sphere with a diameter of 8 in.
3. The area of the great circle of a sphere is 75 in.^{2} What is the surface area of the sphere?
4. Find the radius of a sphere with a volume of 84.82 in.^{3}
5. Find the diameter of a sphere with a surface area of 201.0 in.^{2}
6. Find the volume of a sphere whose surface area is 64 ft^2.
7. A sphere just fits in a cube whose volume is 5,830 in.^{3} What is the volume of the sphere?
8. A hollow spherical metal ball has an outside diameter of 13 in. and is $\frac{1}{2}$ in. thick. Find the volume of the metal in the ball.
9. Find the volume enclosed in the figure.

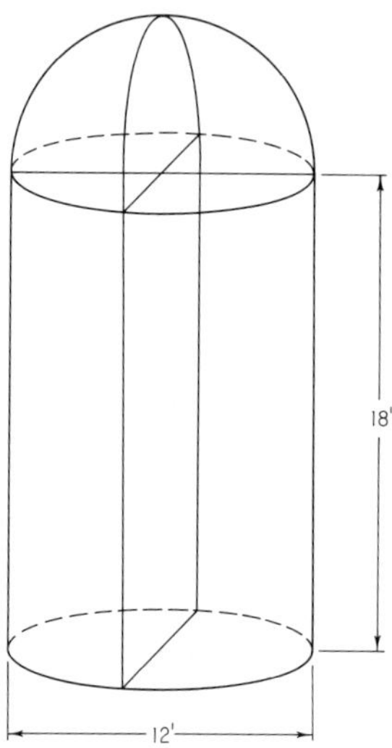

PROBS. 9 and 10.

10. Find the total surface of the figure. Include the lateral surface of the cylinder, the area of the base, and the area of the hemisphere.
11. Find the surface of the torus with an inside diameter of 16 in. and an outside diameter of 18 in.
12. Find the volume of the torus of Prob. 11.

SECTION 4

ADVANCED ALGEBRA FOR TECHNICIANS

CHAPTER 19

EXPONENTS AND RADICALS

19-1 LAWS OF EXPONENTS In earlier chapters we have treated only integral exponents. In this chapter we will introduce fractional exponents in such a way that the laws of exponents (see Sec. 8.7) given for integral exponents will be valid for any rational exponent. For the purpose of review, the fundamental laws for use with integral exponents will be listed again below.

If n is a positive integer, then $a^n = a \cdot a \cdot a \cdot\cdot\cdot$ to n a's. The expression a^n is called the nth *power* of a. The number a is the *base*, and n is the exponent of the number. For all positive integers m and n:

$$a^m \cdot a^n = a^{m+n} \tag{8-1}$$

$$a^m \div a^n = a^{m-n} \qquad \text{if } m > n \text{ and } a \neq 0 \tag{8-2}$$

$$= \frac{1}{a^{n-m}} \qquad \text{if } n > m \text{ and } a \neq 0 \tag{8-2a}$$

$$(a^m)^n = a^{mn} \tag{8-3}$$

$$(a^m b^n)^k = a^{km} b^{kn} \tag{8-4}$$

$$\left(\frac{a^m}{b^n}\right)^k = \frac{a^{km}}{b^{kn}} \qquad \text{if } b \neq 0 \tag{8-5}$$

19-2 ZERO EXPONENT Consider the law for exponents given in Eq. (8-2). If $m = n$, we get

$$\frac{a^n}{a^n} = a^{n-n} = a^0 \qquad a \neq 0$$

But, we know that any number, other than zero, divided by itself is 1. Hence we know that

$$a^0 = 1 \qquad a \neq 0 \tag{19-1}$$

PRINCIPLE *Any number a (other than zero) raised to the zero power is unity.*

$$a^0 = 1$$

Example 19-1

$$(7)^0 = 1 \qquad (-\tfrac{3}{4})^0 = 1 \qquad (x - 3y)^0 = 1$$

19-3 NEGATIVE EXPONENT Suppose we tried to use the above second law for exponents (Eq. 8-2) in which n is larger than $m (n > m)$. In those instances $m - n$ will be negative. Take for example:

$$\frac{a^3}{a^5} = a^{3-5} = a^{-2}$$

But

$$\frac{a^3}{a^5} = \frac{a \cdot a \cdot a}{a \cdot a \cdot a \cdot a \cdot a} = \frac{1}{a \cdot a} = \frac{1}{a^2}$$

Hence a^{-2} must be equivalent to $\frac{1}{a^2}$. Using the above approach, it can be readily shown that

$$a^{-n} = \frac{1}{a^n} \qquad a \neq 0 \tag{19-2}$$

In like manner it can be shown that

$$\frac{1}{a^{-n}} = \frac{1}{1/a^n} = a^n \qquad a \neq 0 \tag{19-3}$$

Any quantity with a negative exponent is equivalent to the reciprocal of that quantity with the corresponding positive exponent.

PRINCIPLE *Any factor of the numerator (denominator) of a fraction may be transferred to the denominator (numerator) if the sign of the exponent is changed.*

The student is cautioned that

$$\frac{a^2 + b^{-3}}{x^{-2}y^3} \neq \frac{a^2 + x^2}{b^3y^3}$$

The fraction on the left above cannot be simplified since b^{-3} is a term of the numerator and is not a factor.

19-4 FRACTIONAL EXPONENTS AND RADICALS We can define fractional exponents in a manner which will satisfy our laws of exponents by noting that

$$a^{1/2} \cdot a^{1/2} = a^{1/2+1/2} = a^1 = a$$

and

$$a^{1/3} \cdot a^{1/3} \cdot a^{1/3} = a^{1/3+1/3+1/3} = a^1 = a$$

We note, then, that $a^{1/2}$ is one of the two equal factors of a; $a^{1/3}$ is one of three equal factors of a. Thus

$$a^{1/2} = \sqrt{a}$$

$$a^{1/3} = \sqrt[3]{a}$$

$$a^{1/q} = \sqrt[q]{a}$$

$a^{1/q}$ is defined as the qth root of a.

Next let us multiply $a^{1/q}$ by itself p times to get

$$\overbrace{a^{1/q} \cdot a^{1/q} \cdot a^{1/q} \cdot \cdot \cdot}^{p \text{ factors}} = a^{1/q+1/q+1/q+\cdots} = a^{p(1/q)} = a^{p/q}$$

It will be noted that $a^{p/q} = (a^p)^{1/q} = (a^{1/q})^p$. $a^{1/q}$ is the pth root of $a^{p/q}$ and $a^{p/q}$ is the pth power of the qth root of a. This can be written

$$a^{p/q} = (\sqrt[q]{a})^p = \sqrt[q]{a^p} \qquad (19\text{-}4)$$

The pth power of the qth root of a number a is equivalent to the qth root of the pth power of the number a.

$(\sqrt{a})^p = \sqrt{a^p}$ if $a \neq 0$, $q \neq 0$

Example 19-2

$8^{2/3} = \sqrt[3]{8^2} = \sqrt[3]{64} = 4$

or

$8^{2/3} = (\sqrt[3]{8})^2 = (2)^2 = 4$

Exercise 19-1

Reduce each expression in Probs. 1 to 24 to its simplest numerical equivalent.

1. $(25)^{1/2}$
2. $(0.16)^{1/2}$
3. $(-8)^{1/3}$
4. $\left(\frac{3}{4}\right)^0$
5. $(32)^{4/5}$
6. $(64)^{2/3}$
7. $\left(\frac{8}{125}\right)^{1/3}$
8. 5^{-2}
9. $(144)^{-1/2}$
10. $32^{-1/5}$
11. $4^{-3/2}$
12. $(-8)^{-2/3}$
13. $\left(\frac{25}{36}\right)^{-1/2}$
14. $\left(\frac{4}{9}\right)^{-3/2}$
15. $3a^0$
16. $(-27)^{-4/3}$
17. $(16^{3/2})^{1/2}$
18. $[(-8)^{2/3}]^{1/2}$
19. $[(-8)^{2/3}]^{-1/2}$
20. $[(-8)^{-2/3}]^{-1/2}$
21. $(17^0)^{1/3}$
22. $[25^{1/2}]^0$
23. $[4^{1/2} + 8^{1/3} - 16^{-1/4}]^0$
24. $[27^{2/3} + 64^{2/3}]^{3/2} - 9^2$

In Probs. 25 to 66 perform all operations and express the answer with positive exponents. Simplify whenever possible.

25. $a^{3/2} \cdot a^{1/2}$
26. $x^{1/3} \cdot x^{5/3}$
27. $x^{1/3} \cdot x \cdot x^{2/3}$
28. $a \cdot a^{1/2}$
29. $y^{1/2} \cdot y^{1/3}$
30. $b^{2/3} \cdot b^{3/2}$
31. $x^{3/4} \div x^{1/4}$
32. $a^{5/6} \div a^{1/3}$
33. $6^{2/3} \div 6^{1/4}$
34. $a^{2/3} \cdot a^{4/3} \cdot a^{-2}$
35. $x^3 \cdot x^0 \cdot x^{-2}$
36. $\dfrac{x^{2/5} \cdot x^{3/10} \cdot x^{-4/15}}{x^{1/2}}$
37. $y^{-2} \div y^{-4}$
38. $z^{-2} \div z$
39. $a \div a^2 \cdot a^3$
40. $(a^{1/3})^6$

41. $(a^{-2}bc^3)^2$

42. $\left(\frac{x}{y}\right)^{-3}$

43. $(-\frac{1}{2}x^{-1/2})^2$

44. $(-\frac{1}{2}x^{-1/2})^{-2}$

45. $\left(\frac{-a^5b^{-10}}{32c^{15}}\right)^{-2/5}$

46. $\left(\frac{8x^6y^3z^{-3}}{a^{-12}}\right)^{-2/3}$

47. $x^{2n-3} \div x^{n-3}$

48. $a^{2n+4} \cdot a^{n-4}$

49. $a^{-1} + b^{-1}$

50. $\frac{x}{y^{-1}} + \frac{y}{x^{-1}}$

51. $(a + b)^{-1}$

52. $\frac{a}{b} - \frac{b^{-1}}{a^{-1}}$

53. $\frac{a}{b^{-1}} - \frac{b}{a^{-1}}$

54. $\frac{x^{-1} + y^{-1}}{x^{-1}}$

55. $xy^{-1} + x^{-1}y$

56. $\frac{1}{a^{-1} + b^{-1}}$

57. $\frac{1}{x^{-2} - y^{-2}}$

58. $\frac{x - y}{x^{-1} - y^{-1}}$

59. $\frac{x^{-2} - y^{-2}}{x^2 - y^2}$

60. $(x^{1/2} + y^{1/2})(x^{1/2} - y^{1/2})$

61. $(a^{1/2} + b^{1/2})^2$

62. $(a^{1/2} - a^{-1/2})^2$

63. $\frac{a^{n+4} - a^{n+1}}{a(a^{n+2})}$

64. $\left(\frac{a^{3n+2}}{a^{2n-2}}\right)^2$

65. $\left(\frac{b^{m+3}}{b^{2m-1}}\right)^2$

66. $\left(\frac{x^{n+3}}{x^{n-3}}\right) \div (x^2)^3$

67. Write each number in scientific notation:
(a) 0.203
(b) 0.000107
(c) 0.00000008312
(d) −0.0537

68. Write each number in decimal notation:
(a) 3.51×10^{-2}
(b) 5.67×10^{-5}
(c) -2.03×10^{-1}
(d) -4.28×10^{-3}

19-5 RATIONAL AND IRRATIONAL NUMBERS *A rational number* is a number which, when reduced to its simplest form, can be expressed as the quotient of two (positive or negative) integers. In general, if a and b are integers and $b \neq 0$, then $\frac{a}{b}$ is a rational number. Any real number that is not rational is called an *irrational number.*

Example 19-3 $8^{2/3}$, $5\frac{2}{3}$, -7, 0.37, $\sqrt{16}$, $\sqrt[3]{-27}$, $\sqrt{49 - 4\sqrt{36}}$ are rational because they can be written as $\frac{4}{1}$, $\frac{17}{3}$, $\frac{-7}{1}$, $\frac{37}{100}$, $\frac{4}{1}$, $\frac{-3}{1}$, $\frac{5}{1}$.

$\sqrt{2}$, $\sqrt[3]{5}$, and $\sqrt{49 + 4\sqrt{36}}$ are irrational numbers because they cannot be reduced to a quotient of two integers. Later in this text we will discuss logarithms and trigonometric functions of angles. Most of them are irrational numbers.

The number $\sqrt{9}$ is a rational number because the square root of 9 is exactly 3; that is, $(3)(3) = 9$. The number $\sqrt{2}$ is irrational because the exact square root cannot be expressed as a common fraction or as a repeating decimal. Every common fraction can be written as a repeating decimal. Conversely, every repeating decimal can be written as an equivalent fraction.

Example 19-4

$\frac{3}{4} = 0.75000\ldots$

$\frac{2}{3} = 0.666\ldots$

$\frac{13}{27} = 0.481481481\ldots$

$0.875000\ldots = \frac{7}{8}$

$0.222\ldots = \frac{2}{9}$

$1.571428571428\ldots = \frac{11}{7}$

While there is no exact rational number which multiplied by itself is equivalent to $\sqrt{2}$, we can still approximate the irrational number $\sqrt{2}$. All irrational numbers can be approximated. If we let the symbol $\approx$ stand for "approximates" or "is approximately equal to," we can write

$\sqrt{2} \approx 1.414$

$\sqrt{3} \approx 1.732$

$\pi \approx 3.1416$

We cannot write the square root of 2 as an exact number as we do for rational numbers; yet we can write the square root of 2 as an exact number by writing $\sqrt{2}$. If $\sqrt{2}$ stands for the exact square root of 2, we must conclude by the laws of algebra and the definition of square root that

$$\sqrt{2} \cdot \sqrt{2} = 2$$

In like manner

$$\sqrt[3]{5} \cdot \sqrt[3]{5} \cdot \sqrt[3]{5} = 5$$

$$\sqrt[n]{a} \cdot \sqrt[n]{a} \cdot \sqrt[n]{a} \ldots \text{ to } n \text{ factors of } \sqrt[n]{a} = a$$

19-6 VOCABULARY OF RADICALS The following definitions will be useful in subsequent study.

The *radical sign* or simply *radical* is the name given to the symbol $\sqrt{\ }$.

The *radicand* is the number or expression under the radical sign.

The small number placed in the notch of the radical sign is called the *index* or *order* of the radical. Thus, for $\sqrt[3]{89}$, the radicand is 89 and the index is 3 (or the radical is of the third order). In writing a square root, it is customary to omit the index 2.

The *coefficient* of a radical is the number prefixed to it to show how many times it is used as an addend in a sum. Thus, 3 is the coefficient of $\sqrt{2}$ for $3\sqrt{2}$. $3\sqrt{2} = \sqrt{2} + \sqrt{2} + \sqrt{2}$.

A *surd* is the indicated irrational number in which the radicand is rational. Thus $\sqrt{6}$, $\sqrt[3]{5}$, and $\sqrt{\frac{4}{3}}$ are surds, while $\sqrt{5 - \sqrt[3]{2}}$ is not a surd because the radicand is not rational.

19-7 LAWS OF RADICALS The definition of a radical and the laws of exponents can be used to develop four important relationships. In each of the following four cases we will assume $a \geqslant 0$, $b > 0$, and m and n are positive whole numbers.

First, we have

$$(\sqrt[n]{a})^n = \sqrt[n]{a^n} = a \tag{19-5}$$

This relationship can be verified by showing that

$$(a^{1/n})^n = (a^n)^{1/n} = a$$

The second important relationship is

$$\sqrt[n]{a}\ \sqrt[n]{b} = \sqrt[n]{ab} \tag{19-6}$$

We can verify this relationship by showing

$$\sqrt[n]{a}\ \sqrt[n]{b} = a^{1/n} \cdot b^{1/n} = (ab)^{1/n} = \sqrt[n]{ab}$$

Conversely, we can verify that $\sqrt[n]{ab} = \sqrt[n]{a}\ \sqrt[n]{b}$ (19-6*a*)

The third relationship is

$$\frac{\sqrt[n]{a}}{\sqrt[n]{b}} = \sqrt[n]{\frac{a}{b}} \tag{19-7}$$

Here we can show that

$$\frac{\sqrt[n]{a}}{\sqrt[n]{b}} = \frac{a^{1/n}}{b^{1/n}} = \left(\frac{a}{b}\right)^{1/n} = \sqrt[n]{\frac{a}{b}}$$

Reversing our steps will show that

$$\sqrt[n]{\frac{a}{b}} = \frac{\sqrt[n]{a}}{\sqrt[n]{b}} \tag{19-7a}$$

The fourth important relationship is

$$\sqrt[cn]{a^{cm}} = \sqrt[n]{a^m} \tag{19-8}$$

or

$$\sqrt[n]{a^m} = \sqrt[cn]{a^{cm}} \tag{19-8a}$$

Equation (19-8) follows from the fact that

$$\sqrt[cn]{a^{cm}} = (a^{cm})^{1/cn} = a^{cm/cn} = a^{m/n} = \sqrt[n]{a^m}$$

19-8 TRANSFORMATION OF RADICALS A radical is said to be *transformed* when its form has been changed without changing its value. The following examples will illustrate the four most common transformations.

Example 19-5 Simplify by removing factors from the radicand:

(a) $\sqrt{75a^3}$

(b) $\sqrt[3]{54x^2y^3}$

SOLUTIONS

(a) $\sqrt{75a^3} = \sqrt{25a^2 \cdot 3a} = \sqrt{25a^2} \cdot \sqrt{3a} = 5a\sqrt{3a}$

(b) $\sqrt[3]{-54x^2y^3} = \sqrt[3]{(-27y^3)(2y^2)} = \sqrt[3]{(-3y)^3} \cdot \sqrt[3]{2y} = -3y\sqrt[3]{2y}$

Example 19-6 Transform the following to surds with coefficient of 1:

(a) $7\sqrt{5}$

(b) $5 \cdot \sqrt[3]{4}$

(c) $2x \cdot \sqrt[5]{4x^2y^3}$.

SOLUTIONS

(a) $7\sqrt{5} = \sqrt{49} \cdot \sqrt{5} = \sqrt{49 \cdot 5} = \sqrt{245}$

(b) $5 \cdot \sqrt[3]{4} = \sqrt[3]{(5)^3} \cdot \sqrt[3]{4} = \sqrt[3]{125 \cdot 4} = \sqrt[3]{500}$

(c) $2x\sqrt[5]{4x^2y^3} = \sqrt[5]{32x^5} \cdot \sqrt[5]{4x^2y^3} = \sqrt[5]{128x^7y^3}$

Example 19-7 Transform to equivalent expressions in which the denominator of the integrand is 1:

(a) $\sqrt[3]{\frac{3}{4}}$

(b) $\sqrt{\dfrac{5a}{8b}}$

SOLUTIONS

(a) $\sqrt[3]{\frac{3}{4}} = \sqrt[3]{\frac{3}{4}} \cdot \sqrt[3]{\frac{2}{2}} = \sqrt[3]{\frac{6}{8}} = \dfrac{\sqrt[3]{6}}{\sqrt[3]{8}} = \dfrac{\sqrt[3]{6}}{2}$ or $\frac{1}{2}\sqrt[3]{6}$

(b) $\sqrt{\dfrac{5a}{8b}} = \sqrt{\dfrac{5a}{8b} \cdot \dfrac{2b}{2b}} = \dfrac{\sqrt{10ab}}{\sqrt{16b^2}} = \dfrac{\sqrt{10ab}}{4b}$

The process of Example 19-7 is called *rationalizing the denominator.*

Example 19-8 Reduce the order of:

(a) $\sqrt[4]{9}$

(b) $\sqrt[6]{4x^2y^4}$

SOLUTION

(a) $\sqrt[4]{9} = [3^2]^{1/4} = 3^{2/4} = 3^{1/2} = \sqrt{3}$

(b) $\sqrt[6]{4x^2y^4} = [(2xy^2)^2]^{1/6} = (2xy^2)^{1/3} = \sqrt[3]{2xy^2}$

Exercise 19-2

Write the expressions of Probs. 1 to 9 in simplest radical form.

1. $k^{1/3}$
2. $y^{2/3}$
3. $a^{-1/2}$
4. $2^{3/2} \cdot a^{5/2}$
5. $4b^{2/3}$
6. $a^2b^{4/3}c^{5/3}$
7. $a^{-1/2}b^{1/2}$
8. $(4a^2 + 9b^2)^{1/2}$
9. $2x^{1/2} - 5y^{1/2}$

Write the expressions of Probs. 10 to 18 with fractional exponents.

10. $\sqrt[3]{5^2}$
11. $\sqrt[4]{y^3}$
12. $\sqrt[5]{16x^3}$
13. $\sqrt[3]{6a^5b^2}$
14. $\sqrt{x^2 + y^2}$
15. $\sqrt[3]{(x^2 + y^2)^2}$
16. $\sqrt{a^3}\ \sqrt[4]{b^3}$
17. $\sqrt{x^3}\ \sqrt[3]{y^2}$
18. $\sqrt{a^3\sqrt{b^5}}$

In Probs. 19 to 36 reduce the radical expressions.

19. $\sqrt{50}$
20. $\sqrt{27}$
21. $\sqrt{72}$
22. $\sqrt{36x^3}$
23. $\sqrt[3]{24}$
24. $\sqrt[3]{48}$
25. $\sqrt{+8a^2}$
26. $\sqrt[3]{-54x^3y}$
27. $\sqrt[4]{32a^4b}$
28. $\sqrt[3]{16a^4b^3c^2}$
29. $\sqrt{\dfrac{4x^2y^3}{9ab^3}}$
30. $\sqrt[3]{\dfrac{16a^4b^3}{27c^3d^2}}$
31. $\sqrt{x^2 + x^3}$
32. $\sqrt{a^2x^2 - a^2x^3}$
33. $\sqrt{8x^2y - 16x^2y^2}$
34. $\sqrt[n]{x^{2n}y^{3n}}$
35. $\sqrt[n-1]{x^n}$
36. $\sqrt[n+1]{a^{n+2}}$

Write the expressions of Probs. 37 to 45 as radicals with the coefficient 1.

37. $2\sqrt{3}$
38. $2\sqrt[3]{3}$
39. $4\sqrt{x}$
40. $a\sqrt{a}$
41. $a\sqrt{5a}$
42. $2a\sqrt{3a}$
43. $b\sqrt[3]{b}$
44. $(x - y)\sqrt{x + y}$
45. $(a + b)\sqrt{\dfrac{a - b}{a + b}}$

In Probs. 46 to 63, rationalize the denominator and remove any perfect powers in the radicand.

46. $\sqrt{\frac{2}{3}}$
47. $\sqrt{\frac{1}{7}}$
48. $\sqrt{\dfrac{1}{a}}$

49. $\sqrt{\dfrac{a}{b}}$

50. $\sqrt{\dfrac{5x}{6y}}$

51. $\dfrac{2}{\sqrt{5}}$

52. $\dfrac{\sqrt{6}}{\sqrt{11}}$

53. $\sqrt[3]{\frac{2}{3}}$

54. $\sqrt[3]{\frac{1}{4}}$

55. $\sqrt[3]{\dfrac{1}{x^2}}$

56. $\sqrt[3]{\dfrac{x^2y^2}{z}}$

57. $\sqrt[3]{\dfrac{2}{3a}}$

58. $\sqrt{\frac{25}{32}}$

59. $\sqrt{\frac{16}{27}}$

60. $\sqrt{\frac{27}{80}}$

61. $\sqrt[3]{\dfrac{128a^4b^3}{x^6b^5}}$

62. $\sqrt[4]{\dfrac{81a^5b^{12}}{a^9c^3}}$

63. $\sqrt[n]{\dfrac{x^{5n}}{y^{n-1}}}$

In Probs. 64 to 69, transform to radicals of the same order.

64. $\sqrt{5}, \sqrt[3]{4}$

65. $\sqrt{2}, \sqrt[3]{5}, \sqrt[4]{3}$

66. $\sqrt{x}, \sqrt[3]{y}, \sqrt[5]{z}$

67. $\sqrt[3]{\dfrac{x^2}{2}}, \sqrt[4]{\dfrac{y^3}{3}}$

68. $\sqrt{\frac{2}{3}}, \sqrt[3]{\frac{3}{4}}$

69. $\sqrt[3]{\frac{3}{4}}, \sqrt[6]{\frac{4}{5}}$

19-9 ADDITION AND SUBTRACTION OF RADICALS Two radicals which have the same radicand and the same index are called *like* or *similar radicals.* Such radicals may be added or subtracted by combining their coefficients. The algebraic sum or difference of unlike radicals can only be indicated.

Example 19-9

$$\sqrt{3} + 5\sqrt{3} - 4\sqrt{3} = 2\sqrt{3}$$

Example 19-10

$$\begin{aligned} &5\sqrt{12} - \sqrt{48} + 7\sqrt{27} - 4\sqrt{\tfrac{1}{3}} = \\ 5\sqrt{4}\ \sqrt{3} - &\sqrt{16}\ \sqrt{3} + 7\sqrt{9}\ \sqrt{3} - 4\sqrt{\tfrac{1}{9}}\ \sqrt{3} = \\ &10\sqrt{3} - 4\sqrt{3} + 21\sqrt{3} - \tfrac{4}{3}\sqrt{3} = \tfrac{77}{3}\sqrt{3} \end{aligned}$$

Exercise 19-3

Transform, if necessary, and combine similar terms.

1. $6\sqrt{2} - 4\sqrt{2} + 3\sqrt{2}$
2. $4\sqrt{5} - 3\sqrt{5} - 7\sqrt{5}$
3. $3\sqrt{12} - 4\sqrt{27} + 5\sqrt{3}$
4. $4\sqrt{3} - 5\sqrt{27} + 6\sqrt{48} - 3\sqrt{12}$
5. $5\sqrt{98} - 4\sqrt{72} + 3\sqrt{8}$
6. $3\sqrt{45} - 7\sqrt{20} - \sqrt{80}$
7. $7\sqrt{125} - 2\sqrt{63} - 5\sqrt{45} + 4\sqrt{175}$
8. $4\sqrt{\frac{3}{10}} + 5\sqrt{\frac{15}{32}} - 7\sqrt{\frac{5}{6}}$
9. $a\sqrt{y} + \sqrt{a^2y} + 3a\sqrt{y}$
10. $3\sqrt{\frac{2}{9}} - \sqrt[4]{\frac{1}{4}} + \sqrt{\frac{8}{25}}$
11. $7\sqrt{\frac{4}{3}} - 5\sqrt{\frac{3}{4}} - \sqrt{\frac{1}{12}}$
12. $3\sqrt[3]{2} - 5\sqrt[3]{16} + 7\sqrt[3]{24}$
13. $4\sqrt[3]{48} - 3\sqrt[3]{64} - 5\sqrt{\frac{1}{36}}$
14. $2\sqrt[3]{6} - 5\sqrt[3]{-48} + 4\sqrt[6]{36}$
15. $\sqrt[3]{-108} + 3\sqrt[3]{32} + 4\sqrt[3]{-\frac{1}{2}}$
16. $5\sqrt[4]{4} - 6\sqrt[6]{8} + 7\sqrt[8]{16}$
17. $2\sqrt{18x} + 3\sqrt{8x} - 4\sqrt{50x}$
18. $\sqrt[4]{81a^5} - \sqrt[4]{ab^8} + \sqrt{25a^3c^4}$
19. $\sqrt[3]{\frac{a^2}{b}} + \sqrt[3]{\frac{a}{b^2}} + \sqrt[3]{a^2b^2}$
20. $\sqrt{(x-y)^3} - \sqrt{x^3 - x^2y} + \sqrt{xy^2 - y^3}$

19-10 MULTIPLICATION OF RADICALS The product of two radicals of the same order can be found directly by applying the identity $\sqrt[n]{a}\ \sqrt[n]{b} = \sqrt[n]{ab}$. The product of two radicals of different order can be found in the same way after they have been transformed to radicals of the same order.

Example 19-11 Find the product $(-2\sqrt{6})(3\sqrt{2})$.

SOLUTION

$$(-2\sqrt{6})(3\sqrt{2}) = -6\sqrt{12} = -6\sqrt{4}\sqrt{3} = -12\sqrt{3}.$$

Example 19-12 Find the product $(\sqrt{5})(\sqrt[3]{4})$.

SOLUTION

$$(\sqrt{5})(\sqrt[3]{4}) = \sqrt[6]{5^3}\ \sqrt[6]{4^2} = \sqrt[6]{5^3 \cdot 4^2} = \sqrt[6]{2{,}000}$$

Example 19-13 Find the product $(2\sqrt{3} - 5\sqrt{2})(7\sqrt{3} + 6\sqrt{2})$.

SOLUTION

$$\begin{aligned}(2\sqrt{3} - 5\sqrt{2})(7\sqrt{3} + 6\sqrt{2}) &= 14\sqrt{9} - 35\sqrt{6} + 12\sqrt{6} - 30\sqrt{4} \\ &= 42 - 35\sqrt{6} + 12\sqrt{6} - 60 \\ &= -18 - 23\sqrt{36}\end{aligned}$$

Exercise 19-4

Multiply and simplify if possible:

1. $\sqrt{3}\ \sqrt{4}$
2. $\sqrt{18}\ \sqrt{6}$
3. $2\sqrt{5} \cdot 3\sqrt{7}$
4. $7\sqrt{5}\ \sqrt{45}$
5. $\sqrt{2a}\ \sqrt{3a}$
6. $\sqrt{x}\ \sqrt{x^7}$
7. $\sqrt{15}\ \sqrt{\frac{7}{3}}$
8. $\sqrt{\dfrac{x}{y}}\ \sqrt{\dfrac{y}{x}}$
9. $\sqrt[3]{4}\ \sqrt[3]{9}$
10. $\sqrt{15}\ \sqrt{20}$
11. $(5\sqrt{24})(2\sqrt{45})$
12. $(2\sqrt{75})(-7\sqrt{6})$
13. $(\sqrt{49x})(\sqrt{21x^2})$
14. $\sqrt[3]{12}\ \sqrt[3]{90}$
15. $(4\sqrt[3]{5})(2\sqrt[3]{4})$
16. $(3\sqrt{4})^2$
17. $(\sqrt{x} + 2)^2$
18. $(2\sqrt{a} - 3\sqrt{b})^2$
19. $(\sqrt{k} - 1)^2$
20. $(\sqrt{x} + \sqrt{y})^2$
21. $(2\sqrt{a} - 3)^2$
22. $(\sqrt{a} + \sqrt{b})(\sqrt{a} - \sqrt{b})$
23. $(\sqrt{2} - \sqrt{3})(\sqrt{2} + \sqrt{3})$
24. $(\sqrt{x} - 3)(\sqrt{x} + 4)$
25. $(\sqrt{5} - 2\sqrt{3})(\sqrt{5} + \sqrt{3})$

19-11 DIVISION OF RADICALS The quotient of two radicals is usually best simplified by the process of rationalizing the denominator. Sometimes it is not necessary to rationalize the denominator, however, as will be seen in Example 19-14 below.

A conjugate radical is used in rationalizing a binomial denominator (see Example 19-16). $\sqrt{a} + \sqrt{b}$ and $\sqrt{a} - \sqrt{b}$ are binomials that differ only in the sign of the second term. They are called *conjugates* or are said to be conjugate to each other.

Example 19-14 Divide $15\sqrt{72}$ by $3\sqrt{18}$.

SOLUTION

$$\frac{15\sqrt{72}}{3\sqrt{18}} = 5\sqrt{\frac{72}{18}} = 5\sqrt{4} = 10$$

Example 19-15 Divide $4\sqrt{3}$ by $5\sqrt{2}$.

SOLUTION

$$\frac{4\sqrt{3}}{5\sqrt{2}} = \frac{4\sqrt{3}}{5\sqrt{2}} \cdot \frac{\sqrt{2}}{\sqrt{2}} = \frac{4\sqrt{6}}{5 \cdot 2} = \frac{2\sqrt{6}}{5}$$

Example 19-16 Divide $(5\sqrt{3} - 7\sqrt{2})$ by $(2\sqrt{3} + 3\sqrt{2})$.

SOLUTION

$$\frac{5\sqrt{3} - 7\sqrt{2}}{2\sqrt{3} + 3\sqrt{2}} = \frac{5\sqrt{3} - 7\sqrt{2}}{2\sqrt{3} + 3\sqrt{2}} \cdot \frac{2\sqrt{3} - 3\sqrt{2}}{2\sqrt{3} - 3\sqrt{2}}$$

$$= \frac{30 - 29\sqrt{6} + 42}{12 - 18}$$

$$= \frac{72 - 29\sqrt{6}}{-6}$$

$$= -12 + \frac{29}{6}\sqrt{6}$$

Exercise 19-5

Perform the indicated divisions and simplify the quotients:

1. $\sqrt{6} \div \sqrt{3}$
2. $5\sqrt{28} \div 3\sqrt{7}$
3. $3\sqrt{12} \div 5\sqrt{3}$
4. $7\sqrt{27} \div 7\sqrt{3}$
5. $20 \div 15\sqrt{5}$
6. $6\sqrt{45} \div 2\sqrt{5}$
7. $\sqrt[3]{16} \div \sqrt[3]{2}$
8. $7\sqrt[3]{20} \div 5\sqrt[3]{4}$
9. $\sqrt{\frac{3}{2}} \div \sqrt{\frac{2}{3}}$
10. $\sqrt{\frac{27}{7}} \div \sqrt{\frac{2}{15}}$
11. $\sqrt{a^2b} \div \sqrt{ab^2}$
12. $8 \div \sqrt{xy}$
13. $8\sqrt{\frac{2}{3}} \div 5\sqrt{\frac{10}{33}}$
14. $\sqrt[3]{4} \div \sqrt[3]{2}$
15. $\dfrac{9}{2\sqrt{3}}$
16. $\dfrac{3\sqrt{5}}{2\sqrt[3]{6}}$
17. $\dfrac{8\sqrt[3]{2}}{2\sqrt{3}}$
18. $\dfrac{\sqrt[6]{a^2b^5c^4}}{\sqrt[4]{a^3bc^2}}$
19. $\sqrt[3]{-\frac{1}{4}} \div \sqrt{\frac{1}{2}}$
20. $5\sqrt[4]{\frac{1}{27}} \div \sqrt{\frac{1}{3}}$
21. $\sqrt[3]{81a^5b^4} \div \sqrt[3]{3a^2b}$
22. $\sqrt[3]{\frac{25}{108}} \div \sqrt[3]{\frac{7}{4}}$
23. $\dfrac{4}{\sqrt{3}+\sqrt{2}}$
24. $\dfrac{5}{\sqrt{3}-\sqrt{2}}$
25. $\dfrac{3}{\sqrt{5}-\sqrt{2}}$
26. $\dfrac{\sqrt{10}}{\sqrt{5}-2\sqrt{2}}$
27. $\dfrac{2+\sqrt{3}}{5-\sqrt{3}}$
28. $\dfrac{\sqrt{5}+3}{\sqrt{5}-4}$
29. $\dfrac{\sqrt{5}+\sqrt{3}}{\sqrt{5}-\sqrt{3}}$
30. $\dfrac{\sqrt{2}+\sqrt{5}}{\sqrt{2}-\sqrt{5}}$
31. $\dfrac{2\sqrt{3}+5\sqrt{2}}{\sqrt{3}+\sqrt{2}}$
32. $\dfrac{5\sqrt{3}-2\sqrt{2}}{\sqrt{6}+\sqrt{24}}$
33. $\dfrac{3\sqrt{7}-2\sqrt{6}}{4\sqrt{7}-5\sqrt{6}}$
34. $\dfrac{8\sqrt{6}+3\sqrt{5}}{7\sqrt{6}+2\sqrt{5}}$

19-12 EQUATIONS INVOLVING RADICALS An equation in which the unknown quantity appears under a square radical sign may often be reduced to an equivalent linear equation. In solving such an equation, it is usually convenient to proceed as follows:

1 If the equation contains only one square radical, isolate the radical to one side of the equation by transposition; then square both members, and solve the resulting equation.

2 If the equation contains more than one square radical transpose the terms so that one square radical is the only term on one side of the equation; then square both sides. If a radical remains, repeat the procedure, and solve the resulting equation.

Since the operation of squaring both sides of an equation may introduce extraneous roots, it is of utmost importance to test the values obtained for the unknown by substituting each value in the *original equation.*

Example 19-17 Solve the equation $\sqrt{4x+1}+3=2\sqrt{x-5}$.

SOLUTION

$$\sqrt{4x+1}+3=2\sqrt{x-5}$$

Square both members: $4x+1+6\sqrt{4x+1}+9=4(x-5)$

Transpose and simplify: $6\sqrt{4x+1}=-30$

or $\sqrt{4x+1}=-5$

Square both members again: $4x+1=25$

$$4x=24$$

$$x=6$$

CHECK

$\sqrt{4(6)+1}+3\neq 2\sqrt{6-5}$

Hence 6 is *not* a root. The equation has no root.

Example 19-18 Solve the equation $\sqrt{16y+33}-2\sqrt{y+1}=\sqrt{4y+13}$.

SOLUTION

$$\sqrt{16y+33}-2\sqrt{y+1}=\sqrt{4y+13}$$

Square both members: $16y+33-4\sqrt{(16y+33)(y+1)}+4(y+1)=4y+13$

Transpose and simplify: $16y+24=4\sqrt{16y^2+49y+33}$

or $4y+6=\sqrt{16y^2+49y+33}$

Square both members: $16y^2+48y+36=16y^2+49y+33$

$$y=3$$

CHECK

$$\sqrt{16(3)+33}-2\sqrt{3+1}\stackrel{?}{=}\sqrt{4(3)+13}$$

$$\sqrt{48+33}-2\sqrt{3+1}\stackrel{?}{=}\sqrt{12+13}$$

$$9-4\stackrel{?}{=}5$$

$$5=5$$

Then, 3 is a root of the equation.

Exercise 19-6

Solve and check by substitution:

1. $\sqrt{5x-4}=4$
2. $\sqrt{x+4}=7+\sqrt{x+11}$
3. $\sqrt{a+9}=3-\sqrt{a+12}$
4. $2\sqrt{y+11}-\sqrt{9y+4}=\sqrt{y-4}$
5. $\sqrt{3x+28}=\sqrt{3x-5}+3$
6. $\sqrt{7x+8}+\sqrt{7x-31}=13$
7. $\sqrt{x+18}-\sqrt{x-18}=3\sqrt{3}$
8. $\sqrt{a+20}+\sqrt{a-4}=4\sqrt{2}$
9. $\sqrt{9x+28}+3\sqrt{x+12}=10$
10. $\sqrt{x+3}=\sqrt{x}+\sqrt{3}$
11. $\sqrt{4x+1}+\sqrt{16x-7}=2\sqrt{x-1}$
12. $\sqrt{2x-\sqrt{x-3}}=\sqrt{1+2x}$
13. $\sqrt{5x-\sqrt{4x-3}}=\sqrt{5x-3}$
14. $\dfrac{\sqrt{y}-2}{\sqrt{y}+3}=\dfrac{\sqrt{y}+1}{\sqrt{y}-4}$
15. A loading chute delivers cartons from a warehouse to a freight-car door 25 ft below. If the chute is 32 ft long, how far from the wall will the bottom of the chute be?
16. A formula commonly used to get resonant frequency of an electric circuit is $f=\dfrac{1}{2\pi}\sqrt{\dfrac{1}{LC}}$. Express L in terms of f and C.
17. The period T, in seconds, of a pendulum of length l, in meters, is given by the formula $T=2\pi\sqrt{\dfrac{l}{g}}$. Find the value of the acceleration of gravity g if $T=4.04$ seconds when $l=4.04$ m.

19-13 IMAGINARY NUMBERS It is always possible in the real number system to raise any real number to an integral power. However, the inverse operation with real numbers is not always possible. In the real number system it is impossible to extract an even ordered root of a negative number.

Whether a real number is positive or negative, its square is always positive. The square of 0 is 0. Note that the square of a real number is never negative. Thus, if we restrict ourselves to real numbers only, the symbols $\sqrt{-1}$, $\sqrt{-2}$, $\sqrt{-3}$, $\sqrt{-4}$, $\sqrt{-5}$, and so on, are meaningless.

The square roots of negative quantities occur frequently, not only in theoretical mathematics, but also in practical applications. Hence, it becomes desirable to give them meaning. This can be done only by extending the number system. The larger system of numbers that results is called the *complex number system.*

We note that the unit of the real number system is the number 1. To enlarge our system we introduce a new unit, $\sqrt{-1}$, which is usually designated by j (or i) and is defined by the property

$$j^2 = (\sqrt{-1})^2 = -1 \tag{19-9}$$

Then, for any positive real number N we have

$$\sqrt{-N} = \sqrt{-1}\sqrt{N} = j\sqrt{N} \tag{19-10}$$

That is, the square root of any negative number can be expressed as the product of a real number and the number j. The number $\sqrt{-N}$ is called a pure imaginary number. The term "imaginary" is unfortunate, since imaginary numbers do exist and have great significance in the study of electrical systems and other branches of engineering practice.

19-14 TRANSFORMATIONS OF IMAGINARIES In operating with imaginary numbers we will require that they obey the same basic laws of operation that real numbers obey. In general, imaginaries are considered simpler when they have been transformed so that (1) the values of the radicands have been made a positive integral and as small as possible, and so that (2) no radical appears in the denominator of any fraction.

Example 19-19 Simplify:

(a) $\sqrt{-4}$
(b) $-\sqrt{-5}$
(c) $\sqrt{-\frac{3}{4}}$
(d) $\dfrac{\sqrt{-7}}{\sqrt{-2}}$

SOLUTION

(a) $\sqrt{-4} = \sqrt{4}\,\sqrt{-1} = 2j$
(b) $-\sqrt{-5} = -\sqrt{5}\,\sqrt{-1} = -\sqrt{5}\,j$
(c) $\sqrt{\dfrac{-3}{4}} = \dfrac{\sqrt{-3}}{\sqrt{4}} = \dfrac{\sqrt{3}j}{\sqrt{4}} = \dfrac{\sqrt{3}}{2}j$
(d) $\dfrac{\sqrt{-7}}{\sqrt{-2}} = \dfrac{\sqrt{-7}\,\sqrt{-1}}{\sqrt{-2}\,\sqrt{-1}} = \dfrac{\sqrt{7}\overset{1}{\cancel{j}}}{\sqrt{2}\underset{1}{\cancel{j}}} = \dfrac{\sqrt{7}\,\sqrt{2}}{\sqrt{2}\,\sqrt{2}} = \dfrac{\sqrt{14}}{2}$

If j is raised to successive positive integral powers, we get

$$\begin{aligned} j &= \sqrt{-1} & j^5 &= j^4 \cdot j = (1)j = j \\ j^2 &= -1 & j^6 &= j^4 \cdot j^2 = (1)(-1) = -1 \\ j^3 &= j^2 \cdot j = (-1)j = -j & j^7 &= j^4 \cdot j^3 = (1)(-j) = -j \\ j^4 &= j^2 \cdot j^2 = (-1)(-1) = 1 & j^8 &= j^4 \cdot j^4 = (1)(1) = 1 \end{aligned}$$

and so on. It will be noted that successive integral powers of j give a cyclic arrangement of the numbers j, -1, $-j$, and 1.

Example 19-20

$\sqrt{-4}\ \sqrt{-9} = (j\sqrt{4})(j\sqrt{9}) = (2j)(3j) = 6j^2 = 6(-1) = -6$

Example 19-21

$\sqrt{-a}\ \sqrt{-b} = (j\sqrt{a})(j\sqrt{b}) = j^2ab = -ab$

Certain relationships valid for real numbers are not valid for imaginary numbers. For instance (see page 313) if a and b are both real,

$\sqrt{a}\ \sqrt{b} = \sqrt{ab}$

but if a and b are both imaginary

$\sqrt{a}\ \sqrt{b} = -\sqrt{ab}$

In Example 19-20, for instance,

$\sqrt{-4}\ \sqrt{-9} \neq \sqrt{36} = 6$

Thus

NOTE *Before operating on any imaginary number* $\sqrt{-N}$ where $N > 0$, *the number must first be changed to the form* $j\sqrt{N}$.

Example 19-22 Simplify $(j^3 - 2j + 1)(j^4 + 3j^3 - 2j^2 - 2)$.

SOLUTION

$$\begin{aligned}(j^3 - 2j + 1)(j^4 + 3j^3 - 2j^2 - 2) &= \\ [(-j) - 2j + 1][(1) + 3(-j) - 2(-1) - 2] &= \\ (1 - 3j)(1 - 3j) &= \\ (1 - 6j + 9j^2) &= \\ 1 - 6j + 9(-1) &= -8 - 6j\end{aligned}$$

Example 19-23 Simplify $\sqrt{-20} - 2\sqrt{-45} + 4\sqrt{-80}$.

SOLUTION

$$\begin{aligned}\sqrt{-20} - 2\sqrt{-45} + 4\sqrt{-80} &= \\ \sqrt{4(5)(-1)} - 2\sqrt{9(5)(-1)} + 4\sqrt{16(5)(-1)} &= \\ \sqrt{4}\ \sqrt{5}\ \sqrt{-1} - 2\sqrt{9}\ \sqrt{5}\ \sqrt{-1} + 4\sqrt{16}\ \sqrt{5}\ \sqrt{-1} &= \\ 2j\sqrt{5} - 6j\sqrt{5} + 16j\sqrt{5} &= 12j\sqrt{5}\end{aligned}$$

Exercise 19-7

Express each of the following in its simplest form:

1. j^7
2. j^{11}
3. j^{12}
4. j^{17}
5. $\sqrt{-16}$
6. $\sqrt{-36}$

7. $\sqrt{-8}$
8. $2\sqrt{-25}$
9. $\frac{1}{2}\sqrt{-64}$
10. $\sqrt{-18}$
11. $\sqrt{-75}$
12. $\sqrt{-48}$
13. $\sqrt{\dfrac{-5}{2}}$
14. $\sqrt{-\frac{2}{3}}$
15. $\sqrt{-\frac{20}{3}}$
16. $\sqrt{\dfrac{-7}{4}}$
17. $\dfrac{3}{\sqrt{-5}}$
18. $\dfrac{3}{\sqrt{-7}}$
19. $\dfrac{\sqrt{a}}{\sqrt{-b}}$
20. $\dfrac{a}{\sqrt{-b}}$

Simplify and combine similar terms.

21. $8\sqrt{-2} - \sqrt{-50} - \sqrt{-32}$
22. $\sqrt{-12} - \sqrt{-27} + \sqrt{-147}$
23. $3\sqrt{-8} - 2\sqrt{-8} + 9\sqrt{-72}$
24. $3\sqrt{-25} + 5\sqrt{-4} - 2\sqrt{-9}$
25. $7\sqrt{-20} - 5\sqrt{-80} - 3\sqrt{-45}$
26. $\sqrt{-18} + \sqrt{-\frac{2}{25}} + \sqrt{-\frac{8}{9}}$
27. $3\sqrt{-\frac{8}{3}} - \sqrt{-150} - \frac{1}{4}\sqrt{-\frac{3}{2}}$
28. $\dfrac{22}{\sqrt{-3}} + \dfrac{5}{\sqrt{-2}} + \sqrt{-18} + \sqrt{-75}$
29. $\sqrt{-24} + 2\sqrt{-150} - 3\sqrt{\dfrac{-2}{75}}$
30. $\sqrt{-32} + 7\sqrt{-128} + 2\sqrt{72} - 3\sqrt{18}$

Perform the indicated operations and simplify.

31. $\sqrt{-4}\,\sqrt{-16}$
32. $\sqrt{-9}\,\sqrt{-4}\,\sqrt{-25}$
33. $(2\sqrt{-5}\,)^2$
34. $(-5\sqrt{-6}\,)^2$
35. $(-2\sqrt{-6}\,)(5\sqrt{-15}\,)$
36. $(7\sqrt{-15}\,)(4\sqrt{-12}\,)$
37. $(1 + j)(1 - j)$
38. $(5 + 2j)(3 - 4j)$
39. $(j^4 - 3j^2 + 2j + 1)(j^3 + 5j^2 + 3j - 1)$
40. $(j^4 - j^6 + j^8 - j^{10})(j^9 - j^7 + j^5 - j^3 + j - 1)$

19-15 COMPLEX NUMBERS A complex number is the indicated sum or difference of a real number and an imaginary number. A complex number takes the form of $a + bj$ where a and b are real numbers. The number a is called the real part and bj is called the imaginary part of the complex number. If $a \neq 0$, $b = 0$, the complex number reduces to the real number a. If $a = 0$, $b \neq 0$, the complex number becomes the pure imaginary number bj.

Complex numbers differing only in the sign of their imaginary parts are called *conjugate complex numbers* and either is called the conjugate of the other. Thus, $(a + bj)$ and $(a - bj)$ are conjugate complex numbers. Also, bj and $-bj$ are conjugate complex numbers. A real number is its own conjugate.

19-16 FUNDAMENTAL OPERATIONS WITH COMPLEX NUMBERS The ordinary rules of addition, subtraction, multiplication, and division of real numbers also apply to complex numbers. However, in all operations involving complex numbers, each complex number must first be expressed in $a + bj$ form.

Example 19-24 Simplify $(5 - 3\sqrt{-2})(-1 + 6\sqrt{-2})$.

SOLUTION

$$(5 - 3\sqrt{-2})(-1 + 6\sqrt{-2}) =$$
$$(5 - 3j\sqrt{2})(-1 + 6j\sqrt{2}) =$$
$$-5 + 33j\sqrt{2} - 18j^2(2) =$$
$$-5 + 33j\sqrt{2} + 36 = 31 + 33j\sqrt{2}$$

It should be apparent that (1) the product of two pure imaginary numbers is always a real number, and (2) the product of two conjugate complex numbers is a real number.

Example 19-25 Simplify $(3\sqrt{2} - 5\sqrt{-3})(3\sqrt{2} + 5\sqrt{-3})$.

SOLUTION

$$\begin{aligned}(3\sqrt{2} - 5\sqrt{-3})(3\sqrt{2} + 5\sqrt{-3}) &= (3\sqrt{2} - 5j\sqrt{3})(3\sqrt{2} + 5j\sqrt{3}) \\ &= 9(2) - 25j^2(3) \\ &= 18 + 75 \\ &= 93\end{aligned}$$

Example 19-26 Simplify

$$\frac{5 - 2\sqrt{-3}}{4 + \sqrt{-3}}$$

SOLUTION

$$\frac{5-2\sqrt{-3}}{4+\sqrt{-3}}=\frac{5-2j\sqrt{3}}{4+j\sqrt{3}}\cdot\frac{4-j\sqrt{3}}{4-j\sqrt{3}}$$

$$=\frac{20-13j\sqrt{3}+2j^2(3)}{16-3j^2}$$

$$=\frac{20-13j\sqrt{3}-6}{16+3}$$

$$=\frac{14}{19}-\frac{13j\sqrt{3}}{19}$$

Exercise 19-8

Perform the indicated operations and leave in $a+bj$ form.

1. $(2-\sqrt{-7})(2+\sqrt{-7})$
2. $(7+\sqrt{-8})(7-\sqrt{-8})$
3. $(\sqrt{-25}+\sqrt{-4})(\sqrt{-25}+\sqrt{-4})$
4. $(5-\sqrt{-8})^2$
5. $(7-\sqrt{-6})(5+\sqrt{-6})$
6. $(3+\sqrt{-5})(2-\sqrt{-5})$
7. $(4+2\sqrt{-7})^2$
8. $(\sqrt{5}-2\sqrt{-3})(\sqrt{5}+2\sqrt{-3})$
9. $(5\sqrt{2}+4\sqrt{-3})(5\sqrt{2}-4\sqrt{-3})$
10. $(2\sqrt{-36}-3\sqrt{-49})(2\sqrt{-36}+3\sqrt{-49})$
11. $(\sqrt{x}+j\sqrt{y})(\sqrt{x}-j\sqrt{y})$
12. $(\sqrt{\tfrac{2}{3}}+\sqrt{-\tfrac{3}{4}})(\sqrt{\tfrac{2}{3}}-\sqrt{-\tfrac{3}{4}})$
13. $(5\sqrt{2}-4\sqrt{-3})(3\sqrt{2}-\sqrt{-3})$
14. $(6+2\sqrt{-3})(5-2\sqrt{-3})$
15. $\dfrac{3+\sqrt{-4}}{2+\sqrt{-6}}$
16. $\dfrac{7+\sqrt{-25}}{6-\sqrt{-9}}$
17. $\dfrac{7-\sqrt{-4}}{2+\sqrt{-9}}$
18. $\dfrac{5-\sqrt{-4}}{5-\sqrt{-25}}$
19. $\dfrac{5+\sqrt{-2}}{5-\sqrt{-2}}$
20. $\dfrac{4+2\sqrt{-5}}{5-3\sqrt{-5}}$
21. $\dfrac{\sqrt{3}-\sqrt{-3}}{\sqrt{2}+\sqrt{-2}}$
22. $\dfrac{\sqrt{3}+\sqrt{-6}}{\sqrt{3}-\sqrt{-6}}$

23. $\dfrac{7}{2-\sqrt{-5}}$

24. $\dfrac{5}{\sqrt{3}-\sqrt{2}}$

25. $\dfrac{15}{\sqrt{5}-\sqrt{-3}}$

26. $\dfrac{1}{5+\sqrt{-8}}$

CHAPTER 20

QUADRATIC EQUATIONS

20-1 THE GENERAL QUADRATIC EQUATION IN ONE UNKNOWN The *degree* of an equation in a certain letter is the highest power of that letter that occurs in the equation. Thus $2x^4 - 5x^2y^2 + 3xy^3 - 28 = 0$ is an equation of the fourth degree in x and third degree in y. A rational equation of the second degree in one unknown is called a *quadratic equation.* Every quadratic equation in one variable x can, by applying the proper laws, be arranged in the standard form

$$ax^2 + bx + c = 0 \qquad a \neq 0 \tag{20-1}$$

where a, b, and c are real constants. This form is called the *general quadratic equation* in x. If $b = 0$, the equation is called a *pure quadratic equation.* The letters b or c, or both b and c, can be equal to 0. But if $a = 0$, Eq. (20-1) reduces to a linear equation in x. Therefore, we restrict a to values other than zero.

Example 20-1 Reduce the quadratic equation $(x - 2)^2 + 2(x + 1)^2 - (x - 3)^2 = 12$ to the standard form.

SOLUTION

$$(x-2)^2 + 2(x+1)^2 - (x-3)^2 = 12$$

$$(x^2 - 4x + 4) + 2(x^2 + 2x + 1) - (x^2 - 6x + 9) = 12$$

$$x^2 - 4x + 4 + 2x^2 + 4x + 2 - x^2 + 6x - 9 = 12$$

$$2x^2 + 6x - 15 = 0$$

Here $a = 2$, $b = 6$, and $c = -15$.

20-2 SOLUTION OF PURE QUADRATIC EQUATIONS A pure quadratic equation can be written in the form

$$ax^2 + c = 0 \qquad a \neq 0 \tag{20-2}$$

where a and c are constants.

Equation (20-2) is solved by first isolating x^2 on one side of the equation; the roots are then found by extracting the square roots of both members to obtain

$$x = \pm\sqrt{-\frac{c}{a}}$$

Thus, Eq. (20-2) has two roots with equal absolute values. The two roots are real when a and b are unlike in sign, and are imaginary if a and b are alike in sign.

Example 20-2 Solve $\frac{x}{2}(3x + 8) = 4(x - 9)$.

SOLUTION

$$\frac{x}{2}(3x+8)=4(x-9)$$

Remove parentheses: $$\frac{3x^2}{2}+4x=4x-36$$

Solve for x^2: $$\frac{3x^2}{2}=-36$$

$$x^2=\frac{-2(36)}{3}$$

Solve for x: $$=-24$$

$$x=\pm\sqrt{-24}=\pm 2j\sqrt{6}$$

Exercise 20-1

Write each of equations 1-10 in $ax^2+bx+c=0$ form.

1. $5x^2-7x+3=9-3x$
2. $3x^2-2x-1=5x^2-8x+7$
3. $6x^2-3x+4=(2x-1)^2$
4. $5x^2-(3x-2)^2=4(x+7)$
5. $2x(x-k)-5(x+3k)=2k-9$
6. $(x-\sqrt{5})^2+(x+\sqrt{5})^2=0$
7. $\frac{x+3}{x}-\frac{x}{x-3}=\frac{3}{4}$
8. $\frac{2x}{x-2}-\frac{3x}{x+2}=\frac{5}{6}$
9. $\frac{2x-3}{x+4}-\frac{7}{12}=\frac{x-3}{5x-1}$
10. $\frac{5}{4x^2-4}+\frac{1}{2}-\frac{3x}{x+1}=\frac{6}{x-1}$

Solve each of the following equations for x, y, or z.

11. $x^2=81$
12. $4y^2=9$
13. $9z^2+25=0$
14. $8x^2-5=2x^2$
15. $3ay^2-b=c$
16. $18y^2+64=0$
17. $\frac{2}{3}z^2-2=\frac{1}{4}z^2$
18. $y(y-3)=2+3(5-y)$
19. $5x(x-4)=4x(2x-5)-27$
20. $3y(2y-5)-56=5y(4y-3)$

21. $8z(z+3a)=(4z+3a)^2$

22. $\dfrac{z+1}{3z+4}=\dfrac{2z-3}{z-2}$

23. $\dfrac{3}{x+3}-\dfrac{2}{x-2}-1=0$

24. $\sqrt{y+12}=\sqrt{2y+10}-\sqrt{y-12}$

25. $\sqrt{3z+6}-\sqrt{3z-6}=\sqrt{6z-16}$

20-3 SOLUTION BY FACTORING If it is possible to factor the left member of a quadratic equation in standard form, the equation can be solved by applying the following principle.

PRINCIPLE *If the product of two or more factors is equal to zero, then and only then at least one of the factors is equal to zero.*

Thus, if $A \cdot B = 0$, then either $A = 0$, $B = 0$, or both $A = 0$ and $B = 0$.

Example 20-3 Solve $6x^2 - 20 = 7x$.

SOLUTION

$$6x^2 - 20 = 7x$$

Change to standard form: $6x^2 - 7x - 20 = 0$

Factor: $(2x-5)(3x+4) = 0$

Equate each factor to zero: $2x - 5 = 0 \qquad 3x + 4 = 0$

Solve for x: $x = \frac{5}{2} \qquad x = -\frac{4}{3}$

CHECK

$6(\frac{5}{2})^2 - 20 \overset{?}{=} 7(\frac{5}{2}) \qquad 6(-\frac{4}{3})^2 - 20 \overset{?}{=} 7(-\frac{4}{3})$

$\frac{150}{4} - \frac{80}{4} \overset{?}{=} \frac{70}{4} \qquad \frac{96}{9} - \frac{180}{9} \overset{?}{=} -\frac{28}{3}$

$\frac{70}{4} = \frac{70}{4} \qquad -\frac{84}{9} = -\frac{84}{9}$

Hence the solution to the equation is $x = -\frac{4}{3}$ and $x = \frac{5}{2}$.

Example 20-4 Solve $4y^2 + 28y + 49 = 0$.

SOLUTION

$$4y^2 + 28y + 49 = 0$$

Factor: $(2y+7)(2y+7) = 0$

Equate each factor to zero: $2y + 7 = 0 \qquad 2y + 7 = 0$

Solve for y: $y = -\frac{7}{2} \qquad y = -\frac{7}{2}$

CHECK

$$4(-\tfrac{7}{2})^2 + 28(-\tfrac{7}{2}) + 49 \stackrel{?}{=} 0$$

$$49 - 98 + 49 \stackrel{?}{=} 0$$

$$0 = 0$$

The equation $4y^2 + 28y + 49 = 0$ has two equal roots, $-\frac{7}{2}$ and $-\frac{7}{2}$.

Exercise 20-2

Solve each of the following equations for x, y, or z by the method of factoring.

1. $4x^2 - 9 = 0$
2. $9x^2 - 25 = 0$
3. $16z^2 = 9$
4. $x^2 - 5x = 0$
5. $2y^2 - 8y = 0$
6. $3x^2 = 5x$
7. $y^2 - 6y + 9 = 0$
8. $z^2 - 10z + 25 = 0$
9. $x^2 + 5x + 6 = 0$
10. $z^2 + 3z = 10$
11. $4y^2 + 21 = -19y$
12. $47x + 40 + 12x^2 = 0$
13. $45z^2 + 14 = 53z$
14. $5x^2 + 7x = 6$
15. $24y^2 - 28 = 11y$
16. $88x^2 + 50 = 135x$
17. $21y^2 + 26y + 8 = 0$
18. $108z^2 + 15kz - 28k^2 = 0$
19. $30x^2 - 49ax + 6a^2 = 0$
20. $39y^2 - 62cy - 5c^2 = 0$
21. $9a^2 - x^2 + 6x - 9 = 0$
22. $6z^2 - 14cz + 15az - 35ac = 0$
23. $(y + 2)^2 - (y + 2) - 12 = 0$
24. $4(x - 3)^2 + 7(x - 3) - 2 = 0$
25. $\dfrac{3}{x + 1} + \dfrac{4x - 1}{6x} = \dfrac{7}{3}$
26. $\dfrac{5y - 3}{8y + 2} - \dfrac{7y - 4}{13y - 3} + \dfrac{1}{10} = 0$
27. $\dfrac{1}{z - 1} + \dfrac{1}{3} = \dfrac{3z + 1}{5z - 3}$
28. $\dfrac{3x - 2}{7x + 2} + \dfrac{1}{12} = \dfrac{5x - 4}{11x - 4}$

20-4 SOLUTION BY THE QUADRATIC FORMULA Not all quadratic equations can be solved by the factor method. However, a quadratic formula can be derived which can be used to solve all quadratic equations whether or not a solution can be found by factoring. The formula is obtained by a process known as *completing the square.* Its derivation follows. Let the equation be in standard form.

Given: $$ax^2 + bx + c = 0$$

Subtract c from each member: $$ax^2 + bx = -c$$

Divide each term by a: $$x^2 + \frac{b}{a}x = -\frac{c}{a}$$

Add $\left(\frac{b}{2a}\right)^2$ to each member: $$x^2 + \frac{b}{a}x + \frac{b^2}{4a^2} = \frac{b^2}{4a^2} - \frac{c}{a}$$

Then: $$\left(x + \frac{b}{2a}\right)^2 = \frac{b^2 - 4ac}{4a^2}$$

Extract square roots of each member: $$x + \frac{b}{2a} = \pm\frac{\sqrt{b^2 - 4ac}}{2a}$$

Subtract $\frac{b}{2a}$ from both members: $$x = -\frac{b}{2a} \pm \frac{\sqrt{b^2 - 4ac}}{2a}$$

Or: $$x = \frac{-b \pm \sqrt{b^2 - 4ac}}{2a}$$

RULE *The two roots of the quadratic equation*

$ax^2 + bx + c = 0 \qquad a \neq 0$

in which a, b, and c are real numbers given by the equation

$$x = \frac{-b \pm \sqrt{b^2 - 4ac}}{2a} \tag{20-3}$$

Equation (20-3) is known as the *quadratic formula,* and should be memorized, for it will be used frequently hereafter.

Example 20-5 Solve $6x^2 + 7 = 23x$.

SOLUTION Arrange the equation in standard form.

$6x^2 - 23x + 7 = 0$

Here $a = 6$, $b = -23$, and $c = 7$.

Substitute in Eq. (20-3): $$x = \frac{-(-23) \pm \sqrt{(-23)^2 - 4(6)(7)}}{2(6)}$$

Simplify: $$x = \frac{23 \pm \sqrt{529 - 168}}{12}$$

$$= \frac{23 \pm \sqrt{361}}{12} = \frac{23 \pm 19}{12}$$

Then: $$x = \tfrac{42}{12} = \tfrac{7}{2} \qquad \text{or} \qquad x = \tfrac{4}{12} = \tfrac{1}{3}$$

Example 20-6 Solve $\frac{1}{4}y^2 + \frac{2}{3}y + \frac{3}{2} = 0$.

SOLUTION Clear fractions by multiplying each term by 12.

$3y^2 + 8y + 18 = 0$

Here $a = 3$, $b = 8$, $c = 18$.

Substitute in Eq. (20-3): $x = \dfrac{-8 \pm \sqrt{(8)^2 - 4(3)(18)}}{2(3)}$

$$= \frac{-8 \pm \sqrt{-152}}{6} = \frac{-8 \pm 2j\sqrt{38}}{6}$$

$$= -\frac{4}{3} + j\frac{\sqrt{38}}{3} \text{ or } x = -\frac{4}{3} - j\frac{\sqrt{38}}{3}$$

Example 20-7 Solve $kmx^2 - m^2x = klx - ml$.

SOLUTION Arrange the equation in standard form.

$kmx^2 + (-m^2 - kl)x + ml = 0$

Here $a = km$, $b = -m^2 - kl$, and $c = ml$.

Substitute in Eq. (20-3): $x = \dfrac{m^2 + kl \pm \sqrt{m^4 + 2klm^2 + k^2l^2 - 4klm^2}}{2km}$

Simplify: $= \dfrac{m^2 + kl \pm \sqrt{m^4 - 2klm^2 + k^2l^2}}{2km}$

$$= \frac{m^2 + kl \pm (m^2 - kl)}{2km}$$

Then: $x = \dfrac{m}{k}$ or $\dfrac{l}{m}$

Exercise 20-3

Solve each of the following equations by use of the quadratic formula. Express irrational roots in simplest radical form and complex roots in $a + bj$ form.

1. $x^2 - 7x + 3 = 0$
2. $6x^2 - 7x + 2 = 0$
3. $4x^2 - 11x = 20$
4. $15x^2 + 14x = 8$
5. $x^2 + 3x + 4 = 0$
6. $x^2 + 5x + 5 = 0$
7. $x^2 - 4x + 2 = 0$
8. $4x^2 + 3 = 4x$
9. $9x^2 + 6x = 1$
10. $4x^2 + 4 = 7x$
11. $4y^2 + 13 = 8y$
12. $6z^2 + 5 = 13z$
13. $(2x - 1)(x + 3) = 10$
14. $\dfrac{y+1}{y-2} = \dfrac{3y+5}{2y}$
15. $\dfrac{z-2}{z-3} + \dfrac{z-1}{z-2} = \dfrac{7}{2}$
16. $\dfrac{y-1}{y+2} + \dfrac{3y}{y-2} = 5$
17. $(x - 3)(2x + 9) = (3x - 7)(2x + 5)$
18. $(x + 5)^2 - (2x - 3)^2 = 20$

19. $4x^2 + kx - m = 0$
20. $ax^2 + ax - bx + a - b = 0$
21. $abx^2 - b^2x - acx + bc = 0$
22. $3x^2 - 2kx + 2k^2 - 2k - 4x + 10 = 0$
23. $y^2 + 2a^2 - 4y - 6a + 4 = 0$
24. $a^2 + 2z^2 - 4a - 6z + 4 = 0$
25. $T = 2\pi rh + 2\pi r^2$. Solve for r.
26. $x = 2\pi fh - \dfrac{1}{2\pi fc}$. Solve for f.
27. $Lm^2 + Rm + \dfrac{1}{C} = 0$. Solve for m.
28. $s = vt + \frac{1}{2}gt^2$. Solve for t.

20-5 EQUATIONS CONTAINING RADICALS An equation in which the unknown occurs under a radical sign is called a *radical* or *irrational equation.* Such an equation can sometimes be reduced to a linear or quadratic equation by raising both members to the same power until the unknown no longer appears under a radical sign. The derived equation can then be solved by the regular methods discussed thus far. The method of solving radical equations depends upon the following principle.

PRINCIPLE *The root of a given equation is also a root of the derived equation obtained by equating the squares of the members of the given equation.*

However, the converse of this principle is not always true. Since the operation of raising both members to the same power may produce an equation that is not equivalent to the given equation, it is necessary to test all solutions by substituting in the original equation. Any solution not satisfying the original equation is *extraneous* and should be rejected. The student should recall that $\sqrt{N}$ is used to mean only the positive value of the root of N.

Example 20-8 Solve the equation $\sqrt{3x^2 - 2} - 2x - 1 = 0$.

SOLUTION

Isolate $\sqrt{3x^2 - 2}$:	$\sqrt{3x^2 - 2} = 2x + 1$
Square both members:	$3x^2 - 2 = 4x^2 + 4x + 1$
Add $-4x^2 - 4x - 1$ to both members:	$-x^2 - 4x - 3 = 0$
Multiply each term by -1:	$x^2 + 4x + 3 = 0$
Factor:	$(x + 3)(x + 1) = 0$
Solve for x:	$x + 3 = 0$ or $x + 1 = 0$
	$x = -3$ or $x = -1$

CHECK

$$\sqrt{3(-3)^2-2}-2(-3)-1 \stackrel{?}{=} 0 \qquad \sqrt{3(-1)^2-2}-2(-1)-1 \stackrel{?}{=} 0$$
$$\sqrt{27-2}+6-1 \stackrel{?}{=} 0 \qquad \sqrt{3-2}+2-1 \stackrel{?}{=} 0$$
$$5+6-1 \neq 0 \qquad 1+2-1 \neq 0$$

Then -3 is *not* a root and -1 is *not* a root.

Example 20-9 Solve the equation $\sqrt{4x+3}-\sqrt{8x+5}=\sqrt{2x+1}$.

SOLUTION Equate squares of both members of the equation.

$$4x+3-2\sqrt{(4x+3)(8x+5)}+8x+5=2x+1$$

Add $-12x-8$ to both members: $-2\sqrt{(4x+3)(8x+5)}=-10x-7$

Square both members again: $4(32x^2+44x+15)=100x^2+140x+49$

Simplify: $28x^2+36x+11=0$

Factor: $(2x+1)(14x+11)=0$

Solve for x: $2x+1=0$ or $14x+11=0$

$x=-\frac{1}{2}$ $\qquad$ $x=-\frac{11}{14}$

CHECK

$$\sqrt{4(-\tfrac{1}{2})+3}-\sqrt{8(-\tfrac{1}{2})+5} \stackrel{?}{=} \sqrt{2(-\tfrac{1}{2})+1}$$
$$\sqrt{-2+3}-\sqrt{-4+5} \stackrel{?}{=} \sqrt{-1+1}$$
$$0=0$$

Then, $-\frac{1}{2}$ *is* a root

$$\sqrt{4(-\tfrac{11}{14})+3}-\sqrt{8(-\tfrac{11}{14})+5} \stackrel{?}{=} \sqrt{2(-\tfrac{11}{14})+1}$$
$$\sqrt{-\tfrac{1}{7}}-\sqrt{-\tfrac{9}{7}} \stackrel{?}{=} \sqrt{-\tfrac{4}{7}}$$
$$\frac{j\sqrt{7}}{7}-\frac{3j\sqrt{7}}{7} \neq \frac{2j\sqrt{7}}{7}$$

Thus, we know that $-\frac{11}{14}$ *is not* a root.

Example 20-10 Solve the equation $\sqrt{3x+3}+\sqrt{6x-1}=\sqrt{33x-2}$.

SOLUTION Equate the square of both members of the equation.

$$3x+3+2\sqrt{(3x+3)(6x-1)}+6x-1=33x-2$$

Simplify: $2\sqrt{18x^2+15x-3}=24x-4$

Divide both members by 2: $\sqrt{18x^2+15x-3}=12x-2$

Square both members again: $18x^2+15x-3=144x^2-48x+4$

Add $-18x^2-15x+3$ to both members: $0=126x^2-63x+7$

Or: $126x^2-63x+7=0$

Factor: $7(3x-1)(6x-1)=0$

Solve for x: $3x-1=0$ or $6x-1=0$

$x=\frac{1}{3}$ $\qquad$ $x=\frac{1}{6}$

CHECK

$\sqrt{3(\frac{1}{3})+3}+\sqrt{6(\frac{1}{3})-1} \stackrel{?}{=} \sqrt{33(\frac{1}{3})-2}$

$\sqrt{4}+\sqrt{1}=\sqrt{9}$ and $\frac{1}{3}$ *is* a root of the equation.

$\sqrt{3(\frac{1}{6})+3}+\sqrt{6(\frac{1}{6})-1} \stackrel{?}{=} \sqrt{33(\frac{1}{6})-2}$

$\sqrt{\frac{7}{2}}+\sqrt{0}=\sqrt{\frac{7}{2}}$ and $\frac{1}{6}$ *is* a root of the equation.

Exercise 20-4

Solve the following equations.

1. $\sqrt{x-1}=3$
2. $\sqrt{2x+3}=11$
3. $\sqrt{3x+6}=\sqrt{5x-2}$
4. $\sqrt{5x+1}-\sqrt{6x-6}=0$
5. $\sqrt{8x+5}-\sqrt{10x+4}=0$
6. $\sqrt{x^2+3x+4}-\sqrt{-4-3x}=0$
7. $\sqrt{3x+3}=x+1$
8. $x-3=\sqrt{x-3}$
9. $\sqrt{x-7}=7-\sqrt{x}$
10. $\sqrt{3x}+\sqrt{2x+3}=0$
11. $\sqrt{9x+4}+3\sqrt{5x}=2\sqrt{3x+1}$
12. $\sqrt{5x+6}-\sqrt{2x^2+4x-9}=0$
13. $\sqrt{3x^2-8x-7}=2x-5$
14. $\sqrt{3x-3}+\sqrt{2x-8}=\sqrt{2x+1}$
15. $4+\sqrt{x^2-4x-1}=\sqrt{x^2+2x+1}$
16. $\sqrt{8x+7}-\sqrt{x+2}=\sqrt{5x+1}$
17. $\sqrt{3x+7}-1=\sqrt{2x+3}$
18. $\sqrt{x^2+3x+7}+2=\sqrt{x^2-3x+9}$
19. $\sqrt{3x+\sqrt{2x+8}}=4$
20. $\sqrt{\sqrt{3x+1}+\sqrt{x}}=1$
21. $\sqrt{\sqrt{2x+4}+\sqrt{2x}}=1$
22. $\sqrt{\sqrt{2x-5}+\sqrt{2x}}=\sqrt{5}$

20-6 EQUATIONS IN QUADRATIC FORM An equation is said to be in *quadratic form* if it can be written as $au^2+bu+c=0$, where $a \neq 0$ and u is some expression of the unknown.

Example 20-11

$2x^4-3x^2+7=0$ is quadratic in x^2

$x^{2/3}-2x^{1/3}-3=0$ is quadratic in $x^{1/3}$

$4\sqrt{2x+1}+4\sqrt[4]{2x+1}-3=0$ is quadratic in $\sqrt[4]{2x+1}$

Such equations can often be solved by replacing an expression involving the unknown by a single letter. The solutions of such equations are illustrated in the following examples.

Example 20-12 Solve the equation $x^4-5x^2+6=0$.

SOLUTION Let $u=x^2$.

Then: $u^2-5u+6=0$

Factor: $(u-2)(u-3)=0$

Solve for u: $u=2$ or $u=3$

Substitute x^2 for u: $x^2=2$ or $x^2=3$

Solve for x: $x=\pm\sqrt{2}$ or $x=\pm\sqrt{3}$

CHECK

$(\pm\sqrt{2})^4 - 5(\pm\sqrt{2})^2 + 6 \stackrel{?}{=} 0 \qquad (\pm\sqrt{3})^4 - 5(\pm\sqrt{3})^2 + 6 \stackrel{?}{=} 0$

$4 - 10 + 6 \stackrel{?}{=} 0 \qquad 9 - 15 + 6 \stackrel{?}{=} 0$

$0 = 0 \qquad 0 = 0$

Then, $\pm\sqrt{2}$ and $\pm\sqrt{3}$ are roots of the equation.

Example 20-13 Solve the equation $2(2x + 3) - 7\sqrt{2x + 3} + 3 = 0$.

SOLUTION Let $u = \sqrt{2x + 3}$.

Then: $2u^2 - 7u + 3 = 0$

Factor: $(2u - 1)(u - 3) = 0$

Solve for u: $2u - 1 = 0$ or $u - 3 = 0$

$u = \frac{1}{2}$ or $u = 3$

Substitute $\sqrt{2x+3}$ for u: $\sqrt{2x+3} = \frac{1}{2}$ or $\sqrt{2x+3} = 3$

Solve for x: $2x + 3 = \frac{1}{4}$ or $2x + 3 = 9$

$x = -\frac{11}{8}$ or $x = 3$

CHECK

$2(-\frac{11}{4} + \frac{12}{4}) - 7\sqrt{-\frac{11}{4} + \frac{12}{4}} + 3 \stackrel{?}{=} 0$

$\frac{1}{2} - \frac{7}{2} + 3 \stackrel{?}{=} 0$

$0 = 0$

$2(6 + 3) - 7\sqrt{6 + 3} + 3 \stackrel{?}{=} 0$

$18 - 21 + 3 \stackrel{?}{=} 0$

$0 = 0$

Hence, $-\frac{11}{8}$ and 3 are roots of the equation.

Example 20-14 Solve the equation $y^{-6} + 7y^{-3} - 8 = 0$.

SOLUTION Let $u = y^{-3}$.

Then: $u^2 + 7u - 8 = 0$

Factor: $(u + 8)(u - 1) = 0$

Equate factors to zero: $u + 8 = 0$ or $u - 1 = 0$

Substitute y^{-3} for u: $y^{-3} + 8 = 0$ or $y^{-3} - 1 = 0$

Solve for y: $1 + 8y^3 = 0$ or $1 - y^3 = 0$ or $y^3 - 1 = 0$

$(2y + 1)(4y^2 - 2y + 1) = 0$ or $(y - 1)(y^2 + y + 1) = 0$

$y = -\frac{1}{2};\ \frac{1}{4}(1 \pm j\sqrt{3})$ or $y = 1;\ \frac{1}{2}(-1 \pm j\sqrt{3})$

It can be shown that each of the six numbers above satisfy the original equation and are the required roots.

Example 20-15 Solve the equation $(x^2 - x)^2 + 2(x^2 - x) - 8 = 0$.

SOLUTION Let $u = (x^2 - x)$.

Then: $u^2 + 2u - 8 = 0$

Factor: $(u + 4)(u - 2) = 0$

Equate each factor to zero: $u + 4 = 0$ or $u - 2 = 0$

Replace u by $x^2 - x$: $x^2 - x + 4 = 0$ or $x^2 - x - 2 = 0$

Solve for x: $x = \dfrac{1 \pm \sqrt{1 - 16}}{2}$ $\qquad (x - 2)(x + 1) = 0$

$= \frac{1}{2} \pm \dfrac{j\sqrt{15}}{2}$ $\qquad x = -1, 2$

The four numbers above can be checked to show that they are the set of roots which satisfy the original equation.

Exercise 20-5

Solve the following equations.

1. $x - 4\sqrt{x} - 21 = 0$
2. $x - \sqrt{x} - 12 = 0$
3. $x^4 - 2x^2 - 15 = 0$
4. $x^4 - 11x^2 + 24 = 0$
5. $x^{-2} + x^{-1} - 20 = 0$
6. $x^{-2} + 7x^{-1} - 8 = 0$
7. $x^{-4} - 12x^{-2} + 32 = 0$
8. $x^{-4} - 10x^{-2} - 24 = 0$
9. $x^{2/3} - 6x^{1/3} + 8 = 0$
10. $x^{2/3} + 6x^{1/3} + 5 = 0$
11. $x^{1/2} - x^{1/4} - 6 = 0$
12. $x^{1/2} - 3x^{1/4} - 4 = 0$
13. $x^{-6} - 64 = 0$
14. $x^{-6} - 7x^{-3} - 8 = 0$
15. $(x^2 - 2)^4 - 5(x^2 - 2)^2 + 4 = 0$
16. $4(x^2 + 1)^4 + 15(x^2 + 1)^2 - 4 = 0$
17. $(x^2 - 2x)^2 - 2(x^2 - 2x) = 3$
18. $(x^2 - 5x)^2 + x^2 - 5x - 12 = 0$
19. $(x^2 - 2)^2 + 14 = 9(x^2 - 2)$
20. $(x^2 - x)^2 - 8(x^2 - x) + 12 = 0$
21. $(2x^2 - x)^2 + 60 = 16(2x^2 - x)$
22. $\left(\dfrac{x + 2}{x}\right)^2 - \left(\dfrac{x + 2}{x}\right) - 2 = 0$
23. $2\left(\dfrac{2x}{x + 1}\right)^2 - 5\left(\dfrac{2x}{x + 1}\right) = 3$
24. $12\left(\dfrac{x + 1}{x - 1}\right)^2 + 7\left(\dfrac{x + 1}{x - 1}\right) - 12 = 0$

25. $2x^2 - x + \sqrt{2x^2 - x - 3} = 5$

26. $x^2 + 2x + 3 - 2\sqrt{x^2 + 2x + 6} = 0$

20-7 APPLIED PROBLEMS The conditions of many applied problems are such that the conditions may be expressed by quadratic equations in one unknown. Each quadratic equation has two solutions. Both solutions may have a reasonable meaning when applied to a stated problem, but more often only one of the solutions has any real significance. Thus, in solving applied problems, we must reject any solutions which have absurd meanings.

Example 20-16 A boat traveling at full throttle travels 24 mi down a river and back again in 6 hr. If the rate of the river flow is 3 mph, find the rate the boat would travel at full throttle in still water.

SOLUTION Let $r =$ rate of boat in still water in mph. Then $r + 3 =$ rate of boat downstream in mph and $r - 3 =$ rate of boat upstream in mph. Since

$$\text{time} = \frac{\text{distance}}{\text{rate}}$$

$$\frac{24}{r+3} = \text{time in hours to travel 24 mi downstream}$$

and $$\frac{24}{r-3} = \text{time in hours to travel 24 mi upstream}$$

Then $$\frac{24}{r+3} + \frac{24}{r-3} = 6$$

$$24(r - 3) + 24(r + 3) = 6(r^2 - 9)$$

$$24r - 72 + 24r + 72 = 6r^2 - 54$$

$$6r^2 - 48r - 54 = 0$$

$$6(r + 1)(r - 9) = 0$$

$$r = -1, 9$$

The boat would travel 9 mph in still water.

Example 20-17 *ABCD* is a square. $AB = 20$ ft. What must be the width of x which will cause the shaded area to be doubled?

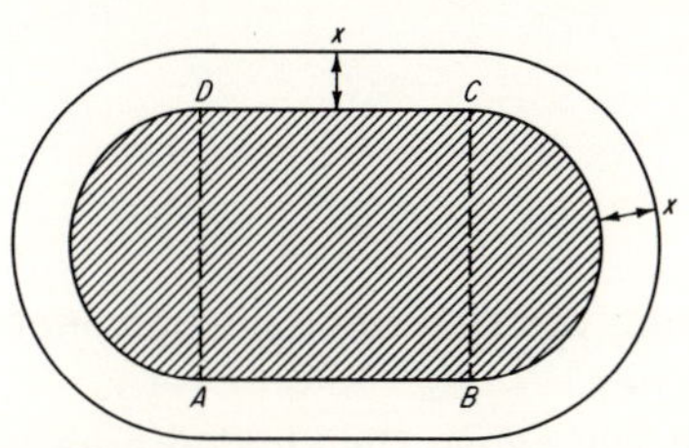

EXAMPLE 20-17.

SOLUTION

$\pi(10 + x)^2 + 20(20 + 2x) = 2[\pi(10)^2 + 20(20)]$

$100\pi + 20\pi x + \pi x^2 + 400 + 40x = 200\pi + 800$

$\pi x^2 + 20(\pi + 2)x - 100\pi - 400 = 0$

$$x = \frac{-20(\pi + 2) \pm \sqrt{400(\pi^2 + 4\pi + 4) - 4\pi(-100\pi - 400)}}{2\pi}$$

$$= \frac{-10\pi - 20 \pm 10\sqrt{2\pi^2 + 8\pi + 4}}{\pi}$$

≈ 5.90 ft

Exercise 20-6

1. The width of a rectangle is 12 ft shorter than its length. The area of the rectangle is 448 ft^2. Find the dimensions of the rectangle.
2. A 96-ft-high tree broke during a windstorm so that the top of the tree was 48 ft from the foot of the tree. What was the height of the part of the tree left standing?

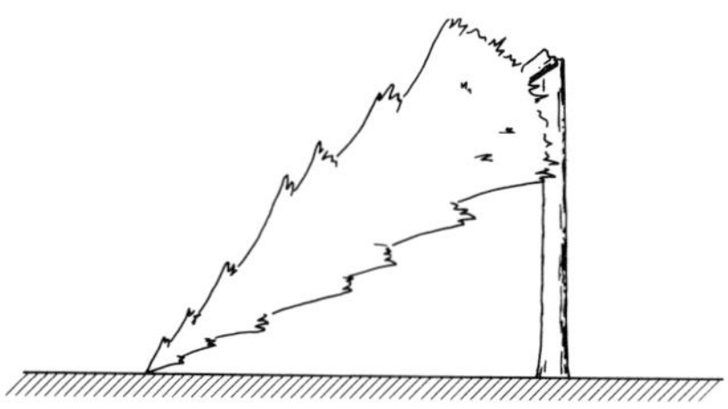

PROB. 2.

3. The cross-sectional area of the L-beam is $41\frac{15}{16}$ in.^{2}. Find the thickness x.

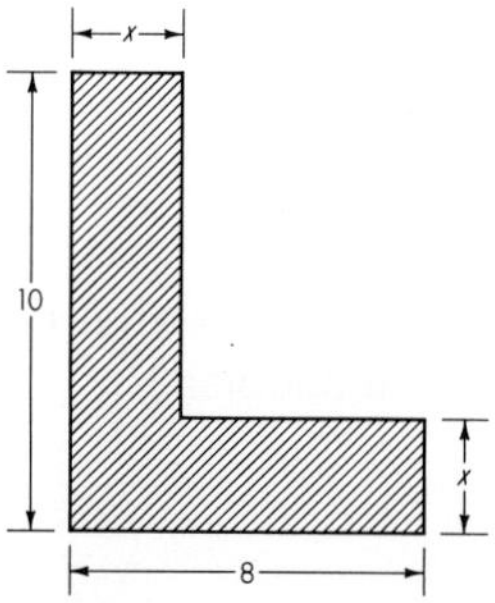

PROB. 3.

4. Find the radius of the sphere whose volume in cubic inches equals the surface area in square inches.
5. Find the value of x which will cause the shaded area to be doubled.

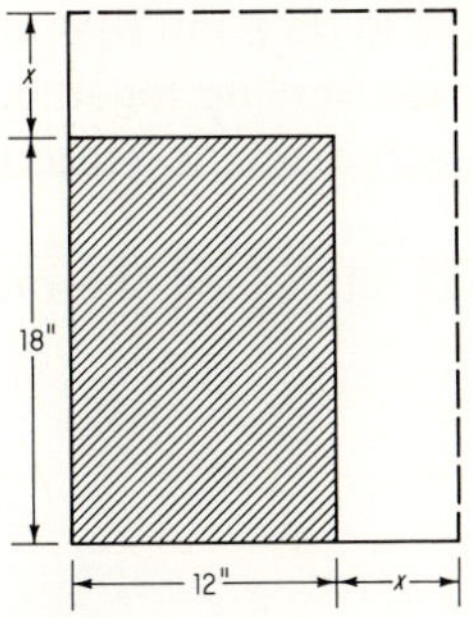

PROB. 5.

6. A bin is to be built to have a volume of 1,200 ft^3. It must be 8 ft deep and 5 ft longer than it is wide. What must be the dimensions of the bin?
7. Circle O is tangent to circle O'. Both circles are tangent to $\overrightarrow{BA}$ and $\overrightarrow{BC}$. Find x.

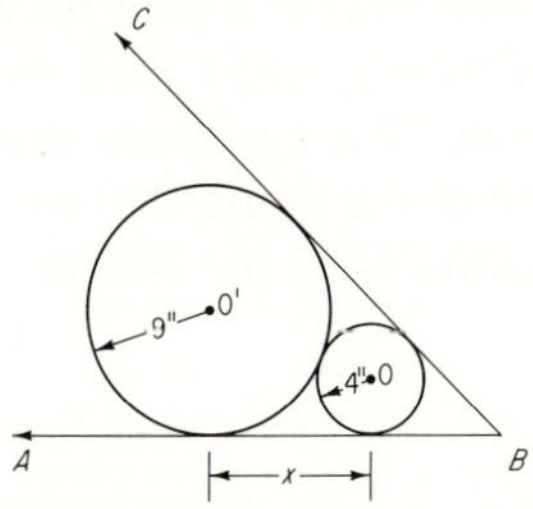

PROB. 7.

8. Find the dimensions of the rectangle whose area is 945 ft^2 and whose perimeter is 124 ft.
9. The altitude of a triangle is 6 in. longer than its base and the area is 108 $in.^2$. Find the lengths of the altitude and the base.
10. The width of a rectangle is $\frac{2}{3}$ of its length. If the dimensions are reduced 6 ft, the area of the rectangle is diminished 324 ft^2. Find the original dimensions of the rectangle.
11. *Given:* $ABCD$ is a trapezoid; $AB = 45$ in.; $BC = 13$ in.; $CD = 24$ in.; $AD = 20$ in. *Find:* The altitude h.

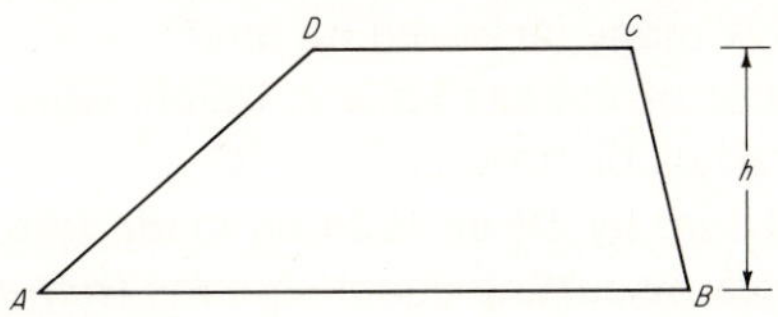

PROB. 11.

12. It takes machine B 3 hr longer to do a job than machine A requires to do the same job. Both machines working together can finish the job in 3 hr 36 min. How long will each machine require to do the job by itself?
13. Circle O is tangent to circle O' at T. Find the radius r. (*Hint:* $\overline{O'O}$ will pass through T.)

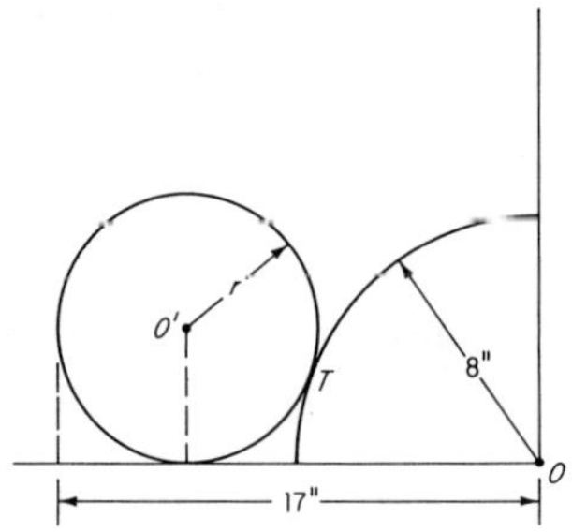

PROB. 13.

14. A group of fishermen decided to rent a boat for \$160 a day. However, when it came time to go fishing, eight men did not show up. Consequently, it cost each man \$1 a day more than they had originally planned. What was the size of the group that went fishing?
15. *Given:* Area of rectangle $OABC = 30$ in.^{2} Radius $OD = 10$ in. *Find:* OA and AB.

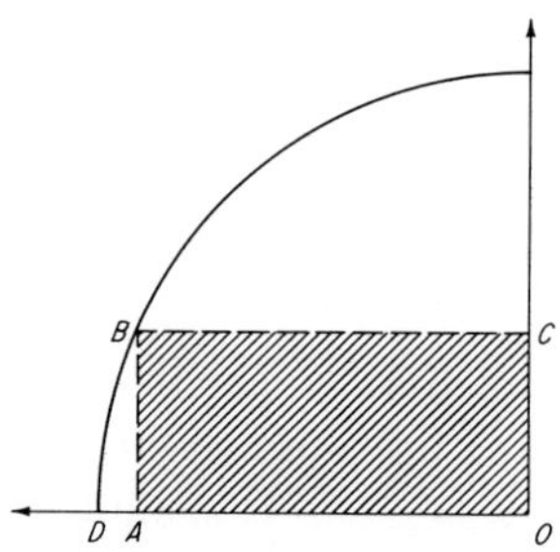

PROB. 15.

16. A retailer paid \$400 for a number of hydraulic jacks. He sold them for a profit of \$2 each. He found that when he had 10 jacks left he had collected from his sales as much as he had invested in the purchase of the jacks. How many jacks did he buy?
17. What is the radius of the gas balloon which, when $\frac{9{,}500}{3}\pi$ ft^3 of gas escapes, has its radius diminished by 5 ft?
18. A piece of tin 12 in. by 18 in. is to be made into an open box with a base 112 in.^{2} area by cutting equal squares from the four corners and then bending up the edges. What will be the volume of the box?

19. Circle O'' is resting on congruent circles O and O'. What is the radius of circle O''?

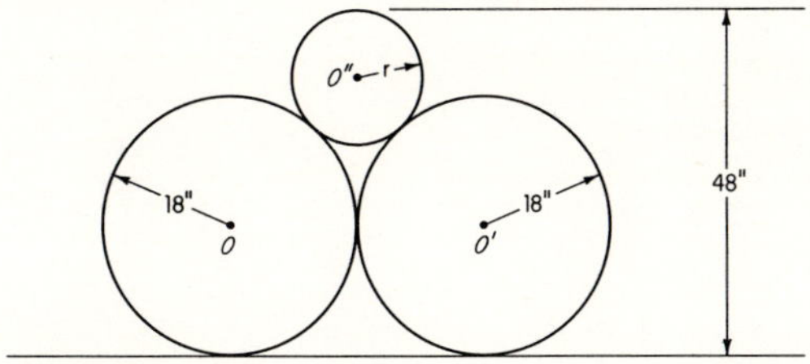

PROB. 19.

20. Find the value of x.

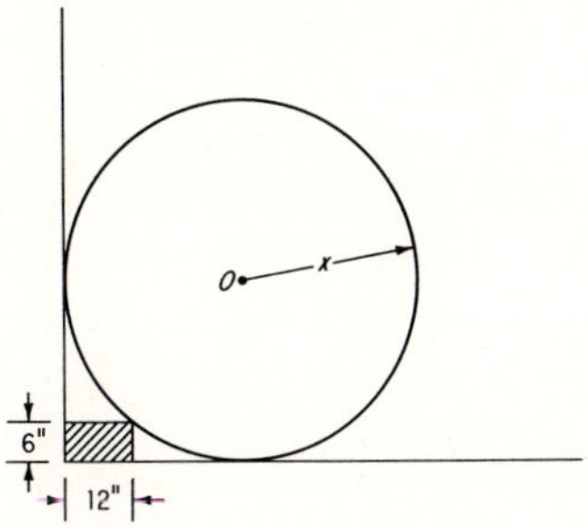

PROB. 20.

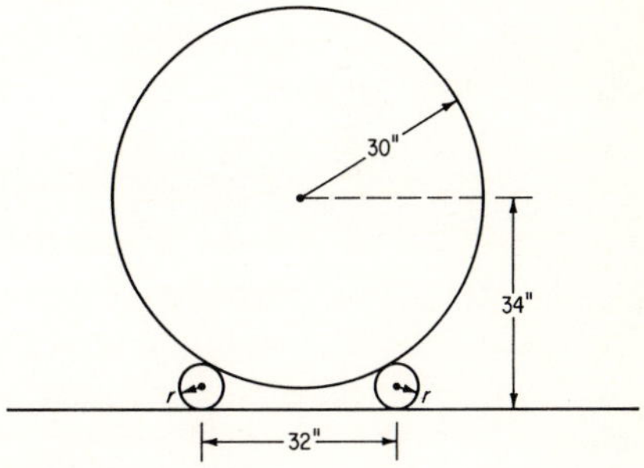

PROB. 21.

21. Find the value of r (the length of the radius of the small $\cong$ circles).

CHAPTER 21

BAR, CIRCLE, AND LINE GRAPHS

21-1 PURPOSES OF GRAPHS A graph is a type of picture that shows the relationship existing between quantities. Graphs serve many purposes. A graph usually provides a quick way of "seeing" the relationship which may exist between sets of numbers, and for this reason graphs are used extensively in the presentation of physical principles and data of technology. Laboratory work frequently calls for the preparation or interpretation of graphs. We shall discuss the bar, circle, and line graphs in this chapter.

21-2 TYPES OF GRAPHS As a simple form of *bar graph* consider Fig. 21-1, which portrays the 1972 oil reserves of the leading oil countries of the

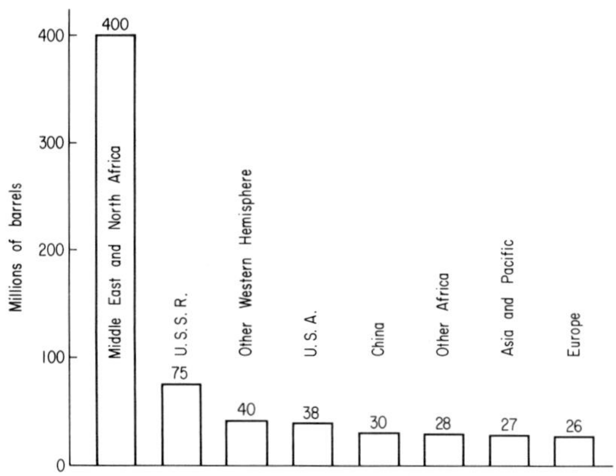

FIG. 21-1. Bar graph showing oil reserves in 1972.

TABLE 21-1 Enrollment in Federally Aided Vocational Classes in 1970 (Data for Fig. 21-2)

ITEM	NUMBER	PERCENT	CENTRAL ANGLE
Agriculture	853,000	10	36
Trades and industry	1,906,000	23	83
Home economics	2,570,000	31	112
Distributive occupations	529,000	6	21
Health occupations	198,000	2	7
Technical education	272,000	3	11
Office occupations	2,112,000	25	90
	8,440,000	100	360

world. The bar graph is used for showing the relative magnitudes of discrete data.

The *circle graph* is used for showing the relative sizes or magnitudes of various parts of a whole category. The circle graph is illustrated in Fig. 21-2. Table 21-1 shows the information needed to construct the graph.

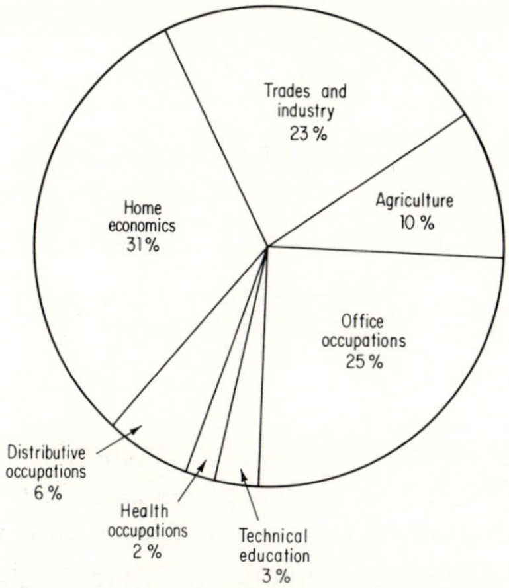

FIG. 21-2. Circle graph showing the number of federally aided vocational and technical students in 1970 in U.S.

As an example, the percentage of students enrolled in the trades and industry classes was found as follows:

$$\frac{1,906,000}{8,440,000} \times 100 = 23 \text{ percent}$$

The size of the sector for the trades and industry was found by taking 23 percent of 360°:

$$\frac{23}{100} \times 360° = 83°$$

Other percentages and sector sizes were found in the same manner.

A short study of the *broken-line graph* of Fig. 21-3 should give a clear picture of the average monthly rainfall for the years 1931–1960 inclusive at Fresno, California. Peak months and dry months are clearly shown. Many hundreds of words would be necessary to present the rainfall record as clearly and as simply as the graph does.

Figure 21-4 shows a line graph that has been "smoothed" so that a flowing curve results. In this figure the maximum output in horsepower is plotted against the speed in revolutions per minute (rpm) at which a certain auto engine is tested. The general shape of the curve is typical of all

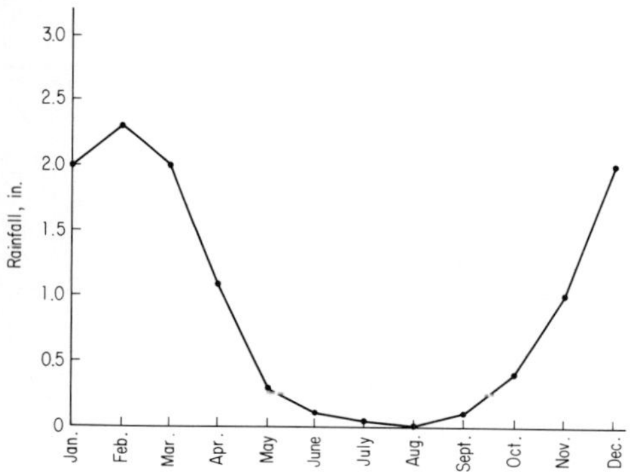

FIG. 21-3. Broken-line graph showing average rainfall at Fresno, California, for 1931–1960 inclusive.

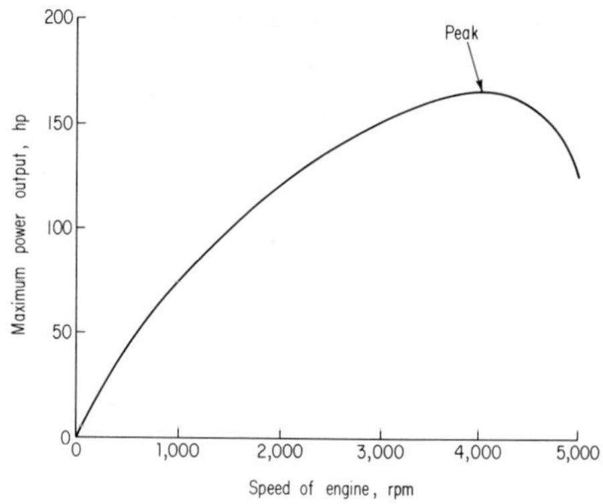

FIG. 21-4. Power-output–engine speed curve for a test automobile engine.

engines of the same model when run under identical conditions. Note that this engine "peaks" at 4,000 rpm and that power output drops off sharply thereafter. In general, when plotting line graphs for physical data, the quantity whose variation is of the most immediate interest is usually plotted along the vertical axis, since its "ups and downs" are thus made more evident. Every graph should have a clear, well-phrased title and each axis must be carefully labeled so there will be no doubt about the units or quantities being plotted.

Exercise 21-1

In Probs. 1 to 5, make a bar graph of the given data.

1. The world electric generating capacity as of Jan. 1, 1969 in millions of kilowatts is given in the following table.

USA	311	Canada	36
USSR	143	France	34
United Kingdom	60	Italy	30
Japan	53	Mainland China	16
West Germany	47	India	14

2. The highest altitudes in thousands of feet for each of the continents are:

Asia	Mount Everest, Nepal-Tibet	29
South America	Mount Aconcagua, Argentina	23
North America	Mount McKinley, Alaska	20
Africa	Kilimanjaro, Tanzania	19
Europe	Mount Elbrus, USSR	19
Antarctica	Vinson Massif	17
Australia	Mount Kosciusko, South Wales	7

3. According to the U.S. Department of Labor, Bureau of Labor Statistics, the purchasing power of the U.S. dollar (using 1967 = $100 as a standard) was:

1940	$2.38
1945	1.92
1950	1.39
1955	1.25
1960	1.13
1965	1.06
1970	0.86

4. The per capita public debt of the United States since 1870 follows:

1870	$ 61.16
1880	41.60
1890	17.80
1900	16.60
1910	12.41
1920	228.23
1930	131.51
1940	367.48
1950	1,696.67
1960	1,584.70
1970	1,923.12

5. The average rise and fall of tides in various cities in the United States are given on page 352. *Source:* National Ocean Survey.

Baltimore, MD	1 ft 1 in.	New York, NY	4 ft 6 in.
Boston, MA	9 ft 6 in.	Philadelphia, PA	5 ft 11 in.
Colon, Panama Canal Zone	1 ft 1 in.	Portland, ME	9 ft 0 in.
Eastport, ME	18 ft 2 in.	San Francisco, CA	4 ft 0 in.
Galveston, TX	1 ft 4 in.	Seattle, WA	7 ft 7 in.
New London, CT	2 ft 7 in.	Tampa, FL	2 ft 10 in.

In Probs. 6 to 9, make a circle graph of the given data. Accompany the graphs with a table similar to Table 21-1.

6. It takes exactly 62 lb of tin, 15 lb of zinc, 11 lb of aluminum, 8.3 lb of lead, 2.5 lb of copper and 1.2 lb of antimony to make 100 lb of sterling aluminum solder.
7. The population of the world in 1971 by continents was as follows:

Africa	344,000,000	Asia (excluding USSR)	2,056,000,000
N. America	321,000,000	Europe (excluding USSR)	462,000,000
S. America	190,000,000	USSR	243,000,000
Oceania	19,000,000		

8. In a recent year the federal government estimated that the source of each budgeted dollar would be:

Individual income taxes	46¢
Corporation income taxes	19¢
Social insurance taxes	23¢
Excise taxes	8¢
Other	4¢

9. In a recent year the federal government proposed to spend each dollar of the federal budget as follows:

National defense	41¢
Social insurance trust fund	22¢
Education and other major social programs	12¢
Veterans	4¢
Interest	6¢
Debt reduction	2¢
International	2¢
Other	11¢

In Probs. 10 to 13, construct broken-line graphs for the given data.

10. The average monthly temperature in degrees Fahrenheit for 1931–1960 for Denver, Colorado is given below:

Jan.	29	July	73
Feb.	32	Aug.	72
Mar.	36	Sept.	63
Apr.	46	Oct.	51
May	56	Nov.	38
June	67	Dec.	32

11. The average monthly rainfall in inches at Memphis, Tennessee over a period of thirty years is given below:

Jan.	6.1	July	3.5
Feb.	4.7	Aug.	3.0
Mar.	5.1	Sept.	2.8
Apr.	4.6	Oct.	2.7
May	4.2	Nov.	4.4
June	3.7	Dec.	4.9

Which appears to be the driest month in Memphis?

12. The price of a share of stock in dollars for the EBS corporation for each month of the year is given below:

Jan.	$15\frac{1}{8}$	July	$22\frac{3}{4}$
Feb.	$18\frac{3}{8}$	Aug.	$20\frac{5}{8}$
Mar.	$19\frac{1}{4}$	Sept.	$27\frac{1}{4}$
Apr.	20	Oct.	$25\frac{3}{4}$
May	$24\frac{7}{8}$	Nov.	$22\frac{3}{8}$
June	$21\frac{5}{8}$	Dec.	$24\frac{1}{4}$

13. The refinery runs of crude oil in thousands of barrels a day are given for both the Western Hemisphere and the Eastern Hemisphere. Plot both sets of data on the same graph. Label your graphs. What do you see as a trend in the relative runs? *Source:* Standard Oil Co. of California.

YEAR	WESTERN HEMISPHERE	EASTERN HEMISPHERE
1963	596	391
1964	698	425
1965	743	464
1966	780	511
1967	774	580
1968	805	650
1969	805	755
1970	849	892
1971	952	947

14. Stopping distances (including reaction time) on dry pavement at various speeds is shown in the following table. Construct a "smooth line" graph for the data. From the graph, determine the stopping distance for a speed of 45 mph. (Data is from General Motors Corp.)

SPEED, MPH	STOPPING DISTANCE, FT
20	40
30	77
40	128
50	190
60	263
70	346

15. The following data was developed experimentally in a physics laboratory. It relates the distances a body fell from a rest position and the time the body fell. From a smooth line graph, determine approximately how long it will take the body to fall 100 meters.

ELAPSED TIME, SEC	DISTANCE FALLEN, METERS
0	0
1	4.9
2	19.6
3	44.1
4	78.4
5	122.5

16. The following results were obtained by a student in a physics laboratory experiment in which he measured the pressures and volumes of a given quantity of air. Draw a smooth line graph of the data.

VOLUME, IN.^{3}	400	340	300	260	240	220	200	185
PRESSURE, LB/IN.^{2}	30	35	40	45	50	55	60	65

17. From the following data draw a smooth curve relating the density of water and the temperature of the water. At what temperature does water have its greatest density? (Let the vertical scale range from 0.99700 to 1.00000 gm/cm^3.)

TEMPERATURE, °C	DENSITY, GM/CM^3
0	0.99987
2	0.99997
4	1.0000
6	0.99997
10	0.99988
15	0.99912
20	0.99823
25	0.99705

18. Draw a smooth line graph for the population growth of the United States indicated in the table.

YEAR	U.S. POPULATION, MILLIONS	YEAR	U.S. POPULATION, MILLIONS
1790	3.9	1890	62.9
1800	5.3	1900	76.0
1810	7.2	1910	92.0
1820	9.6	1920	105.7
1830	12.9	1930	122.8
1840	17.1	1940	131.7
1850	23.2	1950	150.7
1860	31.4	1960	179.3
1870	38.6	1970	203.2
1880	50.2		

21-3 RECTANGULAR (OR CARTESIAN) COORDINATES The extension of the relationship between algebra and geometry is made possible by locating points in a plane or in space by sets of ordered pairs or triplets of numbers called rectangular (or cartesian) coordinates. To locate a point in a plane, two perpendicular lines $X'X$ and $Y'Y$ intersecting at 0 (see Fig. 21-5) are used.

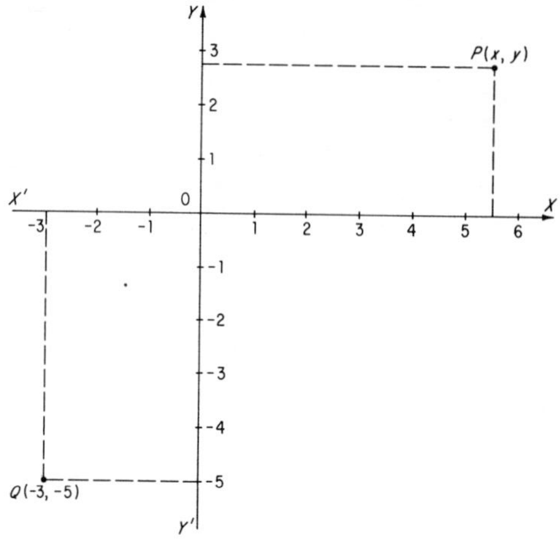

FIG. 21-5.

These lines are called the *coordinate axes.* It is customary to draw line $X'X$, called the x axis, horizontal, with displacements to the right of 0 matched with positive real numbers, and displacements to the left of 0 matched with negative real numbers. Line $Y'Y$, called the y axis, is vertical, with displacements above 0 matched with positive real numbers, and displacements below 0 matched with negative real numbers. The point 0 is called the *origin.* The position of a point P in the plane of the axes is determined if its distance from the y axis (parallel to $\overleftrightarrow{X'X}$) and its distance from the x axis (parallel to $\overleftrightarrow{Y'Y}$) are known. The distance of a point from the y axis is called the *x distance* or *abscissa* of the point. The distance of the point from the x axis is called the *y distance* or *ordinate* of the point. The abscissa and the ordinate taken together in that order are called the *rectangular* or *cartesian coordinates* (or simply *coordinates*) of P.

The coordinates of a point are an ordered pair of real numbers in which the x coordinate is the first number of the pair and the y coordinate is the second. The coordinates of point Q in Fig. 21-5 are $(-3, -5)$. If the coordinates are variable or unknown, the point's coordinates are written (x, y) or $P(x,y)$. It should be clear that (2,3) and (3,2) are different ordered pairs. It is true that $(a,b) = (c,d)$ if, and only if, $a = c$ and $b = d$.

The coordinates of a point depend on the choice of the scale used along with the coordinate axes. If you halve the length of a unit segment in a given coordinate system, you double the coordinates of each point in the system. If you double the length of the unit segment, the coordinate of each point will be halved. There is exactly one pair of real numbers assigned to each point in a given system. Conversely, if (a,b) is any ordered pair of real numbers, there is exactly one point P in a given system which has (a,b) as its coordinates.

21-4 QUADRANTS The rectangular coordinate axes divide a plane into four regions, called quadrants. These quadrants are numbered I, II, III, IV, as shown in Fig. 21-6. The *first quadrant* or *quadrant I* is the set of all points

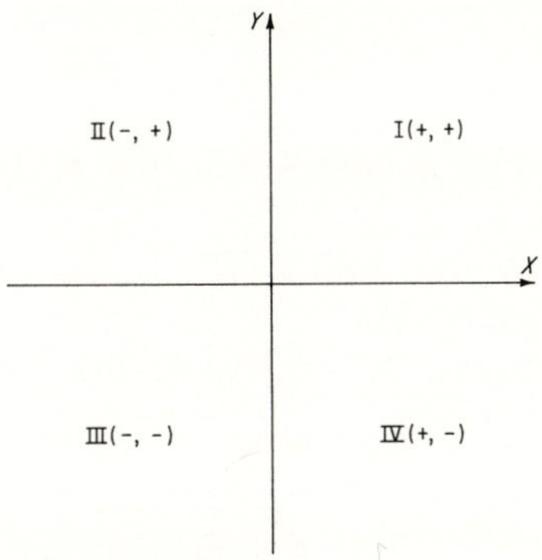

FIG. 21-6. Quadrants.

whose x and y coordinates are both positive. The *second quadrant* or *quadrant II* is the set of all points whose x coordinate is negative and whose y coordinate is positive. The *third quadrant* or *quadrant III* is the set of all points whose x and y coordinates are both negative. The *fourth quadrant* or *quadrant IV* is the set of all points whose x coordinate is positive and whose y coordinate is negative.

The process of locating a point when its coordinates are given is called *plotting* or *graphing.* Plotting is facilitated by the use of coordinate or graph paper whose parallel lines divide the paper into small squares.

Example 21-1 Plot the points $A(-2,2)$, $B(3,-1\frac{1}{2})$, $C(-3,-2\frac{1}{2})$.

SOLUTION (See figure, page 358.)

Example 21-2 Draw a line segment with end points $A(3,5)$ and $B(-2,-4)$.

SOLUTION (See figure, page 358.)

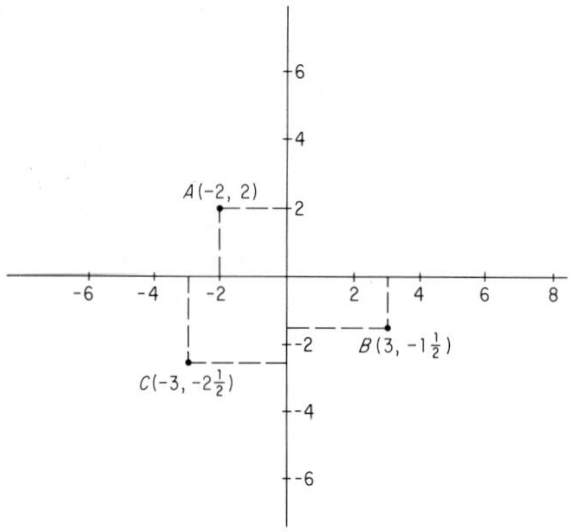

EXAMPLE 21-1.

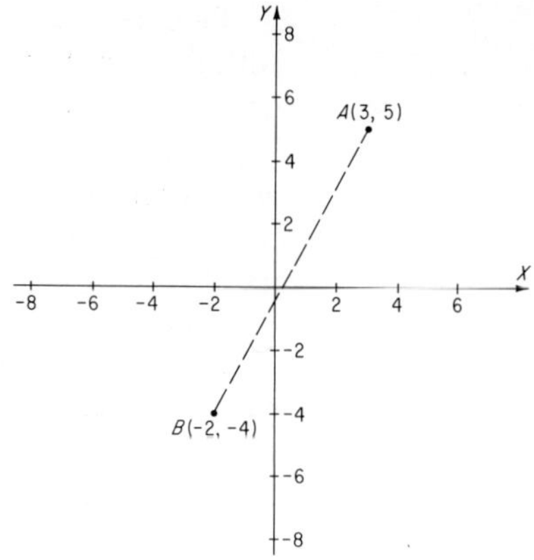

EXAMPLE 21-2.

Exercise 21-2

1. Plot each of the following points: $A(3,0)$, $B(0,4)$, $C(-2,0)$, $D(0,-1)$, $E(-3,5)$, $F(-\frac{1}{2},4\frac{1}{2})$, $G(5,\pi)$.
2. Plot each of the following points: $R(-1,-1)$, $S(0,-4)$, $T(2,-3)$, $U(\sqrt{2},-3)$, $V(\sqrt{3},0)$, $W(2\frac{1}{2},-3\frac{1}{2})$.
3. In the accompanying figure, $OU = 1$. Name the coordinates of the points shown:

	A	B	C	D	E	P_1	P_2	P_3	P_4
x coordinate									
y coordinate									

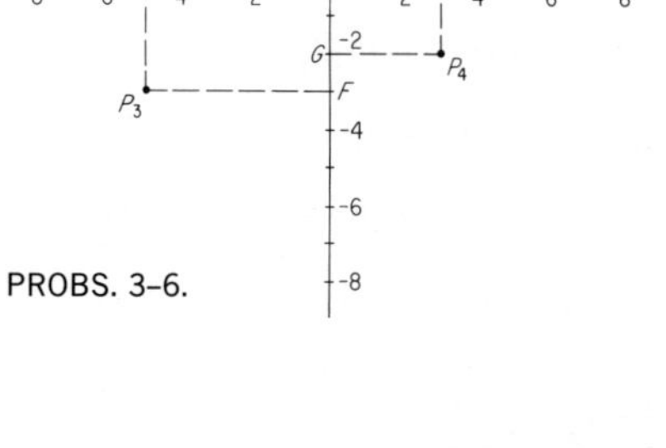

PROBS. 3–6.

4. Using the figure of Prob. 3, but letting $0U = \frac{1}{2}$, complete the following table:

	A	B	C	D	E	P_1	P_2	P_3	P_4
x coordinate									
y coordinate									

5. Using the figure of Prob. 3, but letting $0U = 3$, complete the following table:

	A	B	C	D	E	P_1	P_2	P_3	P_4
x coordinate									
y coordinate									

6. Using the figure for Prob. 3, if $0U = 1$, find: (a) $0A$; (b) UC; (c) BU; (d) $0D$; (e) AB; (f) CD; (g) P_3D; (h) EP_1; (i) HP_2; (j) BP_2; (k) FP_3.
7. Draw $\triangle ABC$ if the vertices are at $A(2,3)$, $B(-4,1)$, $C(2,-3)$.
8. Plot $\triangle STQ$ whose vertices are at $S(-4,-4)$, $T(-2,2)$, $Q(3,-2)$.
9. Describe the set of points whose abscissa always equals 2.
10. Describe the set of points whose ordinate is always -3.
11. Describe the set of points whose abscissa is always numerically equal to its ordinate.
12. Describe the set of points whose abscissa is always numerically equal to the negative of its ordinate.
13. What is the abscissa of a point if it lies on the y axis?
14. What is the ordinate of a point if it lies on the x axis?
15. Plot the set of all points (x,y) for which x and y are integers which satisfy the following conditions:
    (a) $x = 2$, $-1 \leqq y \leqq 4$. (Read "y is greater than or equal to -1 and y is also less than or equal to 4.")
    (b) $y = -3$, $-2 < x < 5$.
    (c) $-2 < x \leqq 0$, $-1 \leqq y < 4$.
    (d) $x = 2$, $y = 3$.
16. Plot the set of points (x,y) which satisfy the following conditions:

(a) $x \geqq 1, y \geqq 3$. (Read "x is greater than or equal to 1 and y is greater than or equal to 3.")
(b) $x \leqq 2, y \leqq 5$.
(c) $x \geqq 4, -5 \leqq y \leqq -2$.

17. Find the length of the line segment from $A(-4,3)$ to $B(2,3)$.
18. Find the length of the line segment from $C(2,-6)$ to $D(-3,-6)$.
19. Find the length of the line segment from $P(x_1,y_1)$ to $Q(x_1,y_2)$.
20. Find the length of the line segment from $R(x_1,y_1)$ to $T(x_2,y_1)$.
21. Find the coordinates of the midpoint of segment AB whose end points are at $A(2,2)$ and $B(7,2)$.
22. Find the coordinates of the midpoint of $\overline{CD}$ (segment CD) whose end points are at $C(-3,4)$ and $D(-3,-6)$.
23. Find the coordinates of the midpoint of $\overline{EF}$ whose end points are at $E(x_1,y_1)$ and $F(x_2,y_1)$.
24. Find the coordinates of the midpoint of $\overline{GH}$ whose end points are at $G(x_1,y_1)$ and $H(x_1,y_2)$.

21-5 THE DISTANCE FORMULA The distance between two points in a plane can be obtained by applying the theorem of Pythagoras. In Fig. 21-7, let

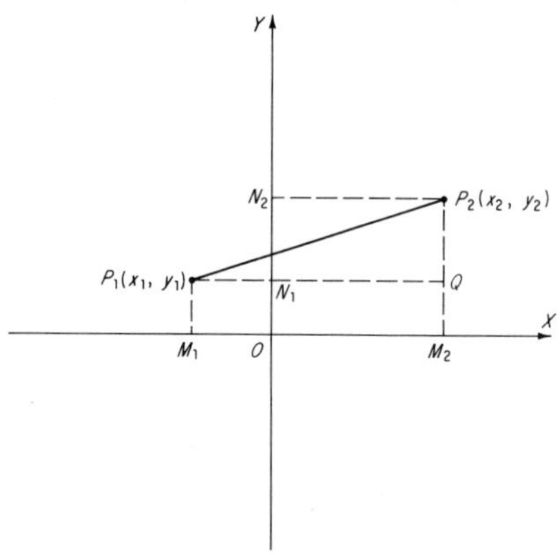

FIG. 21-7.

$P_1(x_1,y_1)$ and $P_2(x_2,y_2)$ be two points such that $\overleftrightarrow{P_1P_2}$ is not parallel to either the x axis or the y axis. Through P_1, draw $\overleftrightarrow{P_1Q}$ parallel to the x axis, and through P_2, draw $\overleftrightarrow{P_2Q}$ parallel to the y axis, thus forming right triangle P_1QP_2. Segment P_1P_2 is the hypotenuse of the triangle. By the Pythagorean theorem, $(P_1P_2)^2 = (P_1Q)^2 + (QP_2)^2$ or $P_1P_2 = \sqrt{(P_1Q)^2 + (QP_2)^2}$. But $P_1Q = M_1M_2 = x_2 - x_1$, and $QP_2 = N_1N_2 = y_2 - y_1$. Using the property of substitution, we get

$$P_1P_2 = \sqrt{(x_2 - x_1)^2 + (y_2 - y_1)^2}$$

Since $(x_2 - x_1)^2 = (x_1 - x_2)^2$ and $(y_2 - y_1)^2 = (y_1 - y_2)^2$, we can state the following formula:

THE DISTANCE FORMULA *For any two points $P_1(x_1,y_1)$ and $P_2(x_2,y_2)$,*

$$P_1P_2 = \sqrt{(x_2 - x_1)^2 + (y_2 - y_1)^2} = \sqrt{(x_1 - x_2)^2 + (y_1 - y_2)^2} \qquad (21\text{-}1)$$

The student should convince himself that the distance formula will also work if the segment is perpendicular to either axis.

Example 21-3 Find the length of the line segment whose end points are at $A(2,-5)$ and $B(-3,-7)$.

SOLUTION Let A correspond to P_1 and B to P_2 in formula given in Eq. (21-1). Then

$$\begin{aligned} AB &= \sqrt{[2-(-3)]^2 + [(-5)-(-7)]^2} \\ &= \sqrt{25+4} \\ &= \sqrt{29} \end{aligned}$$

21-6 MIDPOINT OF A SEGMENT Quite frequently in the study of mathematics it is necessary to determine the coordinates of the midpoint of a line segment. In Fig. 21-8, let $M(x_M,y_M)$ be the midpoint of the line segment

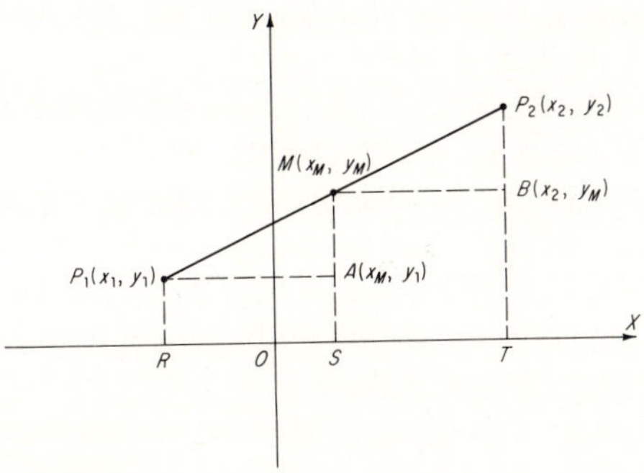

FIG. 21-8.

joining $P_1(x_1,y_1)$ and $P_2(x_2,y_2)$. If through P_1, M, and P_2 we draw lines parallel to the coordinate axes, two congruent right triangles will be formed. (Can you prove them congruent?) We know that $P_1A = MB$. But $P_1A = x_M - x_1$, and $MB = x_2 - x_M$. Therefore,

$$x_M - x_1 = x_2 - x_M$$

Adding $x_M + x_1$ to both members,

$$x_M + x_M = x_1 + x_2$$

Then dividing by 2, $x_M = \frac{1}{2}(x_1 + x_2)$. Similarly, by setting $AM = BP_2$, we get $y_M = \frac{1}{2}(y_1 + y_2)$.

THE MIDPOINT FORMULA *M is the midpoint of $\overline{P_1P_2}$ if, and only if,*

$$x_M = \tfrac{1}{2}(x_1 + x_2) \text{ and } y_M = \tfrac{1}{2}(y_1 + y_2) \qquad (21\text{-}2)$$

Example 21-4 Find the length of the median from vertex B of $\triangle ABC$ with the following vertices: $A(3,5)$, $B(-2,6)$, $C(2,-7)$.

SOLUTION

$$x_M = \frac{x_A + x_C}{2} = \frac{3+2}{2} = \frac{5}{2}$$

$$y_M = \frac{y_A + y_C}{2} = \frac{5 + (-7)}{2} = -1$$

$$\begin{aligned} BM &= \sqrt{(x_M - x_B)^2 + (y_M - y_B)^2} \\ &= \sqrt{[\tfrac{5}{2} - (-2)]^2 + [-1 - (6)]^2} \\ &= \sqrt{\tfrac{81}{4} + 49} \\ &= \tfrac{1}{2}\sqrt{277} \end{aligned}$$

Exercise 21-3

1. Given points $O(0,0)$, $A(2,3)$, $B(4,-2)$, $C(5,0)$, $D(0,-8)$, $E(-1,-1)$. Determine the following lengths: (a) OA, (b) BO, (c) AB, (d) BC, (e) CD, (f) DE, (g) AC, (h) BD, (i) BE.
2. Plot points $R(0,4)$, $S(-3,8)$, $T(1,5)$. Show that $\triangle RST$ is isosceles.
3. What is the perimeter of $\triangle RST$ of Prob. 2?
4. Plot the points $A(-3,-2)$, $B(2,5)$, $C(7,12)$, and prove that they are collinear.
5. Prove that $A(-1,-1)$, $B(5,-1)$, and $C(5,7)$ are vertices of a right triangle. (Assume the converse of the Pythagorean theorem.)
6. Find the area of $\triangle ABC$ of Prob. 5.
7. Prove that $D(-2,2)$, $E(6,7)$, and $F(8,-14)$ are vertices of a right triangle.
8. Find the coordinates of the midpoint of segment AB if $A(4,8)$ and $B(2,6)$ are its end points.
9. Find the coordinates of the midpoint of $\overline{CD}$ if $C(-3,-7)$ and $D(5,3)$ are its end points.
10. Find the coordinates of the midpoints of $\overline{EF}$ if $E(a+b, a-b)$ and $F(-a,b)$, $(a \neq 0)$ are its end points.
11. Prove that $O(3,2)$ is the center of a circle passing through $A(-2,-3)$, $B(4, 9)$, $C(10,3)$.

12. Draw a circle with center at (1,2) and passing through (−9,8). Does the circle pass through (a) $P(7,12)$, (b) $Q(11,-4)$, (c) $R(4,-8)$?
13. Find the coordinates of the midpoint of the hypotenuse of $\triangle DEF$ of Prob. 7. Then show that the midpoint is equidistant from the three vertices of the triangle.
14. Find the lengths of the medians of the triangle whose vertices are at $A(-2,-3)$, $B(6,3)$, and $C(9,-5)$.

CHAPTER 22

SYSTEMS OF EQUATIONS

22-1 LINEAR EQUATIONS IN TWO UNKNOWNS The most general form of an equation of the first degree in x and y is $ax + by + c = 0$, where a and b are not both zero. The graph of a linear equation in two unknowns can be drawn if a set of two ordered pairs which satisfy the equation is known. From the graph, many more ordered pairs which satisfy the equation can then be determined.

Two pairs of numbers commonly used in plotting graphs of linear equations in two unknowns are those indicating where the graph intersects the coordinate axes. The point where the graph of the linear equation $ax + by + c = 0$ intersects the x axis can be determined by letting $y = 0$ in the equation and solving for x. The value $x = -\frac{c}{a}$ is called the x intercept of the graph and $\left(-\frac{c}{a}, 0\right)$ is an ordered pair which satisfies the equation. Likewise, by letting $x = 0$ and solving for y, the y intercept will be found to be $-\frac{c}{b}$ and the graph will be found to pass through the point whose coordinates are the ordered pair $\left(0, -\frac{c}{b}\right)$.

Usually, as a check, a third ordered pair which satisfies the equation is also plotted. This pair can be readily obtained by assigning any other real value to x and solving the equation for the corresponding value of y. The straight line should pass through each of the above determined points.

The graph of every linear equation in x and y is always a straight line. Conversely, every line in a plane is the graph of a linear equation in x and y.

Example 22-1 Plot the graph of the equation $2x - 3y = 18$.

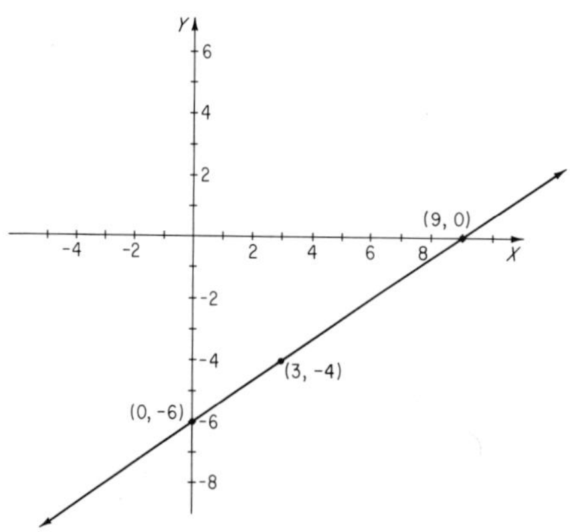

EXAMPLE 22-1.

SOLUTION Let $y = 0$ and solve for x in the equation.

$$2x - 0 = 18$$
$$x = 9 \text{ (the } x \text{ intercept)}$$

Next let $x = 0$ and solve for y.

$$0 - 3y = 18$$
$$y = -6 \text{ (the } y \text{ intercept)}$$

Let $x =$ any other value, say 3, and solve for y.

$$2(3) - 3y = 18$$
$$y = -4$$

Hence, the coordinates of three points which satisfy the equation are (9,0), (0,–6), and (3,–4). We plot these three points and draw straight line AB through them.

22-2 GRAPHIC SOLUTION OF LINEAR EQUATIONS IN TWO UNKNOWNS Often we wish to find the ordered pair that satisfies each of two linear equations in two unknowns. Such equations are called *simultaneous equations.* Any two distinct lines lying in a plane that are not parallel must intersect in one and only one point. The ordered pair which satisfies each of the equations is called the *solution* of the system of equations and can be determined graphically by plotting the two equations on the same set of axes and determining the coordinates of the intersection point.

Example 22-2 Determine graphically the solution to the following system of simultaneous equations.

$$2x - 3y = 12$$
$$3x + \ \ y = \ \ 7$$

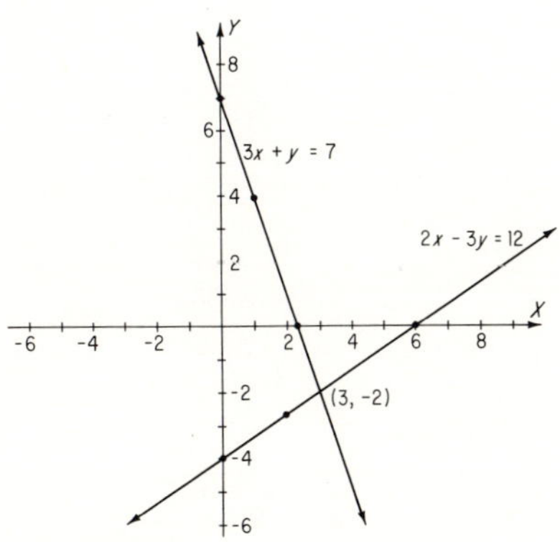

EXAMPLE 22-2.

SOLUTION Prepare for each equation a table of three ordered pairs which satisfy the equation, as follows.

$2x - 3y = 12$

x	0	6	2
y	-4	0	$-\frac{8}{3}$

$3x + y = 7$

x	0	$\frac{7}{3}$	1
y	7	0	4

Plotting the graphs of the two equations, we find the point of intersection. The solution is the ordered pair (3,−2).

When the graphs of two equations intersect in a single point, the equations are said to be consistent and independent. If the set of equations describe two parallel lines, there will be no solution to the simultaneous equations. Such equations are termed *inconsistent* equations. If the graphs of two linear equations are two lines that coincide, the equations are *consistent* and *dependent.*

Example 22-3 Solve the following system of equations graphically:

$3x - 2y = 6$

$4y - 6x = 12$

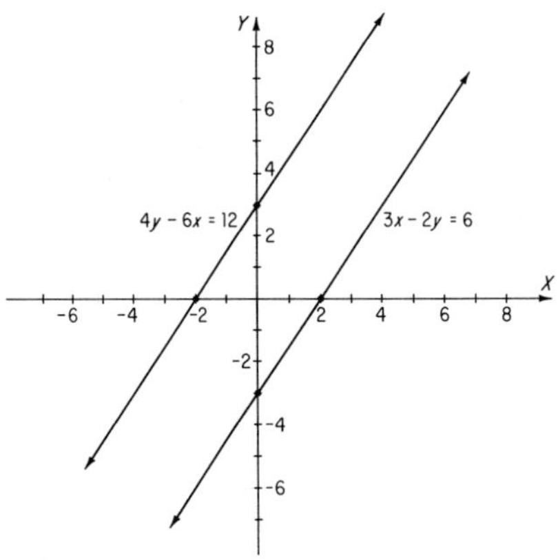

EXAMPLE 22-3.

SOLUTION

$3x - 2y = 6$

x	0	2	1
y	-3	0	$-1\frac{1}{2}$

$4y - 6x = 12$

x	0	-2	2
y	3	0	6

The graphs of the equations are parallel lines and there is no solution to the system. The solution set is a null set.

Example 22-4 Solve graphically the following system of equations:

$4x - 3y = 12$

$6y - 8x = -24$

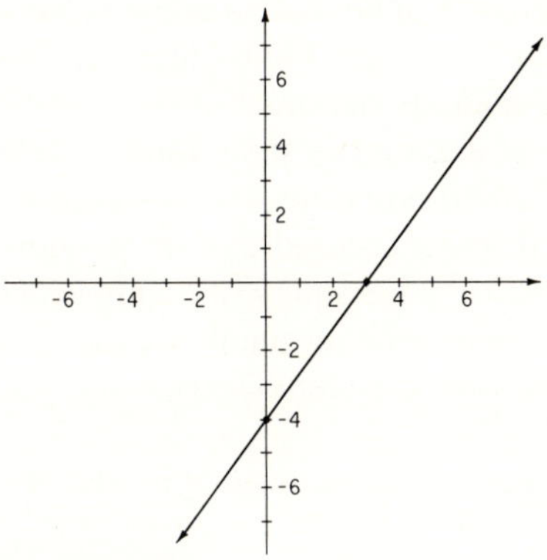

EXAMPLE 22-4.

SOLUTION

$4x - 3y = 12$

x	0	3	−2
y	−4	0	$-6\frac{2}{3}$

$6y - 8x = -24$

x	0	3	−2
y	−4	0	$-6\frac{2}{3}$

The graphs are identical lines and the system has no unique solution. Any ordered pair of numbers which satisfies the one equation will also satisfy the other.

Exercise 22-1

Solve each of the following systems of linear equations graphically:

1. $x + y = 5,\ 2x - y = -2$
2. $2x - y = 0,\ y - x = 3$
3. $2x - 3y = 0,\ 6x + 2y = -22$
4. $x + 5y = 10,\ 2x - 4y = 6$
5. $4x - 3y = -7,\ 2y - x = 3$
6. $5x + 5y = 5,\ 3y - 3x = 15$
7. $3x - 4y = 12,\ 8y - 6x - 24 = 0$
8. $2x = 3y,\ 4x - 6y = 12$
9. $x - 8y = 20,\ 2x + 4y = 0$
10. $5x - 4y = 12,\ 7x - 7y = 21$
11. $6x + 4y = -12,\ 5x + 2y = -10$
12. $2x + 6y = 6,\ x + 3y = 3$
13. $10x - 5y = -10,\ 4y - 2x = -4$
14. $3x + 5y = 0,\ x + y = 1$
15. $3x + y = 8,\ 5x - 3y = -3$
16. $x - 2y = 4,\ 4y - 2x = -8$
17. $3x - 4y = 11,\ 2x + y = 11$
18. $2x = 3y,\ 7y - 5x + 5 = 0$

22-3 ALGEBRAIC SOLUTION OF LINEAR SYSTEMS IN TWO UNKNOWNS The results obtained by the graphic solution of a pair of linear equations in two unknowns are usually only approximate. Often exact solutions are required. Exact solutions of two consistent and independent equations can always be obtained by replacing the given pair of equations by proper equivalent equations which can be combined to yield an equation in just one unknown. The value of the single unknown can be found by methods previously studied. If the value for the one unknown be substituted wherever the unknown appears in either of the original equations, the resulting equation can then be solved for the second unknown.

The student will recall that adding the same term to both members of an equation will yield an equivalent equation. Two equations in two unknowns are said to be *equivalent* if every ordered pair of numbers which satisfies one satisfies the other. Thus, $x - y = 3$ and $x = y + 3$ are equivalent equations.

Another operation which yields an equivalent equation is multiplying each term of the equation by the same constant. Thus, $x - y = 3$ and $2x - 2y = 6$ are equivalent equations.

Two systems of equations are called *equivalent* if each of the systems have identical solutions.

We will now show, by means of examples, how to solve a system of two linear equations in two unknowns by the method of *elimination by addition or subtraction.*

Example 22-5 Solve the following system algebraically:

$$3x - 4y = 13 \qquad (1)$$

$$x + 2y = 1 \qquad (2)$$

SOLUTION

Multiply Eq. (2) by 2: $\quad 2x + 4y = 2 \qquad (3)$

Add corresponding members of Eqs. (1) and (3):

$$\begin{array}{r} 3x - 4y = 13 \\ \hline 5x \quad = 15 \end{array} \qquad (4)$$

Solve Eq. (4) for x: $\quad x = 3$

Substitute $x = 3$ in Eq. (2): $\quad 3 + 2y = 1 \qquad (5)$

Solve Eq. (5) for y: $\quad 2y = -2$

$$y = -1$$

Hence the solution of the given system is $x = 3, y = -1$, also written $(3,-1)$.

CHECK Substitute $x = 3, y = -1$ in Eqs. (1) and (2).

$$3(3) - 4(-1) \stackrel{?}{=} 13 \qquad 3 + 2(-1) \stackrel{?}{=} 1$$

$$9 + 4 \stackrel{?}{=} 13 \qquad 3 - 2 \stackrel{?}{=} 1$$

$$13 = 13 \qquad 1 = 1$$

Example 22-6 Solve the following system algebraically:

$2x + 3y = 7$ (1)

$5x + 4y = 14$ (2)

SOLUTION

Multiply Eq. (1) by 5: $10x + 15y = 35$ (3)

Multiply Eq. (2) by 2: $10x + 8y = 28$ (4)

Subtract Eq. (4) from Eq. (3): $7y = 7$ (5)

Solve Eq. (5) for y: $y = 1$

Substitute $y = 1$ in Eq. (1): $2x + 3(1) = 7$ (6)

Solve Eq. (6) for x: $x = 2$

Hence the solution of the given equation is $x = 2, y = 1$, also written (2,1).

CHECK Substitute $x = 2$, $y = 1$, in Eqs. (1) and (2).

$$2(2) + 3(1) \stackrel{?}{=} 7 \qquad 5(2) + 4(1) \stackrel{?}{=} 14$$
$$4 + 3 \stackrel{?}{=} 7 \qquad 10 + 4 \stackrel{?}{=} 14$$
$$7 = 7 \qquad 14 = 14$$

Example 22-7 Solve the following system algebraically:

$5x - 7y + 11 = 0$ (1)

$2x + 4y = 15$ (2)

SOLUTION

Add -11 to both members of Eq. (1): $5x - 7y = -11$ (3)

Multiply Eq. (3) by 4: $20x - 28y = -44$ (4)

Multiply Eq. (2) by 7: $14x + 28y = 105$ (5)

Add Eqs. (4) and (5): $34x = 61$ (6)

Solve Eq. (6) for x: $x = \frac{61}{34}$

Substitute $x = \frac{61}{34}$ in Eq. (2): $2(\frac{61}{34}) + 4y = 15$ (7)

Solve Eq. (7) for y: $\frac{61}{17} + \frac{68}{17}y = \frac{255}{17}$

$$\frac{68}{17}y = \frac{194}{17}$$
$$y = \frac{194}{17} \cdot \frac{17}{68} = \frac{97}{34}$$

Hence the solution of the system is $x = \frac{61}{34}$, $y = \frac{97}{34}$, also written in the form $(\frac{61}{34}, \frac{97}{34})$.

CHECK Check these values in Eqs. (1) and (2).

$$5(\tfrac{61}{34}) - 7(\tfrac{97}{34}) + 11 \stackrel{?}{=} 0 \qquad 2(\tfrac{61}{34}) + 4(\tfrac{97}{34}) \stackrel{?}{=} 15$$
$$\tfrac{305}{34} - \tfrac{679}{34} + \tfrac{374}{34} \stackrel{?}{=} 0 \qquad \tfrac{61}{17} + \tfrac{194}{17} \stackrel{?}{=} 15$$
$$0 = 0 \qquad 15 = 15$$

A second method of solving systems of equations is the *substitution* method. In this method we first use one of the given equations to express one unknown (say, y) explicitly in terms of the other variable. The resulting expression is substituted for the same unknown in the other equation. This procedure yields a single equation in one unknown, which can usually be solved for the unknown. The remaining steps are the same as that used in the addition or subtraction method.

Example 22-8 Solve the following system algebraically:

$$5x - 7y + 11 = 0 \quad (1)$$

$$2x + 4y = 15 \quad (2)$$

SOLUTION

Solve Eq. (2) for y in terms of x: $$y = \frac{15 - 2x}{4} \quad (3)$$

Substitute Eq. (3) in Eq. (1): $$5x - 7\left(\frac{15 - 2x}{4}\right) + 11 = 0 \quad (4)$$

Solve Eq. (4) for x: $$20x - 105 + 14x + 44 = 0$$

$$34x = 61$$

$$x = \tfrac{61}{34}$$

Substitute $x = \frac{61}{34}$ in Eq. (2): $$2(\tfrac{61}{34}) + 4y = 15 \quad (5)$$

Solve Eq. (5) for y: $$122 + 136y = 510$$

$$136y = 388$$

$$y = \tfrac{388}{136} = \tfrac{97}{34}$$

CHECK Check as in Example 22-7.

Exercise 22-2

Solve each of the following systems of linear equations algebraically and check the results.

1. $2x - y = 5$
   $3x + y = 5$
2. $x - 3y = 7$
   $x - 5y = 13$
3. $3x - 2y = 10$
   $2x + 3y = 11$
4. $x - 2y = 3$
   $2x - 3y = 8$
5. $3x + 4y = 24$
   $5x - 3y = 11$
6. $4x + 7y = 6$
   $6x - 3y = 36$
7. $2x - 3y + 2 = 0$
   $4x - 5y = 0$
8. $3x - 8y = 0$
   $x - 4y + 4 = 0$
9. $6x - 5y = 20$
   $2x - 3y = 8$
10. $7x - 3y = 9$
    $2x - 5y = 44$
11. $3x + y = 5$
    $6x - 7y + 26 = 0$
12. $5x - y = 40$
    $17x - 16y + 116 = 0$

13. $9x - 4y = 5$
$15x + 8y = 1$

14. $15x - 16y = 2$
$9x + 4y + 9 = 0$

15. $12x - 24y = 51$
$16x + 12y + 1 = 3$

16. $9x + 2y = 5$
$y = 3x - \frac{5}{2}$

17. $8x - 15y + 20 = 0$
$3y - 20x = 27$

18. $2x = 3y + 15$
$2x = 7y - 8$

19. $5x = 4y - 11$
$5x = 3y + 7$

20. $3y = 11x - 10$
$3y = 5x + 8$

21. $\frac{x}{5} + \frac{y}{4} = 6$
$\frac{x}{5} - \frac{y}{4} = 0$

22. $\frac{x}{4} + \frac{y}{6} = 7$
$\frac{x}{10} - \frac{y}{12} = 1$

23. $\frac{6}{x} + \frac{9}{y} = 6$
$\frac{4}{x} + \frac{3}{y} = 3$

24. $\frac{15}{x} - \frac{12}{y} = 7$
$\frac{3}{x} + \frac{6}{y} + \frac{7}{5} = 0$

25. $2(x + 4) + 3(y - 1) = 16$
$5(x - 3) - 4(y + 2) = -76$

26. $3(x - 2) - 5(y - 7) + 21 = 0$
$7(x + 5) - 4(y + 2) + 36 = 0$

27. $\frac{1}{2}(x + 6) + \frac{1}{3}(3y + 9) = 5$
$\frac{1}{5}(2x - 8) - \frac{1}{4}(2y - 6) = 3$

28. $\frac{1}{3}(2x - 5) - \frac{1}{5}(4y - 3) + 10 = 0$
$7x + 8y = 21$

22-4 ALGEBRAIC SOLUTION OF LINEAR EQUATIONS IN MORE THAN TWO UNKNOWNS The procedure for finding the algebraic solution of any number of linear equations in the same number of unknowns is an immediate extension of the procedures for solving a system of two linear equations in two unknowns. The procedure is to eliminate one unknown after another.

Let us consider three equations in three unknowns, say, x, y, and z, as representative of this type. If we eliminate the same variable, say, z, from two pairs of the equations, the problem is reduced to a system of two linear equations with the same two variables. These two equations can be solved by the methods of Sec. 22-3. The solution of the two derived equations can be substituted in any of the three original equations to solve for the third unknown.

The above method can be used to solve three or more equations only if there is a unique solution to the system. If there is not a unique solution, the system will probably have an unlimited number of solutions, or the equations will be inconsistent. These types of systems will be recognized as the algebraic solution is carried out.

Example 22-9 Solve the following system:

$$2x - 2y - z = 5 \quad (1)$$
$$3x + y + z = 4 \quad (2)$$
$$x - 3y + 2z = -7 \quad (3)$$

SOLUTION

Add Eqs. (1) and (2):
$$\begin{array}{r} 2x - 2y - z = 5 \\ 3x + y + z = 4 \\ \hline 5x - y \quad = 9 \end{array} \quad (4)$$

Multiply Eq. (1) by 2: $4x - 4y - 2z = 10$ (5)

Add Eqs. (5) and (3):
$$\begin{array}{r} x - 3y + 2z = -7 \\ \hline 5x - 7y \quad = 3 \end{array} \quad (6)$$

Multiply Eq. (4) by 7: $35x - 7y = 63$ (7)

Subtract Eq. (6) from Eq. (7): $30x = 60$ (8)

Solve Eq. (8) for x: $x = 2$

Substitute $x = 2$ in Eq. (4): $5(2) - y = 9$ (9)

Solve Eq. (9) for y: $y = 1$

Substitute $x = 2$, $y = 1$ in Eq. (2): $3(2) + 1 + z = 4$ (10)

Solve Eq. (10) for z: $z = -3$

Hence the solution of the given system is $x = 2$, $y = 1$, $z = -3$, also written $(2,1,-3)$.

CHECK Substitute these values in Eqs. (1), (2), and (3):

$$2(2) - 2(1) - (-3) \stackrel{?}{=} 5$$
$$4 - 2 + 3 \stackrel{?}{=} 5$$
$$5 = 5$$
$$3(2) + 1 + (-3) \stackrel{?}{=} 4$$
$$6 + 1 - 3 \stackrel{?}{=} 4$$
$$4 = 4$$
$$2 - 3(1) + 2(-3) \stackrel{?}{=} 7$$
$$2 - 3 - 6 \stackrel{?}{=} 7$$
$$-7 = -7$$

Example 22-10 Solve the following system of equations:

$$5x - 2y + 2z = 15 \quad (1)$$
$$8x - 6y + 3z = 7 \quad (2)$$
$$3x - 4y + z = 11 \quad (3)$$

SOLUTION

Multiply Eq. (3) by 2: $6x - 8y + 2z = 22$ (4)

Subtract Eq. (1) from Eq. (4): $x - 6y = 7$ (5)

Multiply Eq. (3) by 3: $9x - 12y + 3z = 33$ (6)

Subtract Eq. (2) from Eq. (6): $\underline{8x - 6y + 3z = 7}$

$x - 6y = 26$ (7)

Equations (5) and (6) are inconsistent. Hence the system has no solution.

Example 22-11 In triangle *ABC* the measure of angle *B* is equal to the sum of twice the measure of angle *A* and the measure of angle *C*. If the measure of angle *A* is increased by 46° and the measure of angle *B* decreased by 36° the two new measures will be equal. What is the measure of each angle of the triangle?

SOLUTION

Let $m\angle A = x$; $m\angle B = y$; $m\angle C = z$ (each measured in degrees). Then our system becomes

$x + y + z = 180$ (1)

$y = 2x + z$ (2)

$x + 46 = y - 36$ (3)

Add $-2x - z$ to both members of Eq. (2):

$$-2x + y - z = 0 \quad (4)$$

Add Eqs. (1) and (4): $\underline{x + y + z = 180}$

$-x + 2y = 180$ (5)

Add $-y - 46$ to both members of Eq. (3): $\underline{x - y = -82}$ (6)

Add Eqs. (5) and (6): $y = 98$

Substitute $y = 98$ in Eq. (3): $x + 46 = 98 - 36$ (7)

Solve Eq. (7) for x: $x = 16$

Substitute $x = 16$, $y = 98$ in Eq. (1): $16 + 98 + z = 180$ (8)

Solve Eq. (8) for z: $z = 66$

Thus the angle measures are: $m\angle A = 16°$, $m\angle B = 98°$, and $m\angle C = 66°$.

CHECK

$$m\angle A + m\angle B + m\angle C \stackrel{?}{=} 180°$$

$$16° + 98° + 66° \stackrel{?}{=} 180°$$

$$180° = 180°$$

$$m\angle B \stackrel{?}{=} 2m\angle A + m\angle C$$
$$98^\circ \stackrel{?}{=} 2(16^\circ) + 66^\circ$$
$$98^\circ = 98^\circ$$
$$m\angle A + 46^\circ \stackrel{?}{=} m\angle B - 36^\circ$$
$$16^\circ + 46^\circ \stackrel{?}{=} 98^\circ - 36^\circ$$
$$62^\circ = 62^\circ$$

Exercise 22-3

Solve and check the following systems of equations

1. $x + 2y - z = 0$
   $2x - y + z = 7$
   $x + 3y - z = 1$
2. $2x + 3y - 4z = -5$
   $x - 2y - 3z = -8$
   $3x + y + z = 3$
3. $5x - y - z = 13$
   $3x + 2y - z = 9$
   $x + y + z = -1$
4. $2x + y + 3z = 6$
   $x + 4y - 2z = -11$
   $x + 5y - 5z = -15$
5. $6x + 5y = -1$
   $3x - y + 2z = 0$
   $2y - 3z = -4$
6. $3x - 4y = 11$
   $5x + 7z = -1$
   $6x - 2y + 7z = 6$
7. $x + 3y - 2z = 7$
   $5x - 2y + 4z = 20$
   $4x - 5y + 6z = 15$
8. $3x - 2y - 2z = 6$
   $x + y + 4z = 5$
   $4x - y + 2z = 10$
9. $2x - 5y + 7z = 7$
   $4x + 3y - 2z = 12$
   $2x + 8y - 9z = 5$
10. $x + 3y - 4z = 8$
    $2x - 5y + 6z = 11$
    $x - 8y + 10z = 3$
11. $5x + 2y - 3z = 59$
    $3x - 4y + 2z = -29$
    $2x + 3y - 4z = 58$
12. $7x - 3y + 5z = 68$
    $4x + 2y - 3z = -18$
    $6x - 7y - 2z = 73$
13. $2x - 9y + 5z = -4$
    $8x + 6y - 7z = -15$
    $4x - 3y - 2z = -10$
14. $10x + 8y - 9z = 8$
    $7x - 6y - 6z = 11$
    $4x - 4y - 9z = 14$
15. If regular gasoline costs 34 cents/gal and premium gasoline costs 39 cents/gal, how many gallons of each will be required to blend 1,000 gal of a gasoline which should sell at 36 cents/gal?
16. How much type metal containing 16 percent tin and how much solder containing 50 percent tin must be mixed to get 120 lb of solder containing 45 percent tin?
17. An agent collected 240 nickels and dimes from a vending machine. The coins had a value of \$18.15. How many coins of each denomination did he have?
18. The following dimensions are given for the accompanying figure: $OA = 9$ in., $O'B = 6$ in., $OO' = 25$ in. Find the length of $\overline{AB}$.

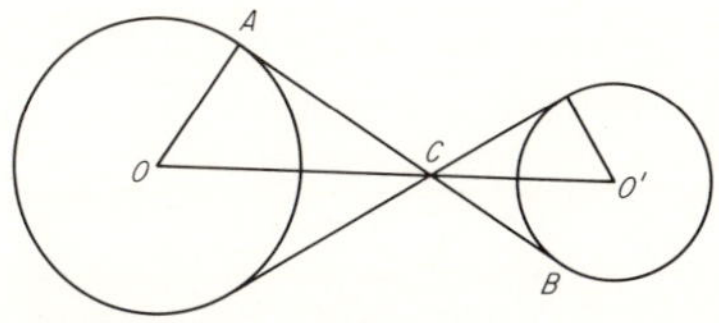

PROB. 18.

19. A man bought 80 acres of land for \$27,500. He paid \$300 an acre for part of the land and \$400 an acre for the rest. How much land did he buy at \$400 an acre?
20. Mr. Smith traveled from town A to town B at a speed of 40 mph and returned from town B to town A at a speed of 60 mph. What was his average speed for the round trip?
21. A and B together can do a job in 2 days; B and C can do it in 4 days; and A and C in $2\frac{2}{5}$ days. Find the number of days for A to do the job alone.
22. A bag of nickels, dimes, and quarters contain 79 coins worth \$6.95. There are five more dimes than quarters. How many of each type of coin are in the bag?
23. A motor boat makes a 50-mi trip upstream in 3 hr, 15 min. The return trip downstream takes 2 hr, 30 min. If the boat maintains a constant speed with respect to the water, and the speed of the river does not vary, find the speed of the river.
24. Mr. Jones has part of \$50,000 loaned at $4\frac{1}{2}$ percent interest and the rest at 6 percent interest. If the income from the money loaned at $4\frac{1}{2}$ percent interest is \$360 more than that at 6 percent, what is the amount of each loan?
25. A merchant bought 15 microphones and 10 speakers for \$181.25. When he found that he had underordered, he bought 9 more microphones and 8 more speakers for \$125.50. Find the price he paid for each microphone and speaker.
26. How many gallons of 40 percent and 70 percent salt solutions should be mixed to get 80 gal of 50 percent salt solution?
27. Three pipes are used to fill a reservoir. The reservoir can be filled by pipes A and B running for 6 hr, or by pipes B and C running for 4 hr, or by A and C running for 8 hr. What length of time would be required for each pipe running alone to fill the reservoir?
28. Three machines A, B, and C operating together can complete a job in 12 hr. If A operates for 4 hr and B for 6 hr, one-third of the job can be done. If B operates for 6 hr and C for 8 hr, one-half of the job can be done. How many hours would it take each machine operating alone to complete the job?

CHAPTER 23
LOGARITHMS

23-1 USE OF LOGARITHMS In Chap. 6 we studied how the slide rule is used to save time and labor in solving problems involving combinations of the operations of multiplication, division, raising to a power, and taking roots of numbers.

Slide rules are based on the principles of the logarithm. Each type of problem we solved in Chap. 6 could also be solved using logarithms. But, whereas the solutions are limited to three-place accuracy when the slide rule is used, the degree of accuracy when logarithms are used is dependent only on the number of significant figures in the table of logarithms. Five- and six-place accuracy is not uncommon in solving problems with logarithms.

In this text we will use a logarithmic table which will only yield four-place accuracy. The objective in this chapter is to learn how logarithms are used in solving problems, not to get high degrees of accuracy in the solutions. The same principles can be used with proper tables to get higher degrees of accuracy.

Students who will continue the study of mathematics beyond this course will find that logarithms have many other uses than those studied in this chapter. Logarithms are still used extensively in describing scientific and engineering phenomena although many of the calculations that until recently could best be solved by logarithmic computations are now solved by electronic computers. A knowledge of logarithmic functions is essential to the study of advanced mathematics, science, and engineering.

23-2 DEFINITION OF A LOGARITHM The concept of a logarithm is directly related to that of an exponent. *A logarithm is an exponent.* (The student might be well advised to review the laws of exponents given in Sec. 19-1.) Since a logarithm is an exponent, it must be an exponent of some number. This number is called the *base.*

DEFINITION *The logarithm of a positive number N to a positive base b ($b \neq 1$) is the exponent x of the power to which the base must be raised to equal that number.*

Thus, if

$$N = b^x \tag{23-1}$$

where $b > 0$ and $b \neq 1$, the exponent x is called the logarithm of N to the base b. This relation is expressed in symbols by

$$x = \log_b N \tag{23-2}$$

It should be emphasized that a logarithm is an exponent. Equations (23-1) and (23-2) yield the same corresponding values for x and N. Equation (23-1) is an exponential equation while Eq. (23-2) is a logarithmic equation. They state the same relations in two different ways.

Example 23-1 The following table shows the two different ways of saying the same thing.

TABLE 23-1 Equivalent Equations

	EXPONENTIAL FORM	LOGARITHMIC FORM
a	$5^3 = 125$	$\log_5 125 = 3$
b	$10^4 = 10{,}000$	$\log_{10} 10{,}000 = 4$
c	$8^{5/3} = 32$	$\log_8 32 = \frac{5}{3}$
d	$10^\circ = 1$	$\log_{10} 1 = 0$
e	$4^{-2} = \frac{1}{16}$	$\log_4 \frac{1}{16} = -2$
f	$16^{-3/4} = \frac{1}{8}$	$\log_{16} \frac{1}{8} = -\frac{3}{4}$

Example 23-2 Find the value of $\log_4 64$.

SOLUTION Let $x = \log_4 64$. Then from the definition of a logarithm we know that $4^x = 64$. But $4^3 = 64$. Then $x = 3$ and $\log_4 64 = 3$.

Example 23-3 Given $\log_4 \frac{1}{32} = y$, find y.

SOLUTION If $\log_4 \frac{1}{32} = y$, then $4^y = \frac{1}{32}$. The exponential equation can be written

$$(2^2)^y = 2^{-5}$$

Hence $$2^{2y} = 2^{-5}$$

and $$2y = -5$$

Hence $$y = -\tfrac{5}{2}$$

Example 23-4 Find the value of N if $\log_{4/9} N = -\frac{3}{2}$.

SOLUTION Write the logarithmic equation in equivalent exponential form:

$$(\tfrac{4}{9})^{-3/2} = N$$

Thus
$$\begin{aligned} N &= [(\tfrac{2}{3})^2]^{-3/2} \\ &= (\tfrac{2}{3})^{-3} \\ &= (\tfrac{3}{2})^3 \\ &= \tfrac{27}{8} \end{aligned}$$

Exercise 23-1

Express each of the following exponential equations as equivalent logarithmic equations.

1. $2^4 = 16$
2. $4^3 = 64$
3. $10^3 = 1{,}000$
4. $10^5 = 100{,}000$
5. $10^{-2} = 0.01$
6. $10^{-3} = 0.001$
7. $8^{2/3} = 4$
8. $216^{1/3} = 6$
9. $8^{-5/3} = \frac{1}{32}$
10. $(\frac{2}{3})^3 = \frac{8}{27}$
11. $b^0 = 1$
12. $a^b = c$

Express each of the following logarithmic equations as equivalent exponential equations.

13. $\log_2 32 = 5$
14. $\log_3 81 = 4$
15. $\log_{64} (8) = \frac{1}{2}$
16. $\log_5 \frac{1}{125} = -3$
17. $\log_{27} 81 = \frac{4}{3}$
18. $\log_7 1 = 0$
19. $\log_9 \frac{1}{27} = -\frac{3}{2}$
20. $\log_m q = p$
21. $\log_{1/16} 32 = -\frac{5}{4}$
22. $\log_{10} 200 = 2.30103$
23. $\log_{10} 6 = 0.77815$
24. $\log_{10} 47.63 = 1.67788$

Find the value of each of the following logarithms.

25. $\log_5 125$
26. $\log_8 2$
27. $\log_9 27$
28. $\log_{10} 10{,}000$
29. $\log_{10} 0.001$
30. $\log_6 1$
31. $\log_2 \frac{1}{32}$
32. $\log_4 \sqrt[3]{4}$
33. $\log_b b^2$
34. $\log_{81} 27$
35. $\log_{12} \frac{1}{144}$
36. $\log_3 \frac{1}{243}$

Find the value of the unknown x, b, or N in each of the following.

37. $\log_6 N = 2$
38. $\log_4 N = 0$
39. $\log_2 N = 5$
40. $\log_b 216 = 3$
41. $\log_2 \frac{1}{8} = x$
42. $\log_6 N = -3$
43. $\log_b 121 = 2$
44. $\log_{5/3} \frac{25}{9} = x$
45. $\log_9 N = -\frac{2}{3}$
46. $\log_{32} 64 = x$
47. $\log_{10} 10^5 = x$
48. $\log_{10} N = -4$

23-3 PROPERTIES OF LOGARITHMS Since logarithms are exponents, the laws relating the use of logarithms are similar to the laws of exponents. In computing with logarithms, we will use certain useful properties of logarithms which can be developed from their exponential analogs.

The following properties of logarithms are used repeatedly in solving multiplication, division, powers and roots problems in this chapter. These properties are also useful in the study of advanced topics of mathematics.

PROPERTY 1 *The logarithm of a product is equal to the sum of the logarithms of the separate factors.*

PROOF Let $M = b^x$ or $x = \log_b M$

and $N = b^y$ or $y = \log_b N$

Hence $MN = (b^x)(b^y) = b^{x+y}$

and in logarithmic form: $\log_b MN = x + y$

Substitute for x and y to get: $\log_b MN = \log_b M + \log_b N$ (23-3)

Similarly it can be proved that

$$\log_b MNP = \log_b M + \log_b N + \log_b P \quad (23\text{-}3a)$$

Example 23-5

$\log_{10} (27.3 \times 80.5) = \log_{10} 27.3 + \log_{10} 80.5$

$\log_3 (5 \times 7 \times 8 \times 9) = \log_3 5 + \log_3 7 + \log_3 8 + \log_3 9$

PROPERTY 2 *The logarithm of a quotient is equal to the logarithm of the numerator minus the logarithm of the denominator.*

The proof of this property is similar to that for Property 1. Thus, we can prove

$$\log_b \frac{M}{N} = \log_b M - \log_b N \tag{23-4}$$

Example 23-6 If $\log_{10} 7 = 0.84510$ and $\log_{10} 2 = 0.30103$, then

$$\begin{aligned}\log_{10} 3.5 &= \log_{10} \tfrac{7}{2}\\ &= \log_{10} 7 - \log_{10} 2\\ &= 0.84510 - 0.30103\\ &= 0.54407\end{aligned}$$

Example 23-7

$$\begin{aligned}\log_2 \frac{32 \times 53}{21 \times 68} &= \log_2 (32 \times 53) - \log_2 (21 \times 68)\\ &= [\log_2 32 + \log_2 53] - [\log_2 21 + \log_2 68]\\ &= \log_2 32 + \log_2 53 - \log_2 21 - \log_2 68\end{aligned}$$

PROPERTY 3 *The logarithm of a power of a number is equal to the product of the exponent of the power and the logarithm of the number.*

PROOF Let $M = b^x$ or $x = \log_b M$. Then

$$M^p = (b^x)^p = b^{px}$$

Written in logarithmic form we get

$$\log_b M^p = px$$

Replacing x by its value above:

$$\log_b M^p = p \log_b M \tag{23-5}$$

Example 23-8

$$\log_3 7^2 = 2 \log_3 7$$

$$\begin{aligned}\log_{10} (5^3)(8^2) &= \log_{10} 5^3 + \log_{10} 8^2\\ &= 3 \log_{10} 5 + 2 \log_{10} 8\end{aligned}$$

PROPERTY 4 *The logarithm of a root of a number is equal to the logarithm of the number divided by the index of the root.*

PROOF Since the root of a number can be expressed as a fractional power,

Eq. (23-5) can be employed to obtain the logarithm of a root of a number. Thus

$$\log_b \sqrt[r]{M} = \frac{1}{r} \log_b M \qquad (23\text{-}6)$$

Example 23-9
$\log_{10} \sqrt[3]{7} = \log_{10} 7^{1/3}$
$= \frac{1}{3} \log_{10} 7$

The student should distinguish between

$\log_b (M + N)$ and $\log_b M + \log_b N$

$\log_b (M - N)$ and $\log_b M - \log_b N$

$(\log_b M)(\log_b N)$ and $\log_b M + \log_b N$

$\dfrac{\log_b M}{\log_b N}$ and $\log_b M - \log_b N$

In none of the above four pairs are the expressions identities.

Since $b^0 = 1$, we apply Eq. (23-1) and Eq. (23-2) to get

$$\log_b 1 = 0 \qquad (23\text{-}7)$$

Similarly from $b^1 = b$, we obtain

$$\log_b b = 1 \qquad (23\text{-}8)$$

If $N = b^x$, then $x = \log_b N$. Replacing x in the exponential equation by its equivalent logarithmic form yields

$$b^{\log_b N} = N \qquad (23\text{-}9)$$

Similarly, replacing N in the logarithmic equation by its equivalent in exponential form yields

$$\log_b b^x = x \qquad (23\text{-}10)$$

The student should verify the following additional properties of logarithms.

1 The number zero has no logarithm.
2 Negative numbers have no real logarithms.

Example 23-10 Express $\log_{10} 2\pi\sqrt{l/g}$ as an algebraic sum of logarithms.

SOLUTION

$\log_{10} 2\pi \sqrt{\dfrac{l}{g}} = \log_{10} 2 + \log_{10} \pi + \log_{10} \left(\dfrac{l}{g}\right)^{1/2}$ By Eq. (23-3)

$= \log_{10} 2 + \log_{10} \pi + \frac{1}{2} \log_{10} \dfrac{l}{g}$ By Eq. (23-5)

$= \log_{10} 2 + \log_{10} \pi + \frac{1}{2}(\log_{10} l - \log_{10} g)$ By Eq. (23-4)

$= \log_{10} 2 + \log_{10} \pi + \frac{1}{2} \log_{10} l - \frac{1}{2} \log_{10} g$

Example 23-11 Express $\log_b 3 + \log_b x + \frac{1}{2}\log_b 5y - \log_b 7 - 3\log_b z$ as a single logarithm.

SOLUTION

$\log_b 3 + \log_b x + \frac{1}{2}\log_b 5y - \log_b 7 - 3\log_b z$

$= \log_b 3x + \frac{1}{2}\log_b 5y - \log_b 7 - 3\log_b z$ By Eq. (23-3)

$= \log_b 3x + \log_b (5y)^{1/2} - \log_b 7 - \log_b z^3$ By Eq. (23-5)

$= \log_b 3\sqrt{5}\,xy^{1/2} - \log_b 7z^3$ By Eq. (23-3)

$= \log_b \dfrac{3\sqrt{5}\,xy^{1/2}}{7z^3}$ By Eq. (23-4)

Exercise 23-2

Find the values of the following:

1. $\log_2 8^{-4}$
2. $\log_3 9^2$
3. $\log_2 4^{1/3}$
4. $\log_5 (125)^{-2}$
5. $\log_4 1$
6. $\log_4 (64)^{2/3}$
7. $\log_3 (3^2 \cdot 3^4)$
8. $\log_b b^{2x-1}$
9. $\log_4 (4^2 \cdot 4^{-5})$
10. $\log_2 (4^2 \cdot 4^{-5})$
11. $\log_e e$
12. $\log_e e^{5x} \cdot e^{-2x}$
13. $\log_{10} \sqrt{10}(10)$
14. $\log_{10} \dfrac{10^2 \cdot 10^{-3}}{10^{-4}}$
15. $2^{\log_2 7}$
16. $10^{\log_{10} 100}$
17. $10^{\log_{10}(x+y)}$
18. $4^{\log_4 64}$
19. $e^{\log_e e^2}$
20. $b^{\log_b 10}$
21. $\log_2 \sqrt[4]{4}$
22. $\log_3 \sqrt[6]{27}$
23. $\log_2 \sqrt[4]{64}$
24. $\log_4 \sqrt[3]{16}$

Expand each of the following logarithms as an algebraic sum of logarithms:

25. $\log_b \pi r^2$
26. $\log_b \dfrac{xy}{z}$
27. $\log_a \frac{2}{3}xy^2$
28. $\log_c \sqrt{2gh}$
29. $\log_e \dfrac{1}{2\pi fC}$
30. $\log_b \dfrac{4\pi^2 R}{T^2}$
31. $\log_c \dfrac{K(t_2 - t_1)\alpha T}{d}$
32. $\log_a \dfrac{1}{2rl}\sqrt{\dfrac{T}{\pi d}}$

Express each of the following as a single logarithm:

33. $\log_b x - \log_b y + \log_b z$
34. $3\log_a x - 2\log_a z$
35. $5\log_c y + \frac{1}{3}\log_c z$
36. $\log_e 4 + 3\log_e r - \log_e 3 + \log_e \pi$
37. $\frac{1}{2}\log_a g + \frac{1}{2}\log_a l - \frac{1}{2}\log_a 2 - \frac{1}{2}\log_a \pi$
38. $\log_e R + \log_e X - \frac{1}{2}\log_e (R^2 + X^2)$
39. $\log_a M - \frac{1}{2}\log_a \left(1 - \dfrac{v}{c}\right)^2$
40. $\frac{1}{3}\log_b (x+a) + \frac{1}{2}\log_b (x+b) - \log_b (x-c) - 2\log_b (x+d)$

Simplify each of the following and indicate whether the statement is true or false:

41. $\dfrac{\log_b M}{\log_b N} = \log_b M - \log_b N$

42. $\dfrac{\log_4 16}{\log_4 \sqrt{4}} = \dfrac{3}{2}$

43. $\log_b \dfrac{AB^2}{2AB} = \log_b \dfrac{B}{2}$

44. $\log_a \dfrac{1}{B} = -\log_a B$

45. $\log_5 5^x = x$

46. $\log_b (x + y) - \log_b y = \log_b x$

47. $\log_c \dfrac{xy}{z} = \dfrac{\log_c x + \log_c y}{\log_c z}$

48. $(\log_b M)^P = P \log_b M$

49. $\log_2 \log_2 2x^0 = 0$

50. $10^{3\log_{10}\sqrt{x}} = x^{2/3}$

51. $\log_e \dfrac{\sqrt{x}}{x} = -\frac{1}{2} \log_e x$

52. $\log_3 \dfrac{3x - 3y}{y - x} = -1$

53. $\log_b \dfrac{A}{3} = \frac{1}{3} \log_b A$

54. $\log_a M - \log_a \dfrac{M}{M - N} = \log_a (M - N)$

55. $\log_c \frac{1}{3}AB = \frac{1}{3}(\log_c A + \log_c B)$

23-4 COMMON LOGARITHMS Any positive number except 1 could be used as a base number in logarithms. There are two systems of logarithms in general use; one is defined by

$$y = \log_{10} x \tag{23-11}$$

and the other, by

$$y = \log_e x \tag{23-12}$$

where e is an irrational number (approximate value 2.7182818).

Logarithms to the base 10 are called *common,* or *Briggs,* logarithms. Common logarithms can be used most effectively for general computation because they are best suited to our decimal numerical system. We will only use common logarithms in this text. Common practice in mathematics texts is to abbreviate $\log_{10} N$ as $\log N$ (with the base 10 understood). We will use this abbreviation hereafter in this text.

The system of logarithms to the base e is referred to as the system of *natural,* or *Naperian,* logarithms. The use of natural logarithms occurs in the solutions of many physical and engineering problems, as well as in many branches of theoretical mathematics.

The values for log x can easily be determined for all values of x that are integral powers of 10:

$\log_{10} 1 = 0$, since $10^0 = 1$

$\log_{10} 10 = 1$, since $10^1 = 10$

$\log_{10} 100 = 2$, since $10^2 = 100$

$\log_{10} 1{,}000 = 3$, since $10^3 = 1{,}000$

and

$\log_{10} 0.1 = -1$, since $10^{-1} = 0.1$

$\log_{10} 0.01 = -2$, since $10^{-2} = 0.01$

$\log_{10} 0.001 = -3$, since $10^{-3} = 0.001$

The common logarithm of a number that is not an integral power of 10 cannot be computed by elementary methods. However, extensive tables have been prepared that enable us to determine a decimal approximation to the common logarithm of any positive number. A table of logarithms is given in Table 1 of the appendixes for values of $\log_{10} x$ where $1 \leqslant x \leqslant 10$. Let us consider how this table is used. As an example, to find $\log_{10} 4.36$, look at the intersection of the row containing 4.3 under x and the column containing 6 to the right of x. Thus,

$$\log_{10} 4.36 = 0.6395$$

Similarly, $\log_{10} 3.97 = 0.5988$ and

$$\log_{10} 4.08 = 0.6107$$

It should be pointed out that $\log_{10} 4.36$ is only approximately equal to 0.6395. The degree of accuracy of the logarithm is inherent in the number of significant figures in the table. Even though logarithms like the above are irrational and cannot be expressed as an exact decimal number, we shall use the = sign and leave the intent up to the reader's understanding.

The approximate values for $\log_{10} x$ for values of x outside the range of Table 1 (that is, for $0 < x < 1$ or $x > 10$) can be determined by representing the number in scientific notation and then applying the properties of logarithms discussed in Sec. 23-3. For example,

$$\begin{aligned}\log_{10} 43.6 = \log_{10} (4.36 \times 10^1) &= \log_{10} 4.36 + \log_{10} 10^1 \\ &= 0.6395 + 1 \\ &= 1.6395\end{aligned}$$

$$\begin{aligned}\log_{10} 436 = \log_{10} (4.36 \times 10^2) &= \log_{10} 4.36 + \log_{10} 10^2 \\ &= 0.6395 + 2 \\ &= 2.6395\end{aligned}$$

$$\begin{aligned}\log_{10} 43{,}600 = \log_{10} (4.36 \times 10^4) &= \log_{10} 4.36 + \log_{10} 10^4 \\ &= 0.6395 + 4 \\ &= 4.6395\end{aligned}$$

Note that the decimal portion of the above logarithms is always equal to 0.6395, and the *integral portion is the exponent of 10 when the number is written in scientific notation.*

Thus, each logarithm can be seen to consist of a whole number (called the *characteristic*) and a nonnegative decimal fraction (called the *mantissa*). The table of values for log x can be considered as a table of mantissas for all $x > 0$.

Again, to find $\log_{10} 4{,}080$, we write

$$\begin{aligned}\log_{10} 4{,}080 &= \log_{10} (4.08 \times 10^3) \\ &= 3.6107\end{aligned}$$

Next, let us consider an example of $\log_{10} x$ for $0 < x < 1$. To find $\log_{10} 0.00436$, we write

$$\begin{aligned}\log_{10} 0.00436 &= \log_{10} (4.36 \times 10^{-3}) \\ &= 0.6395 - 3\end{aligned}$$

Combining terms would give $\log_{10} 0.00436 = -2.3605$. However, now the decimal part of the logarithm no longer is nonnegative. When the characteristic is negative, it will be more convenient to use an equivalent form which will retain a nonnegative decimal part. In the foregoing example, we write

$$\begin{aligned}\log_{10} 0.00436 &= 0.6395 - 3 \\ &= 0.6395 + (7 - 10) \\ &= 7.6395 - 10\end{aligned}$$

Thus, a nonnegative decimal part is maintained just as it is obtained from the table of logarithms. The above logarithm could be written many other ways, such as $6.6395 - 9$ and $11.6395 - 14$. But it is customary to make the negative part a multiple of -10, such as $7.6395 - 10$, $17.6395 - 20$, etc.

23-5 THE ANTILOGARITHM If $\log N = x$, then N is called the antilogarithm (abbreviated antilog) of x. Thus, the antilogarithm of an expression is the number whose logarithm is the given expression. The process of finding an antilogarithm is exactly the reverse of finding the logarithm.

Example 23-12 If $\log N = 2.9154$, find the value of N.

SOLUTION From Table 1 in the Appendix we find that $\log 8.23 = 0.9154$.

Then $2.9154 = 0.9154 + 2$
$= \log 8.23 + \log 10^2$
$= \log (8.23 \times 10^2)$
$= \log 823$

Hence $N = 823$ or the antilog of $2.9154 = 823$

Example 23-13 Find the antilog of $6.7143 - 10$.

SOLUTION From Table 1 we note that $\log 5.18 = 0.7143$.

Then $6.7143 - 10 = 0.7143 - 4$
$= \log 5.18 + \log 10^{-4}$
$= \log (5.18 \times 10^{-4})$
$= \log 0.000518$

Hence the antilog of $6.7143 - 10 = 0.000518$

Exercise 23-3

In Probs. 1 to 12, find the characteristic of the common logarithm for each number.

1. 31
2. 458
3. 1,760
4. 80,900
5. 0.00351
6. 2.76×10^{-4}
7. 0.07000
8. 7,000
9. 3.47×10^6
10. 0.00007
11. 15.70
12. 0.06090

In Probs. 13 to 24, use Table 1 to find the logarithm for each number.

13. 34.5
14. 879
15. 59,700
16. 0.0132
17. 0.0075
18. 0.0000278
19. 2.41×10^8
20. 4,860
21. 0.00006
22. 0.000058
23. 0.0000372
24. 28.5×10^3

In Probs. 25 to 36, use Table 1 to find the antilogarithm of each number.

25. 1.8756
26. 3.7435
27. 7.4829
28. $9.5276 - 10$
29. $7.9809 - 10$
30. $15.7466 - 20$
31. 5.9542
32. $6.1239 - 10$
33. 3.6010
34. $2.0294 - 10$
35. $4.9638 - 10$
36. $4.8597 - 10$

23-6 LINEAR INTERPOLATION Table 1 in the Appendix enables a person to find directly the $\log_{10} x$ when x is given accurate to three significant figures. The same table can be used to find approximations to logarithms for numbers with four-digit numerals. The process will be illustrated with several examples.

Example 23-14 Find log 38.64.

SOLUTION Use Table 1.

$$\begin{array}{ll} & x \qquad\qquad \log x \\ 0.10\left\{\begin{array}{l} \qquad 38.70 \\ 0.04\left\{\begin{array}{l}38.64\\38.60\end{array}\right. \end{array}\right. & \begin{array}{l} 1.5877 \\ \quad ? \\ 1.5866 \end{array} \left.\begin{array}{l} \\ \left.\begin{array}{l}\\ \\ \end{array}\right\} k \end{array}\right\} 0.0011 \end{array}$$

Set up a proportion and solve for k

$$\frac{k}{0.0011} = \frac{0.04}{0.10} = \frac{4}{10}$$

$$k = \frac{4}{10}(0.0011) = 0.0004$$

Hence $\log 38.64 = 1.5866 + 0.0004 = 1.5870$

Example 23-15 Find antilog $8.5639 - 10$.

SOLUTION Use Table 1.

$$\begin{array}{ll} & x \qquad\qquad \log x \\ 0.0010\left\{\begin{array}{l} \qquad 0.03670 \\ c\left\{\begin{array}{l}?\\0.03660\end{array}\right. \end{array}\right. & \begin{array}{l} 8.5647-10 \\ 8.5639-10 \\ 8.5635-10 \end{array} \left.\begin{array}{l} \\ \left.\begin{array}{l}\\ \\ \end{array}\right\} 0.0004 \end{array}\right\} 0.0012 \end{array}$$

$$\frac{c}{0.0010} = \frac{0.0004}{0.0012} = \frac{4}{12}$$

$$c = \frac{4}{12}(0.0010)$$

$$= 0.0003$$

Hence $x = \text{antilog } 8.5639 - 10 = 0.03660 + 0.0003$
$= 0.03663$

Exercise 23-4

In Probs. 1 to 12, find the logarithms (accurate to four decimals) of the following numbers:

1. 6.243	5. 8.004×10^5	9. 0.03946
2. 349.4	6. 31,770	10. 0.0004851
3. 2,785	7. 0.3517	11. 3.829×10^{-3}
4. 12.93	8. 0.002471	12. 1.124×10^{-4}

In Probs. 13 to 24, find the antilogarithms (accurate to four significant figures) of:

13. 0.6566	17. 9.9578	21. 8.8166
14. 1.7129	18. 1.0261	22. 6.3146 − 10
15. 3.9044	19. 9.6109 − 10	23. 5.1589 − 10
16. 2.7828	20. 7.9039 − 10	24. 7.0345 − 10

In Probs. 25 to 33, find the value of x (to four significant figures) if:

25. $\log x = 0.4859$	28. $\log x = 2.1313$	31. $\log x = 7.9311 - 10$
26. $\log x = 1.9050$	29. $\log x = 4.0225$	32. $\log x = 4.4087 - 10$
27. $\log x = 3.3170$	30. $\log x = 9.8478 - 10$	33. $\log x = 18.2078 - 20$

23-7 MULTIPLICATION AND DIVISION WITH LOGARITHMS By using the properties of logarithms (Sec. 23-3), operations of multiplication or division of numbers can be reduced to the simpler operations of addition or subtraction of logarithms.

Example 23-16 Use logarithms to find the product 28.3×0.000591.

SOLUTION Let $x = 28.3 \times 0.000591$. Then applying Eq. (23-3), we can write

$$\log x = \log 28.3 + \log 0.000591$$

Using Table 1, the completed work appears as follows:

$$\begin{aligned} \log 28.3 &= 1.4518 \\ \log 0.000591 &= \underline{6.7716 - 10} \\ \log x &= 8.2234 - 10 \\ x &= 1.67 \times 10^{-2} \end{aligned}$$

Note that no interpolation was needed in this solution. Since the factors had only three significant figures, the product can be determined only to three places.

Example 23-17 Use logarithms to find the value of (4.365)(52.38)(0.008014).

SOLUTION If $x = (4.365)(52.38)(0.008014)$, then $\log x = \log 4.365 + \log 52.38 + \log 0.008014$. From the tables (the student should check the interpolation involved):

$$\begin{aligned} \log 4.365 &= 0.6400 \\ \log 52.38 &= 1.7191 \\ \log 0.008014 &= \underline{7.9038 - 10} \\ \log x &= 10.2629 - 10 \\ x &= 1.832 \qquad \text{(by interpolation)} \end{aligned}$$

Example 23-18 Use logarithms to find the value of $\dfrac{748.3}{0.5516}$.

SOLUTION If $x = \dfrac{748.3}{0.5516}$, we know from Eq. (23-4) that

$$\log x = \log 748.3 - \log 0.5516.$$

Using Table 1 and interpolation we get

$$\begin{aligned} \log 748.3 &= 2.8741 \\ \log 0.5516 &= \underline{9.7416 - 10} \end{aligned}$$

The difference will be easier to obtain if we rewrite the problem as

$$\begin{aligned} \log 748.3 &= 12.8741 - 10 \\ \log 0.5516 &= \underline{\ 9.7416 - 10} \\ \log x &= 3.1325 \\ x &= 1{,}357 \qquad \text{(by interpolation)} \end{aligned}$$

Example 23-19 Use logarithms to find the value of

$$\frac{28.36 \times 0.8019 \times 0.005124}{4.607 \times 0.03272}.$$

SOLUTION Let $x = \dfrac{28.36 \times 0.8019 \times 0.005124}{4.607 \times 0.03272}$. Then

$$\log x = \log(28.36 \times 8.019 \times 0.005124) - \log(4.607 \times 0.03272)$$

The necessary operations and results are outlined below.

$$\begin{aligned} \log 28.36 &= 1.4527 \\ \log 0.8019 &= 9.9041 - 10 \\ \log 0.005124 &= \underline{7.7096 - 10} \\ \log \text{numerator} &= 19.0664 - 20 \\ \log 4.607 &= 0.6634 \\ \log 0.03272 &= \underline{8.5148 - 10} \\ \log \text{denominator} &= 9.1782 - 10 \end{aligned}$$

$\log x = \log \text{numerator} - \log \text{denominator}$

$$\begin{aligned} & 19.0664 - 20 \\ & \underline{\ 9.1782 - 10} \\ \log x &= 9.8882 - 10 \\ x &= 0.7730 \end{aligned}$$

Exercise 23-5

Compute the following by using logarithms:

1. 24.8×83.1
2. 3.95×807
3. 0.0107×4.39
4. 0.000376×25.8
5. $37.6 \times 1.45 \times 0.842$
6. $10.3 \times 0.00936 \times 0.158$
7. 3.572×149.5
8. 40.63×0.005177

9. $0.0008672 \times 13{,}590 \times 0.04135$

10. $647.8 \times 5.129 \times 0.001738$

11. $\dfrac{71.8}{0.0534}$

12. $\dfrac{0.672}{0.0935}$

13. $\dfrac{6{,}294}{27.31}$

14. $\dfrac{0.007913}{0.05555}$

15. $\dfrac{0.003728}{0.04905}$

16. $\dfrac{0.8092 \times 34.72}{56.05}$

17. $\dfrac{5.38 \times 71.6}{22.9}$

18. $\dfrac{0.0527 \times 493}{0.778}$

19. $\dfrac{47.6}{2.98 \times 0.109}$

20. $\dfrac{538}{21.7 \times 32.8}$

21. $\dfrac{58.72 \times 8.231}{26.65}$

22. $\dfrac{0.009204 \times 0.8663}{0.05177}$

23. $\dfrac{39.64}{807.6 \times 0.002213}$

24. $\dfrac{0.07825}{0.3471 \times 9.804}$

25. $\dfrac{0.592 \times 0.818 \times 11.8}{2.64 \times 5.05}$

26. $\dfrac{444 \times 50.3 \times 0.0217}{0.00904 \times 11{,}500}$

27. $\dfrac{4.083 \times 0.7921 \times 21.85}{37.54 \times 6.198}$

28. $\dfrac{2.753 \times 46.29 \times 80.77}{54.73 \times 0.006562}$

29. $\dfrac{4.146 \times 3.758}{0.09782 \times 5.663 \times 4.005}$

30. $\dfrac{0.5903 \times 0.002765}{0.02211 \times 0.3574 \times 0.0002200}$

23-8 ROOTS AND POWERS WITH LOGARITHMS Equations (23-5) and (23-6) can effectively be used to reduce the operations of raising a number to a power or taking the root of a number to multiplying or dividing the logarithm of the base by the exponent of the power or index of the root.

Example 23-20 Compute $(0.05437)^4$.

SOLUTION Let $x = (0.05437)^4$. Then, by Eq. (23-5), we know that $\log x = 4 \log 0.05437$. The solution follows in outline form.

$$\log 0.05437 = 8.7354 - 10$$

$$\log x = 4(8.7354 - 10) = 34.9416 - 40$$

$$= 4.9416 - 10$$

$$x = 0.000008742 \qquad \text{(by interpolation)}$$

Hence $(0.05437)^4 = 8.742 \times 10^{-6}$

Example 23-21 Compute $\sqrt[3]{\dfrac{36.7 \times 0.0549}{0.792}}$.

SOLUTION If we let $x = \sqrt[3]{\dfrac{36.7 \times 0.0549}{0.792}}$, then

$$\log x = \tfrac{1}{3} \log \frac{36.7 \times 0.0549}{0.792}$$
$$= \tfrac{1}{3}[(\log 36.7 + \log 0.0549) - (\log 0.792)]$$

The solution in outline form follows:

$$\begin{aligned} \log 36.7 &= 1.5647 \\ \log 0.0549 &= \underline{8.7396 - 10} \\ \log \text{numerator} &= 10.3043 - 10 \end{aligned} \quad \text{(add)}$$

$$\begin{aligned} \log \text{numerator} &= 10.3043 - 10 \\ \log 0.792 &= \underline{9.8987 - 10} \\ \log \text{radicand} &= 0.4056 \end{aligned} \quad \text{(subtract)}$$

$$\begin{aligned} \log x &= \tfrac{1}{3}(0.4056) \\ &= 0.1352 \\ x &= 1.37 \end{aligned}$$

Example 23-22 Compute $\sqrt[3]{\dfrac{0.005497 \times 0.7605}{2.805 \times 30.14}}$.

SOLUTION Let $x = \sqrt[3]{\dfrac{0.005497 \times 0.7605}{2.805 \times 30.14}}$. Then

$$\log x = \tfrac{1}{3}[(\log 0.005497 + \log 0.7605) - (\log 2.805 + \log 30.14)]$$

$$\begin{aligned} \log 0.005497 &= 7.7401 - 10 \\ \log 0.7605 &= \underline{9.8811 - 10} \\ \log \text{numerator} &= 17.6212 - 20 \end{aligned} \quad \text{(add)}$$

$$\begin{aligned} \log 2.805 &= 0.4479 \\ \log 30.14 &= \underline{1.4791} \\ \log \text{denominator} &= 1.9270 \end{aligned} \quad \text{(add)}$$

$$\begin{aligned} & \ 17.6212 - 20 \\ & \ \underline{1.9270} \\ \log \text{radicand} &= 15.6942 - 20 \end{aligned} \quad \text{(subtract)}$$

In order to avoid a fraction for the characteristic of log x, we will change $15.6942 - 20$ to $25.6942 - 30$. Then

$$\begin{aligned} \log x = \tfrac{1}{3}(15.6942 - 20) - \tfrac{1}{3}(25.6942 - 30) \\ = 8.5647 - 10 \\ x = 0.03670 \end{aligned}$$

Exercise 23-6

Compute by using logarithms.

1. $(3.85)^4$
2. $(2.76)^5$
3. $(0.714)^3$
4. $(0.0912)^4$
5. $(2.735)^3$
6. $(6.341)^4$
7. $(0.9631)^5$
8. $(0.0009247)^3$
9. $(0.01823)^5$
10. $\sqrt{83.74}$
11. $\sqrt[3]{0.04915}$
12. $\sqrt[4]{547{,}300}$
13. $\sqrt{\dfrac{0.05913}{0.002476}}$
14. $\sqrt[3]{\dfrac{27.39}{4.505}}$
15. $\sqrt[3]{\dfrac{0.002705}{0.9806}}$
16. $\sqrt{\dfrac{26.3 \times 4.59}{7.71}}$
17. $\sqrt[3]{\dfrac{374 \times 0.918}{2.56}}$
18. $\sqrt{\dfrac{0.000821}{0.347 \times 0.00595}}$
19. $\sqrt{\dfrac{3.541 \times 87.02}{24.36}}$
20. $\sqrt[3]{\dfrac{569.1 \times 11.53}{8.045}}$
21. $\sqrt{\dfrac{8{,}783 \times 2{,}691}{32.07}}$
22. $\sqrt{\dfrac{0.05937 \times 0.004407}{0.7918 \times 2.653}}$
23. $\sqrt[3]{\dfrac{6.807 \times 1.542}{0.04826 \times 0.002205}}$
24. $\sqrt[4]{\dfrac{0.5909 \times 80.03}{2.747 \times 6.604}}$

SECTION 5

ELEMENTARY TRIGONOMETRY FOR TECHNICIANS

CHAPTER 24

TRIGONOMETRY AND RIGHT TRIANGLES

24-1 THE SUBJECT MATTER OF TRIGONOMETRY The word *trigonometry* is derived from the Greek words for *triangle* and *measure.* Trigonometry was originally a branch of geometry and dealt primarily with the measurement of sides and angles of triangles. Trigonometry furnishes us with a powerful method for computing certain unknown angles and sides of a triangle from the numerical values of other known angles and sides of the triangle. Triangle measurement is a necessary skill for the astronomer, the surveyor, the engineer, the artillery officer, and the navigator of ships and airplanes. It is used in all types of constructions, such as houses, dams, railroad lines, bridges, ships, and airplanes.

Magnitudes of distances and angles may be obtained by direct measurement with the use of instruments such as rulers, steel tapes, transits, sextants, and radio beams. Then from these recorded measurements and formulas derived in the study of trigonometry, other unknown distances and angles can be computed.

Although we restrict ourselves in this text to finding distances and angles without actually measuring them, this is not the only use of trigonometry. The physicist can represent mathematically with the aid of trigonometry certain periodic phenomena such as light, sound, and electricity. Trigonometry has many applications in other branches of mathematics. It is an indispensable tool for calculus, a higher form of mathematics used in scientific study.

24-2 INDIRECT MEASUREMENT Let us suppose that we are assigned the task of determining the height of a flagpole. Let us assume further that it is impossible to scale the pole and measure the height directly. How would we proceed? If the sun's rays strike the pole, a shadow will be cast. The length of this shadow can be measured. If we now place a stick of known height $B'C'$ vertically in the ground, it too will cast a shadow which can be measured. (See Fig. 24-1.)

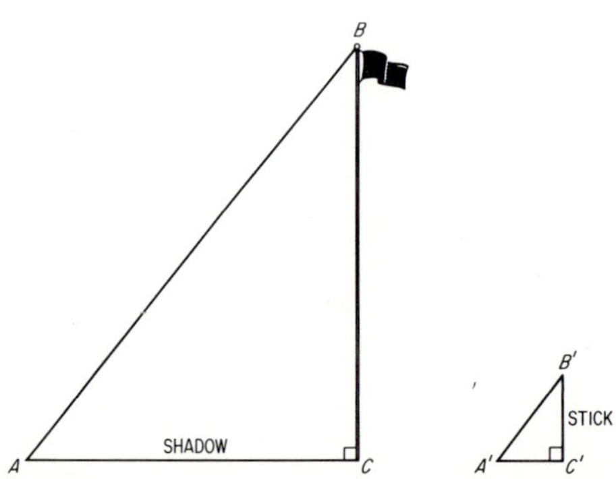

FIG. 24-1.

For all practical purposes, the sun's rays BA and $B'A'$ are parallel. This would make ∠s A and A' congruent and △s ABC and $A'B'C'$ similar. Hence $BC/B'C' = AC/A'C'$. We can measure $B'C'$, AC, and $A'C'$. From these values and the proportion we can compute the height of the flagpole. If, for example, $B'C' = 5$ ft, $AC = 32$ ft, and $A'C' = 4$ ft,

$$\frac{BC}{5} = \frac{32}{4} \qquad \text{or} \qquad BC = \frac{32 \text{ ft}}{4 \text{ ft}} \times 5 \text{ ft} = 40 \text{ ft}$$

It should be clear that we have thus found the height of the flagpole without measuring it directly. This is the method used by Thales, the famous Greek mathematician (about 640–546 B.C.), for calculating the height of an Egyptian pyramid.

Consider a second problem. It is desired to determine roughly the distance between two points C and B on opposite sides of a river. (See Fig. 24-2.) A possible solution follows. From C run a line CA 100 ft long (or any

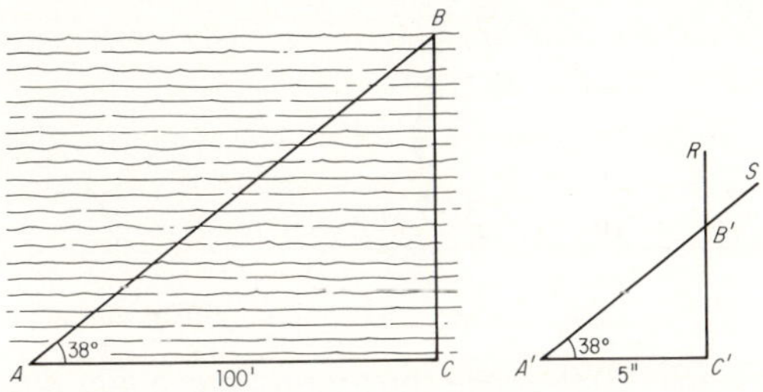

FIG. 24-2.

convenient length) at right angles to BC. By means of a protractor or another angle-measuring device, determine the size of angle CAB. Let us suppose the angle measures 38°.

We next draw to any convenient scale on a piece of paper a triangle similar to $\triangle ACB$. Let us assume we draw $A'C' = 5.0$ in. At C' we construct a line $C'R \perp$ to $A'C'$. With the protractor we draw $\angle C'A'S$ with measure $= 38°$. This will make $\triangle A'C'B' \sim \triangle ACB$. With a ruler we now can measure $B'C'$. Let us assume that $B'C'$ measures 3.9 in.

We then can solve for CB,

$$\frac{CB}{C'B'} = \frac{CA}{C'A'} \qquad \text{or} \qquad CB = 3.9 \text{ in.} \times \frac{100 \text{ ft}}{5.0 \text{ in.}} = 78 \text{ ft}$$

The accuracy of this method depends upon the precision of the instruments. Obviously, if only a protractor is used to measure the angle, not very accurate results can be expected. Today, angles can be measured with a high degree of accuracy by using an instrument called the transit. Distances are measured by steel tapes which are graduated to $\frac{1}{100}$ of a foot. We will learn in the next sections how to compute desired distances more accurately by using mathematical formulas and tables rather than resorting to graphical solutions for which protractor and ruler have to be used.

Exercise 24-1

Solve the following exercises graphically by means of a protractor and ruler. The student will find the solutions easier to obtain if he will use a ruler graduated to tenths of an inch and a protractor on which each degree is marked.

1. The shadow of a pine tree is 60 ft long at the time when the sun's rays make an angle of 40° with the horizontal. Find the height of the tree.

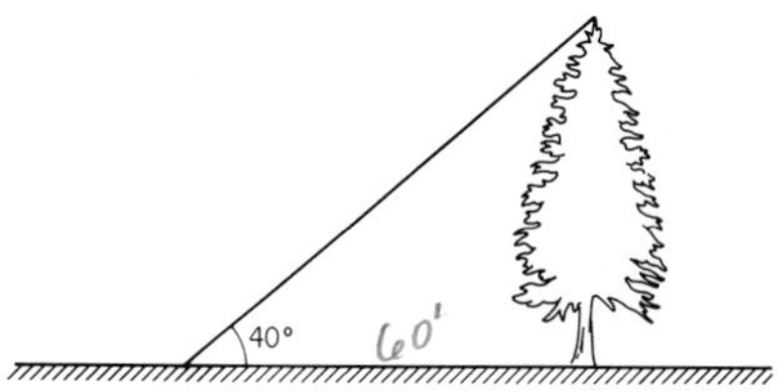

PROB. 1.

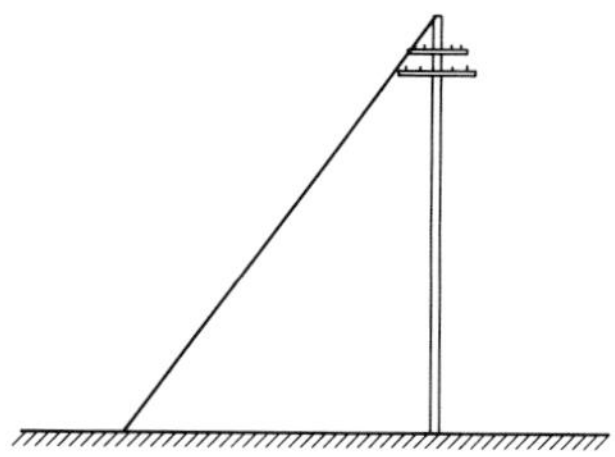

PROB. 2.

2. A telephone pole 20 ft above the ground has a guy wire anchored 15 ft from the pole. What angle does the guy wire make with the ground?
3. From the top of a cliff 124 ft high, the line of sight to a boat makes an angle of 25° with the horizontal (called the *angle of depression*). How far is the boat from the foot of the cliff?

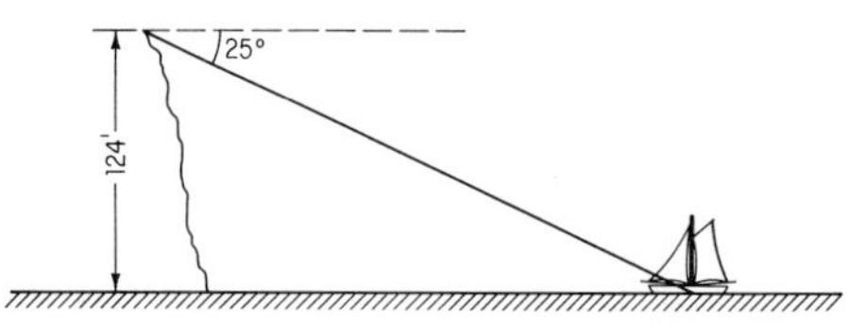

PROB. 3.

4. Find the length of a line drawn from the top of the cliff to the ship in Prob. 3.
5. A ladder 28 ft long rests against a wall. The foot of the ladder is 6 ft from the wall. What angle does the ladder make with the wall?

6. Find the distance RS between two points across a canyon, if $\overline{RT}$ is measured 2,500 ft, $m\angle R = 80°$, $m\angle T = 50°$.

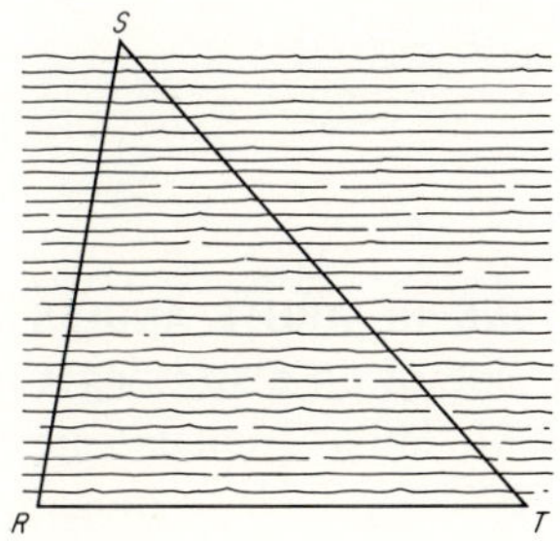

PROB. 6.

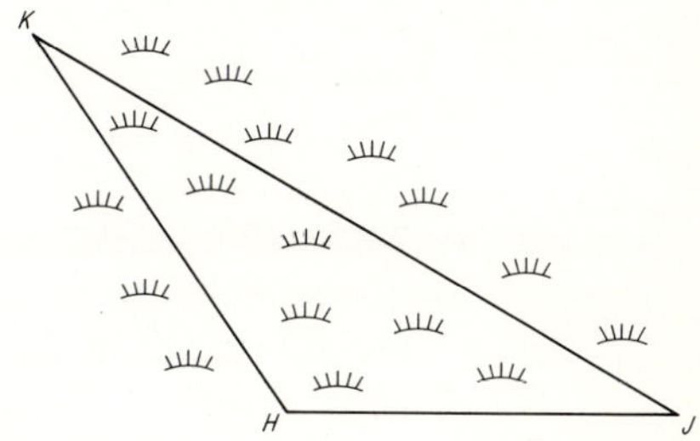

PROB. 7.

7. Find the distance KJ between two points across a marsh, if $HJ =$ 540 ft, $HK = 620$ ft, and $m\angle JHK = 125°$.
8. If the angle of elevation of the sun is 42°, how long a shadow will a man 5 ft 9 in. tall cast on level ground?
9. A surveyor is hired to determine the vertical height PT of a small mountain. With transit he measures the $\angle TSP$ at the base of the mountain to be 46°. He then moves (along horizontal line) to a point R, 1,000 ft from S, and measures $\angle SRP = 27°$. From his measurements, determine the height PT.

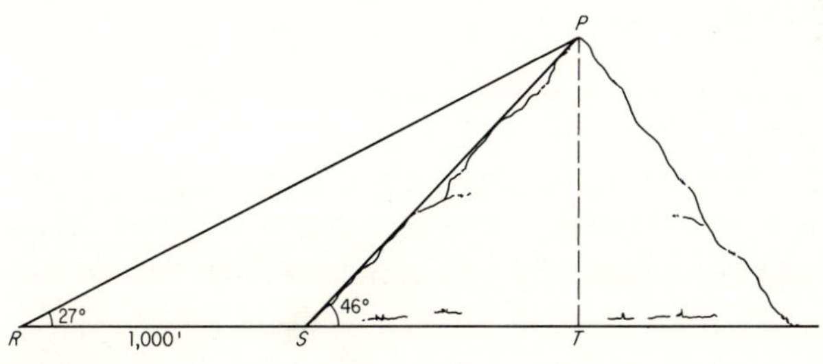

PROB. 9.

10. A monument BC is located on the shore of a small lake. From a point A on the opposite shore, $\angle BAC$ is measured equal to 32°. At a point 200 ft further along the direction BA, $\angle BDC$ is measured equal to 20°. Find (a) the height BC of the monument; (b) the width AB across the lake.

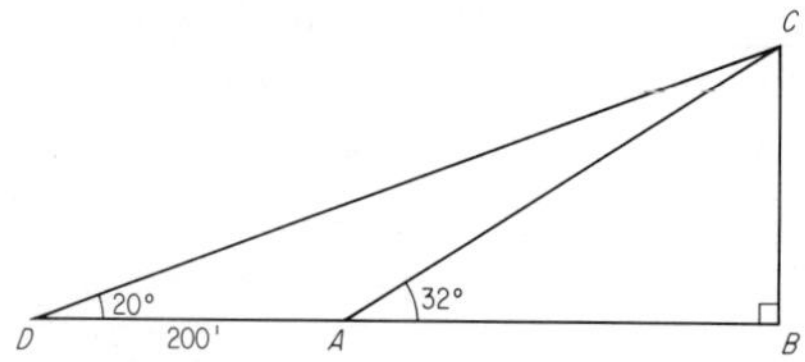

PROB. 10.

24-3 TRIGONOMETRIC FUNCTIONS OF ACUTE ANGLES If lines are drawn from various points along one side of an acute angle perpendicular to the second side of the angle, a set of similar triangles are formed. Thus in Fig. 24-3, $\triangle ABC \sim \triangle AB'C' \sim \triangle AB''C''$ . . . . We know from Sec. 16-4 that

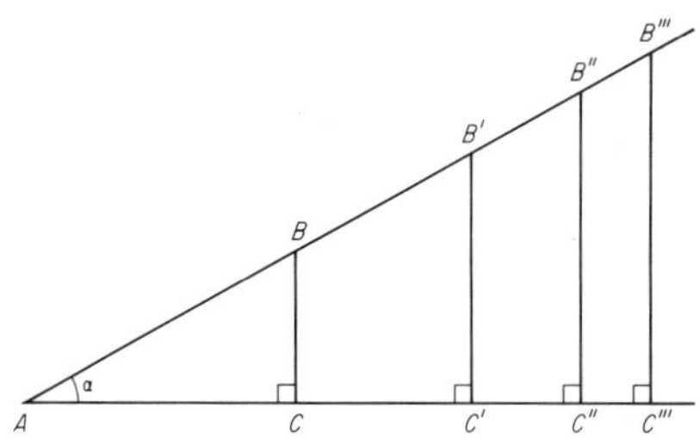

FIG. 24-3.

ratios of sides of similar triangles are equal. Thus

$$\frac{BC}{AB} = \frac{B'C'}{AB'} = \frac{B''C''}{AB''} \cdots \qquad \text{or} \qquad \frac{BC}{AC} = \frac{B'C'}{AC'} = \frac{B''C''}{AC''} \cdots$$

The values of these ratios depend upon the size of $\angle\alpha$ and for this reason are called "functions of $\angle\alpha$."

In the study of trigonometry it is customary to label the sides opposite angles by lowercase letters corresponding to the capital letters used to designate the angles (see Fig. 24-4). The right angle is usually given the capital

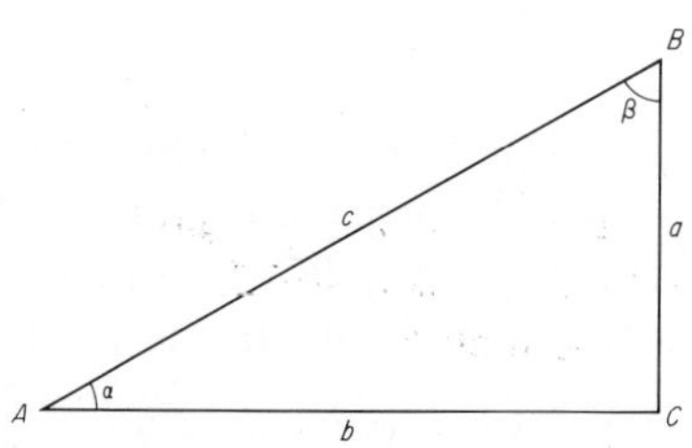

FIG. 24-4.

letter C. This will then result in the hypotenuse being labeled with a lowercase c and the legs with a and b. The acute angles are often labeled with Greek letters. In $\triangle ABC$ side a is opposite to A and adjacent to B, and side b is opposite to B and adjacent to A.

There are six possible ratios that can be formed from three sides of any right triangle. We will be concerned with three of them in this chapter. These ratios are given names. The names and their definitions follow.

$\frac{a}{c}$ = sine of $\angle A$ (abbreviated $\sin A$)

$\frac{b}{c}$ = cosine of $\angle A$ (abbreviated $\cos A$)

$\frac{a}{b}$ = tangent of $\angle A$ (abbreviated $\tan A$)

In like manner we may write:

$$\sin A = \frac{a}{c} = \frac{\text{opposite}}{\text{hypotenuse}} \tag{24-1}$$

$$\cos A = \frac{b}{c} = \frac{\text{adjacent}}{\text{hypotenuse}} \tag{24-2}$$

$$\tan A = \frac{a}{b} = \frac{\text{opposite}}{\text{adjacent}} \tag{24-3}$$

The value for the functions of an acute angle can be roughly determined by drawing with protractor and ruler a right triangle with an angle equal to the given acute angle. Suppose we wish to find the functions of 32°. Let us draw $\triangle ABC$ with $m\angle A = 32°$ and side AC any convenient length, say, 10 in. (Fig. 24-5). With a ruler we then can measure side a and hypotenuse c. Let

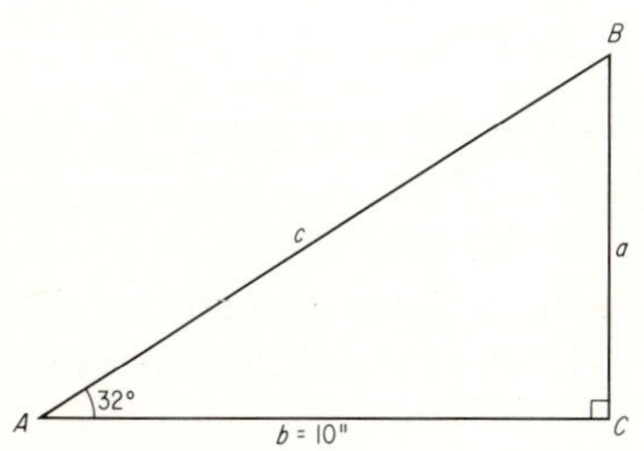

FIG. 24-5.

us assume a measures 6.2 in. and c measures 11.8 in. Our functions then will be:

$$\sin A = \frac{6.2}{11.8} = 0.53$$

$$\cos A = \frac{10}{11.8} = 0.85$$

$$\tan A = \frac{6.2}{10} = 0.62$$

It should be clear that the functions of an angle change as the angle changes. Consider the tangent function. On ruled paper (Fig. 24-6), draw triangles with varying acute angles and a horizontal side equal to 10-unit divisions. In the figure, $\overline{AG}$, $\overline{AF}$, $\overline{AE}$, $\overline{AD}$, and $\overline{AB}$ make with $\overline{AC}$ angles of 20°, 30°, 40°, 50°, and 60°, respectively. By dividing the measure of the side opposite the given acute angle by the measure of the adjacent side, we should get the results (approximately) shown in the table of Fig. 24-6. It will be noted

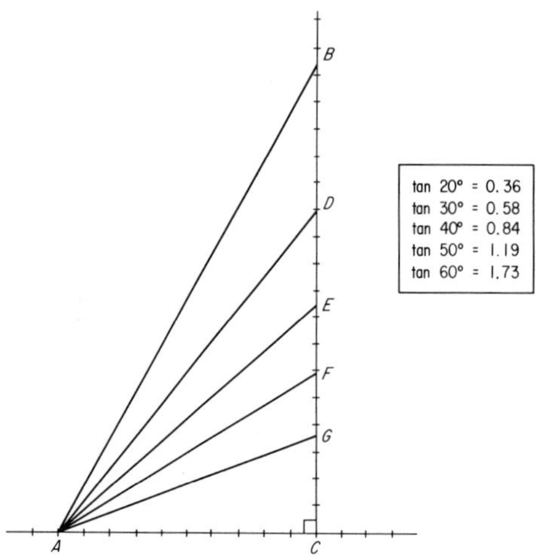

FIG. 24-6.

that as the angle increases, the tangent increases. If the angle is very small, the tangent is small. The tangent of 45° is 1, since we then have an isosceles right triangle with the side opposite equal to the side adjacent. When the angle approaches 90° the hypotenuse is nearly parallel to the side opposite the acute angle. This will make the side opposite very large. Thus, the tangent of an angle near 90° is very large.

24-4 TRIGONOMETRIC TABLES We could find the functions of other angles by the graphic method. However, the procedure would be tedious and the results would be, at best, only crude approximations of the functions. Fortunately, it is not necessary to do this, since these ratios have been computed by higher mathematics more accurately for very many angles. In Table 2 of the appendix, the values of the functions for acute angles containing a whole number of degrees are given accurately to four decimals.

Example 24-1 Compute, accurate to four decimals, the value of tan 67.3°.

SOLUTION Using Table 2, we can write:

$$\begin{array}{cc} & m\angle\alpha & & \tan\alpha & \\ 1.0\left\{\begin{array}{l} \\ 0.3\left\{\begin{array}{l} \\ \\ \end{array}\right. \end{array}\right. & \begin{array}{c} 68.0 \\ 67.3 \\ 67.0 \end{array} & & \begin{array}{c} 2.4751 \\ ? \\ 2.3559 \end{array} & \left.\begin{array}{l} \\ \left.\begin{array}{l} \\ \\ \end{array}\right\}c \end{array}\right\}0.1192 \end{array}$$

Setting up a proportion between corresponding numbers we interpolate

$$\frac{c}{0.1192}=\frac{0.3}{1.0} \qquad \text{or} \qquad \begin{aligned} c &= 0.3(0.1192) \\ &= 0.03576 \\ &\approx 0.0358 \end{aligned}$$

$$\text{Hence} \quad \begin{aligned} \tan\alpha &= 2.3559 + 0.0358 \\ &= 2.3917 \end{aligned}$$

Example 24-2 Determine, to the nearest tenth of a degree, $m\angle\alpha$ if $\sin\alpha =$ 0.6740.

SOLUTION From Table 2, we write

$$1.0\left\{\begin{array}{l} \sin 43.0 = 0.6820 \\ c\left\{\begin{array}{l} \sin\alpha \quad = 0.6740 \\ \sin 42.0 = 0.6691 \end{array}\right. \end{array}\right. \left.\begin{array}{l} \\ \left.\begin{array}{l} \\ \\ \end{array}\right\}0.0049 \end{array}\right\}0.0129$$

Setting up a proportion between corresponding numbers, we get

$$\frac{c}{1.0}=\frac{0.0049}{0.0129} \qquad \text{or} \qquad c \approx 0.4$$

$$\text{Then} \quad \begin{aligned} m\angle\alpha &= 42.0^\circ + 0.4^\circ \\ &= 42.4^\circ \end{aligned}$$

Exercise 24-2

In Probs. 1 to 9, use Table 2 and interpolate to find the value of each of the following trigonometric functions:

1. sin 28.7°
2. cos 16.8°
3. tan 37.6°
4. cos 78.1°
5. tan 59.2°
6. sin 84.3°
7. tan 2.5°
8. sin 36.4°
9. cos 45.9°
10. sin 53.6°
11. cos 11.9°
12. tan 80.4°

In Probs. 13 to 24, use Table 2 and interpolate to find the $m\angle\alpha$ accurate to the tenth of a degree.

13. $\cos\alpha = 0.3157$	17. $\sin\alpha = 0.8337$	21. $\tan\alpha = 0.6775$
14. $\tan\alpha = 1.0109$	18. $\tan\alpha = 5.5413$	22. $\tan\alpha = 1.6103$
15. $\sin\alpha = 0.9808$	19. $\sin\alpha = 0.4387$	23. $\sin\alpha = 0.9340$
16. $\cos\alpha = 0.4531$	20. $\cos\alpha = 0.8482$	24. $\cos\alpha = 0.5074$

In Probs. 25 to 32, find to the nearest degree the size of ϕ for each of the figures.

25.
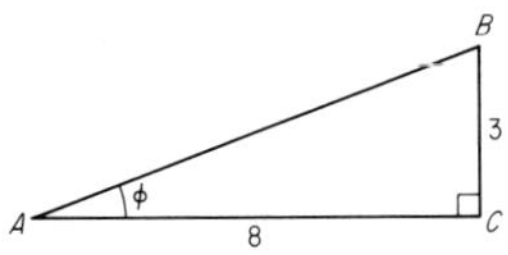

29.
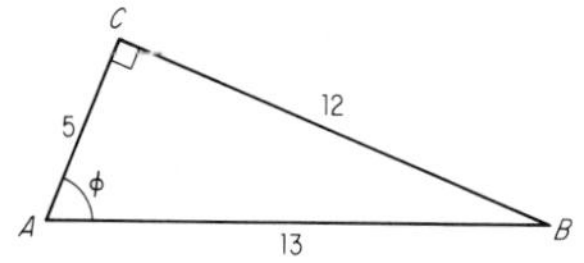

26.
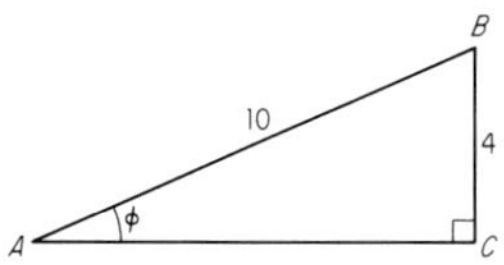

30.
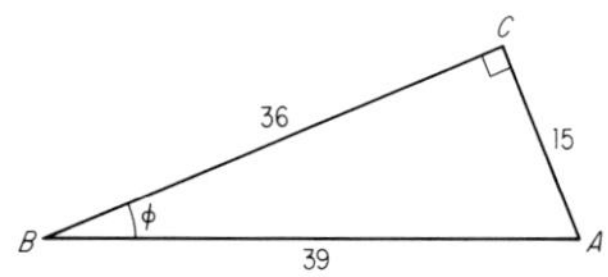

27.
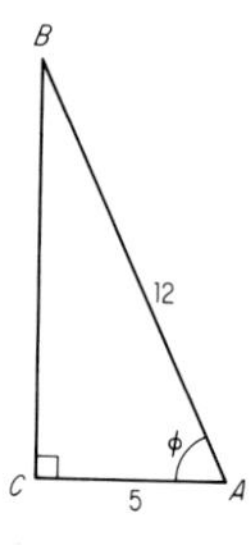

31.
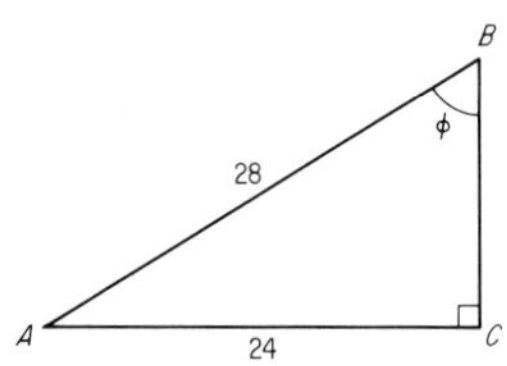

28.
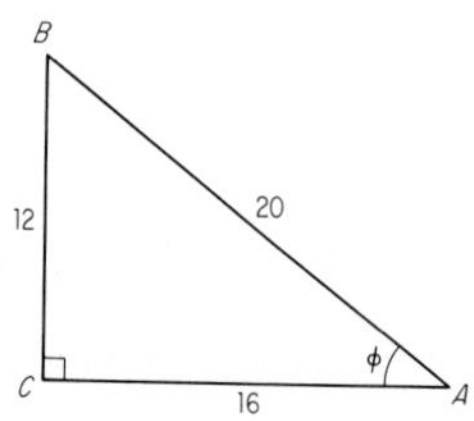

32.
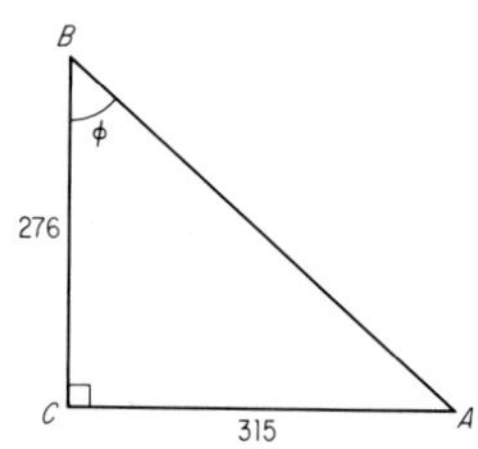

24-5 USE OF TRIGONOMETRIC FUNCTIONS One of the important uses of trigonometry is that of computing unknown angles and sides of a triangle from measured values of other angles or sides of the triangle. With the use of trigonometric functions it is possible to compute any part of a right triangle if any two sides or a side and an acute angle are known. As a general rule, if the angles are measured accurate to the degree, the sides should be measured or computed accurate to three significant places; if the angles are accurate to tenths of a degree, then the sides should be measured or computed accurate to four significant figures. We will follow this practice in the examples and exercises that follow.

Example 24-3

Given: Right $\triangle ABC$ with $b = 23.2$ and $m\angle A = 28°$.
Find: a and c.

SOLUTION

$$\frac{a}{b} = \tan 28°$$

$$a = b \tan 28°$$
$$= 23.2(0.5317)$$
$$= 12.33544 \quad \text{or} \quad 12.3 \text{ (approximately)}$$

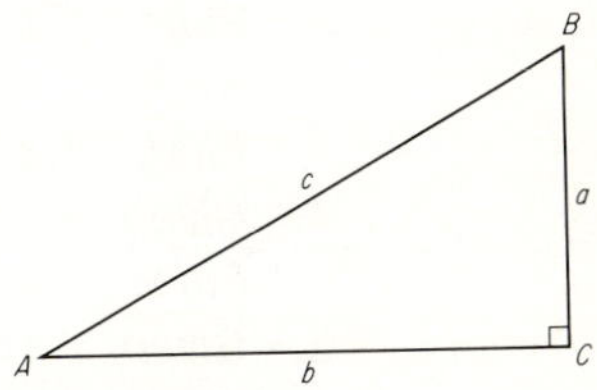

EXAMPLE 24-3.

$$\frac{b}{c} = \cos 28°$$

$$c = \frac{b}{\cos 28°}$$

$$= \frac{23.2}{0.8829}$$

$$= 26.3 \text{ (approximately)}$$

Example 24-4

Given: Right $\triangle RST$ with $s = 15.0$ and $t = 22.0$.
Find: $m\angle R$ and side r.

SOLUTION

$$\cos R = \frac{s}{t}$$

$$= \frac{15.0}{22.0}$$

$$= 0.6818$$

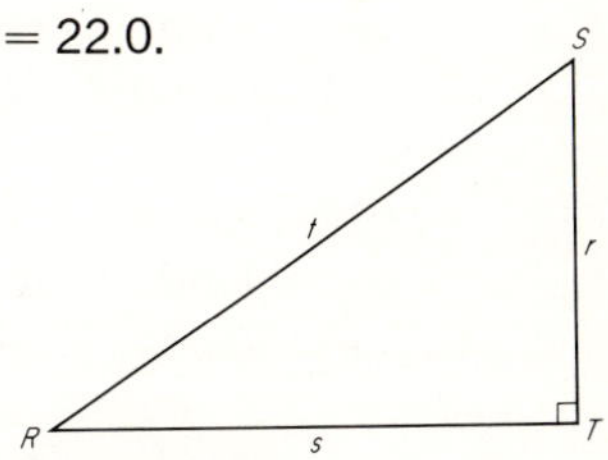

EXAMPLE 24-4.

From Table 2, $m\angle R = 47°$ (accurate to whole degree),

$$\frac{r}{s} = \tan R$$

$$r = s \tan R$$
$$= 15.0(1.0724)$$
$$= 16.1 \text{ (approximately)}$$

Exercise 24-3

In each of the following problems, make your answers have an accuracy that is implied in the given data; i.e., angles in degrees and three-place dimensions of sides of the triangle are consistent. If the angle is given accurate to the tenth of a degree the dimensions of the sides should be given and/or computed accurate to four significant figures. In Probs. 1 to 12, $\triangle ABC$ is a right $\triangle$ and $\angle C$ is the right angle.

1. *Given:* $m\angle A = 54°; b = 15.0$
   Find: a
2. *Given:* $m\angle A = 37°; c = 29.0$
   Find: b
3. *Given:* $m\angle B = 65°; a = 18.0$
   Find: c
4. *Given:* $m\angle B = 23°; c = 0.0280$
   Find: a
5. *Given:* $m\angle A = 67.2°; b = 38.70$
   Find: a
6. *Given:* $m\angle B = 36.5°; a = 2.344$
   Find: b
7. *Given:* $c = 3.80; a = 1.90$
   Find: $m\angle B$
8. *Given:* $b = 0.0340; c = 0.0590$
   Find: $m\angle A$
9. *Given:* $a = 84.0; b = 42.0$
   Find: $m\angle A$
10. *Given:* $a = 0.0820; b = 0.0280$
    Find: $m\angle B$
11. *Given:* $a = 4.804; b = 2.402$
    Find: $m\angle A$
12. *Given:* $b = 35.89; c = 11.24$
    Find: $m\angle A$
13. *Given:* $RT = 15.0; ST = 15.0;$
    $m\angle R = 42°$
    Find: RS

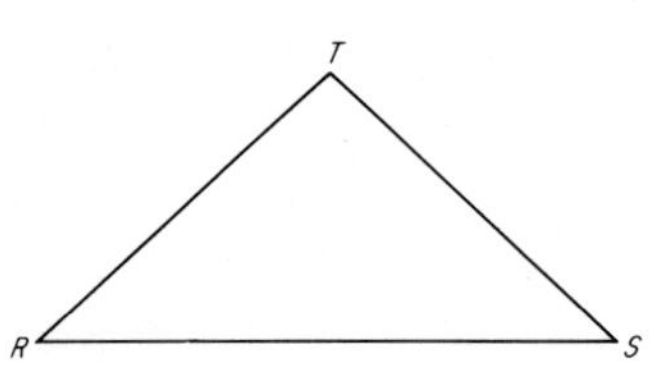

PROB. 13.

14. *Given:* $AC = 28.0$; $m\angle A = 32°$
Find: AD

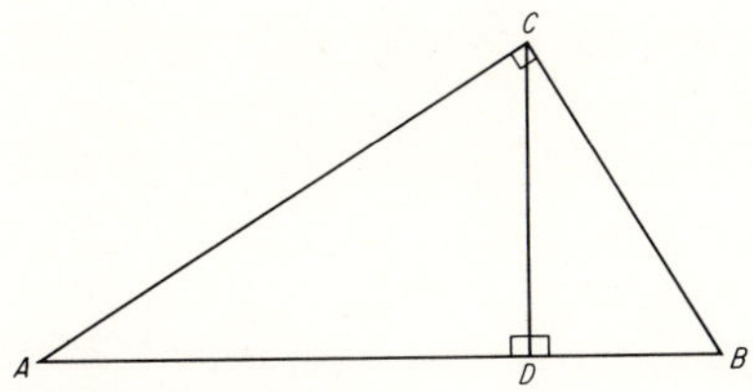

PROBS. 14 and 15.

15. *Given:* $AC = 28.0$;
$m\angle A = 32°$
Find: AB

16. *Given:* $AD = 12.0$;
$AB = 24.0$;
$m\angle A = 60°$
Find: DE

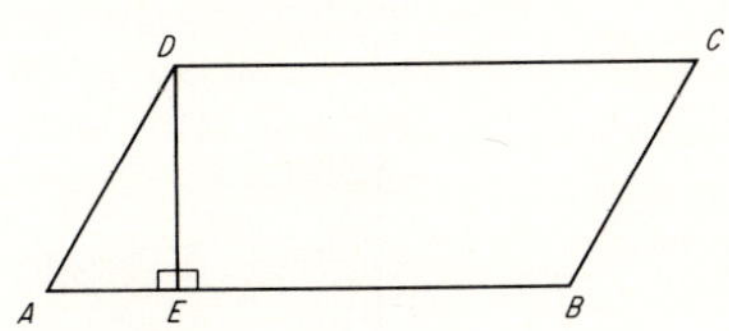

PROB. 16.

17. *Given:* $HK = 18.0$; $JK = 18.0$;
$HJ = 26.0$
Find: $m\angle H$

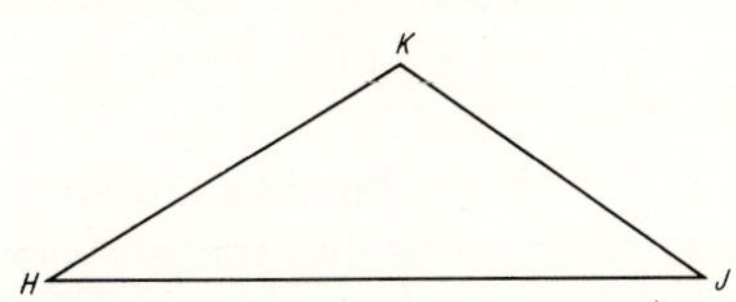

PROBS. 17 and 18.

18. *Find:* The area of $\triangle HJK$ of Prob. 17.

19. *Given:* $\odot O$ with $m\widehat{MN} = 86°$; chord $MN = 654$ ft
Find: radius OM

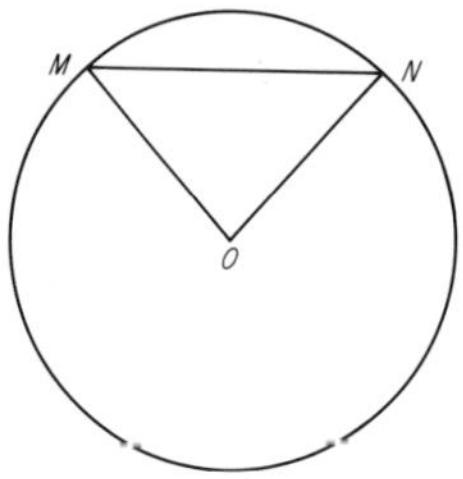

PROB. 19.

20. A railroad track makes an angle of 2° with the horizontal. How many feet will a person have risen after having traveled a mile up the track?
21. Find the area of a regular 10-sided polygon (decagon) inscribed in a circle with a 25-ft radius. (*Hint:* One-half the measure of the central angle $= 18°$.)

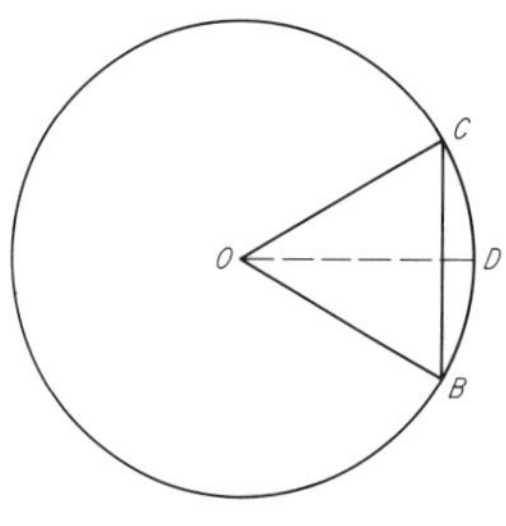

PROB. 21.

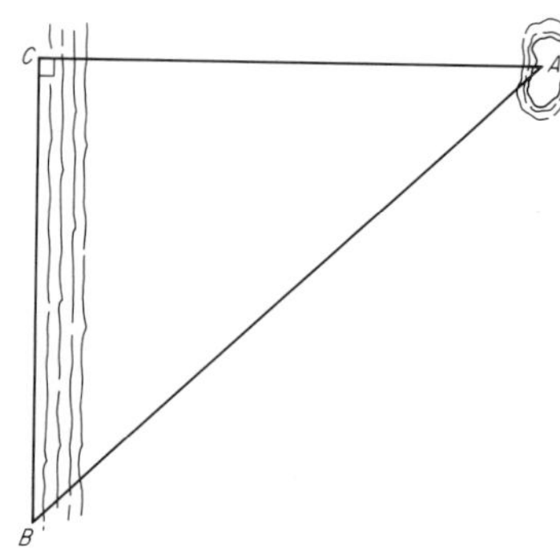

PROB. 22.

22. A surveying party wished to determine the distance between a point C on the shore and a point A on an island. They measured $\overline{CB}$ 485 ft long along the shore at right angle to $\overline{CA}$. They then found $m\angle B = 48°$. What is the distance from C to A?
23. From a lighthouse 216 ft above the ocean, a ship is observed making an angle of depression (with the horizontal) whose measure is 18°. How far is the ship from the observer?
24. A television mast is 186 ft high and one guy wire is 215 ft long. What angle does the guy wire make with the mast? How far is the guy wire anchored from the base of the mast?

25. The angle of elevation of a balloon as observed from a point P has a measure of 46°. If the balloon is directly over a point on the ground 3,850 ft from P, at what height is the balloon?
26. To determine the height of clouds above the earth, a vertical beam of light is sent up to the clouds. An observer 600 ft from the searchlight finds the angle of elevation of the lighted cloud to have a measure of 68°. Find the distance between the earth and the cloud.
27. From the top of a building 325 ft high, the angles of depression to the top and bottom of a monument have measures of 61° and 48° respectively. Find the height of the monument.
28. Find the height KL of a building if $JH = 324$ ft, $m\angle KJL = 49°$, and $m\angle KHL = 36°$.

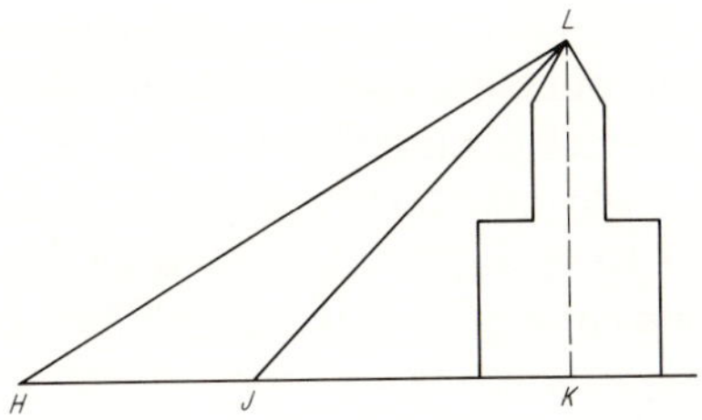

PROB. 28.

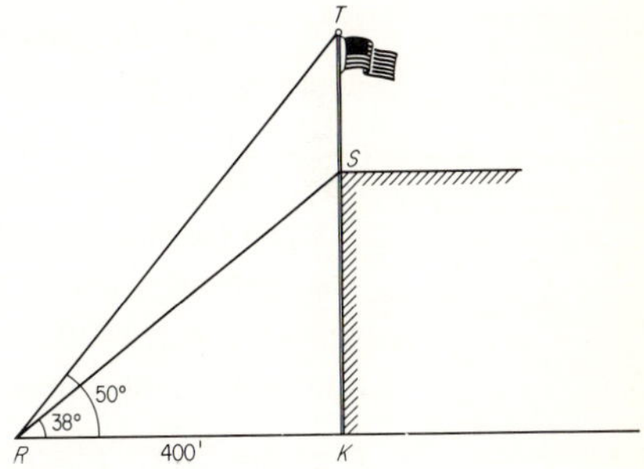

PROB. 29.

29. Find the height ST of the flag on top of the cliff KS if $RK = 400$ ft, $m\angle KRS = 38°$, and $m\angle KRT = 50°$.

24-6 TRIGONOMETRIC SCALES OF A SLIDE RULE Not only can the trigonometric functions of angles be found in tabular form, but the same functions can be determined with a slide rule. Most modern slide rules have three trigonometric scales. These scales are usually on the slide and are identified with the letters S, ST, and T. These three scales are correlated with the C, D, CI, and DI scales. The graduations on the S, ST, and T represent measures of angles. The angles are expressed either in degrees and minutes or in degrees and decimal parts of degrees. In this chapter, we will consider only the rule with angles expressed in degrees and decimal parts of degrees.

24-7 SINE SCALES The S and ST scales are used for finding the sines and cosines of angles. Upon examination of the S scale, two sets of numbers will be found. One set, using black numbers, when read from left to right, covers angles from 5.74° to 90° and is used for finding sines of angles. When read from right to left, using the red numbers, the scale covers angles from 0° to 84.26°, and is used for finding cosines of angles.

Let us consider the black numbers. To find the sine of an angle (between 5.74° and 90°), move the hairline to the angle on the S scale, reading from left to right, the sine of the angle is found under the hairline on the C scale. If the rule is closed during the above manipulations, the sine can also be read on the D scale.

A study of Table 2 will reveal that if an angle has a value less than 90° and greater than 5.74°, the sine of the angle will have a value less than 1.00 and greater than 0.100. Hence, it should be easy to locate the decimal point when using the rule to find sines of such angles in that domain.

Example 24-5 Evaluate sin 34.7°.

SOLUTION Opposite 34.7° on S scale read 569 on the C scale (or on D scale when the rule is closed). Hence $\sin 34.7° = 0.569$.

The ST scale is used to determine the sines of all angles between 0.6° and 5.74°. Angles on the ST scale are paired with the corresponding sines on the C scale (or the D scale if the rule is closed).

A further study of Table 2 will indicate that angles from 0.6° to 5.74° have sines from 0.0100 to 0.100. Here, again, it should be easy to locate the decimal point when finding sines of angles in the 0.6°–5.74° domain.

Example 24-6 Evaluate sin 3.47°.

SOLUTION Opposite 3.47° on the ST scale read 605 on the C scale (or on D scale if rule is closed). Hence, $\sin 3.47° = 0.0605$.

24-8 COSINE OF AN ANGLE Before we can use the slide rule to find cosines of angles, we will need to show how the sines and cosines are related. Consider again Fig. 24-4 and our definitions for the sine and the cosine of an angle. The cosine of an angle was defined as the ratio of the side adjacent the angle to the hypotenuse. Hence, in Fig. 24-4, $\cos \angle B = \frac{a}{c}$. We also know that the sine of an angle is the ratio of the side opposite the angle to the hypotenuse. In Fig. 24-4, sine $\angle A = \frac{a}{c}$. Thus, we have shown that $\cos B = \sin A$. But $\angle A$ and $\angle B$ are complementary angles, and $m\angle A = 90° - m\angle B$. Thus,

$$\cos B = \sin A = \sin(90° - B) \tag{24-4}$$

Because the cosine of an angle is equivalent to the sine of the complement of the angle, we can also use the S scale of the rule to find cosines of angles. To find the cosine of angles from 0° to 84.26°, move the hairline to the angle on the S scale, reading from right to left, the cosine of the angle is found under the hairline on the C scale (or D scale if the rule is closed). Cosines of angles between 0° and 84.26° have a range between 1.000 and 0.100.

Example 24-7 Evaluate cos 52.8°.

SOLUTION Opposite 52.8° on the red S scale read 605 on the C scale (or D scale when the rule is closed). Hence, cos 52.8° = 0.605.

24-9 T (TANGENT) SCALE Tangents of angles are found by matching the T and C (or D on a closed-rule scale). Tangents of angles in the domain from 5.74° to 45° are found by moving the hairline to the angles on the T scale, using the (black) numbers of the scale which increase from left to right, and reading the values of the tangents under the hairline on the C scale. The tangent of any angle in this domain will have a range of values between 0.100 and 1.00.

Example 24-8 Evaluate tan 21.5°.

SOLUTION Opposite 21.5° on the T scale read 394 on the C scale (or D scale if rule is closed). Hence, tan 21.5° = 0.394.

24-10 THE ST SCALE We noted that the ST scale is used to find the sine function of angles between 0.6° and 5.74°. A further study of Table 2 will reveal values for the sines and tangents of angles in the 0.6°–5.74° range are very nearly equal. Within the limits of accuracy of a slide rule, they can be considered equal. Hence, the ST scale can also be used to find tangents of small angles.

Example 24-9 Evaluate tan 1.38°.
SOLUTION Use the ST and C scales to find tan 1.38° = 0.0241.

24-11 TANGENTS OF ANGLES FROM 45° TO 90° Tangents of angles between 45° and 90° have values greater than one. In this domain, the T scale can be used for finding angles up to 84.26°, by reading the red numbers (which increase from right to left). Let us again consider Fig. 24-5. We will need to define a new function, the cotangent. The *cotangent* of an acute angle of a right triangle is defined as the ratio of the adjacent and opposite sides. Thus, we see that the cotangent (abbreviated cot) of an angle is the reciprocal of the tangent of the angle.

$$\tan A = \frac{1}{\cot A} \qquad (24\text{-}5)$$

It can, also, readily be shown that

$$\cot A = \tan B = \tan (90 - A) \tag{24-6}$$

To find the tangent of an angle in the domain between 45° and 84.26°, the hairline is set to the angle on the red T scale, reading from right to left, and the tangent of the angle is under the hairline on the CI scale (or on the DI scale when the rule is closed). In this domain, the tangent function will range from 1.00 to 10.00.

Example 24-10 Evaluate tan 67.4°.

SOLUTION Opposite 67.4° on T *red,* read 240 on the CI scale (or on DI scale if rule is closed). Hence, tan 67.4° = 2.40.

Example 24-11 Given tan x = 1.42. Find x.

SOLUTION Since tan $x > 1.00$, we know that $x > 45°$. Hence, move the hairline over 142 on the CI or DI scale of a closed rule. Read 549 on the red T scale. Hence, $x = 54.9°$.

Example 24-12 Evaluate 5.37 tan 37.4°.

SOLUTION Close the rule. Draw the hairline over 37.4° on the T scale. (You can read tan 37.4° = 0.765 on the D scale.) Draw the right index of the C scale under the hairline. Opposite 537 on the C scale, read 411 on the D scale. Hence, 5.37 tan 37.4° = 4.11.

Exercise 24-4

Evaluate each of the following:

1. sin 28°
2. sin 56°
3. cos 27°
4. cos 61°
5. tan 24°
6. tan 40°
7. sin 3°
8. tan 4°
9. tan 2°
10. sin 2°
11. sin 11.4°
12. sin 63.8°
13. cos 25.9°
14. cos 74.0°
15. cos 47.3°
16. tan 36.7°
17. tan 41.8°
18. sin 3.7°
19. tan 2.81°
20. cos 38.2°
21. cos 67.4°

Find the value of x in each of the following:

22. sin x = 0.500
23. cos x = 0.866
24. cos x = 0.940
25. tan x = 0.249
26. tan x = 0.839
27. cos x = 0.775
28. sin x = 0.208
29. sin x = 0.889
30. tan x = 0.625
31. tan x = 0.424
32. sin x = 0.904
33. sin x = 0.935
34. sin x = 0.0506
35. sin x = 0.0244
36. tan x = 0.0454
37. tan x = 1.54
38. tan x = 1.97
39. tan x = 2.38

Evaluate the following:

40. $27 \sin 60^\circ$
41. $40 \cos 29^\circ$
42. $87 \tan 34^\circ$
43. $\dfrac{54 \sin 27^\circ}{19}$
44. $\dfrac{89 \tan 25^\circ}{34}$
45. $\dfrac{430 \cos 72^\circ}{59}$
46. $11.4 \sin 18.9^\circ$
47. $36.0 \cos 83.2^\circ$
48. $58.3 \tan 37.1^\circ$
49. $\dfrac{65.5 \cos 18.3^\circ}{9.21}$
50. $\dfrac{0.497 \sin 67.3^\circ}{2.55}$
51. $\dfrac{0.835 \tan 40.7^\circ}{0.416}$

APPENDIX 1

GREEK ALPHABET

GREEK ALPHABET

LETTERS		NAMES	LETTERS		NAMES	LETTERS		NAMES
A	α	Alpha	I	ι	Iota	P	ρ	Rho
B	β	Beta	K	κ	Kappa	Σ	σ	Sigma
Γ	γ	Gamma	Λ	λ	Lambda	T	τ	Tau
Δ	δ	Delta	M	μ	Mu	Υ	υ	Upsilon
E	ϵ	Epsilon	N	ν	Nu	Φ	ϕ	Phi
Z	ζ	Zeta	Ξ	ξ	Xi	X	χ	Chi
H	η	Eta	O	o	Omicron	Ψ	ψ	Psi
Θ	θ	Theta	Π	π	Pi	Ω	ω	Omega

APPENDIX 2

SYMBOLS AND ABBREVIATIONS

Symbols and Abbreviations

SYMBOL	MEANING	SYMBOL	MEANING
$\overline{AB}$	segment *AB*	$\lvert x \rvert$	the absolute value of *x*
$\overrightarrow{AB}$	ray *AB*	$\sim$	is similar to
$m\overline{AB}$	the measure of segment *AB*	$\perp$	is perpendicular to
AB	the measure of segment *AB*	$\parallel$	is parallel to
$\angle$	angle	$>$	is greater than
$\angle$s	angles	$<$	is less than
$m\angle ABC$	the measure of angle *ABC*	$\geqslant$	is greater than or equal to
$\widehat{AB}$	arc *AB*	$\leqslant$	is less than or equal to
$=$	is equal to; equals	$\therefore$	therefore
$\neq$	is not equal to	$\ldots$	and so on
$\sqrt{x}$	the nonnegative square root of *x*	$\triangle$	triangle
		$\triangle$s	triangles
▭	rectangle	▱	parallelogram
▭s	rectangles	▱s	parallelograms
$\odot$	circle	$^\circ$	degree
$\odot$s	circles		

SAS — If two triangles have two sides and the included angle of one congruent, respectively, to two sides and the included angle of the other, the triangles are congruent.

ASA — If two triangles have two angles and the included side of one congruent, respectively, to two angles and the included side of the other, the triangles are congruent.

SSS — If two triangles have the three sides of one congruent, respectively, to the three sides of the other, the triangles are congruent.

TABLE 1

COMMON LOGARITHMS

TABLE 1 Common Logarithms

x	0	1	2	3	4	5	6	7	8	9
1.0	.0000	.0043	.0086	.0128	.0170	.0212	.0253	.0294	.0334	.0374
1.1	.0414	.0453	.0492	.0531	.0569	.0607	.0645	.0682	.0719	.0755
1.2	.0792	.0828	.0864	.0899	.0934	.0969	.1004	.1038	.1072	.1106
1.3	.1139	.1173	.1206	.1239	.1271	.1303	.1335	.1367	.1399	.1430
1.4	.1461	.1492	.1523	.1553	.1584	.1614	.1644	.1673	.1703	.1732
1.5	.1761	.1790	.1818	.1847	.1875	.1903	.1931	.1959	.1987	.2014
1.6	.2041	.2068	.2095	.2122	.2148	.2175	.2201	.2227	.2253	.2279
1.7	.2304	.2330	.2355	.2380	.2405	.2430	.2455	.2480	.2504	.2529
1.8	.2553	.2577	.2601	.2625	.2648	.2672	.2695	.2718	.2742	.2765
1.9	.2788	.2810	.2833	.2856	.2878	.2900	.2923	.2945	.2967	.2989
2.0	.3010	.3032	.3054	.3075	.3096	.3118	.3139	.3160	.3181	.3201
2.1	.3222	.3243	.3263	.3284	.3304	.3324	.3345	.3365	.3385	.3404
2.2	.3424	.3444	.3464	.3483	.3502	.3522	.3541	.3560	.3579	.3598
2.3	.3617	.3636	.3655	.3674	.3692	.3711	.3729	.3747	.3766	.3784
2.4	.3802	.3820	.3838	.3856	.3874	.3892	.3909	.3927	.3945	.3962
2.5	.3979	.3997	.4014	.4031	.4048	.4065	.4082	.4099	.4116	.4133
2.6	.4150	.4166	.4183	.4200	.4216	.4232	.4249	.4265	.4281	.4298
2.7	.4314	.4330	.4346	.4362	.4378	.4393	.4409	.4425	.4440	.4456
2.8	.4472	.4487	.4502	.4518	.4533	.4548	.4564	.4579	.4594	.4609
2.9	.4624	.4639	.4654	.4669	.4613	.4698	.4713	.4728	.4742	.4757
3.0	.4771	.4786	.4800	.4814	.4829	.4843	.4857	.4871	.4886	.4900
3.1	.4914	.4928	.4942	.4955	.4969	.4983	.4997	.5011	.5024	.5038
3.2	.5051	.5065	.5079	.5092	.5105	.5119	.5132	.5145	.5159	.5172
3.3	.5185	.5198	.5211	.5224	.5237	.5250	.5263	.5276	.5289	.5302
3.4	.5315	.5328	.5340	.5353	.5366	.5378	.5391	.5403	.5416	.5428
3.5	.5441	.5453	.5465	.5478	.5490	.5502	.5514	.5527	.5539	.5551
3.6	.5563	.5575	.5587	.5599	.5611	.5623	.5635	.5647	.5658	.5670
3.7	.5682	.5694	.5705	.5717	.5729	.5740	.5752	.5763	.5775	.5786
3.8	.5798	.5809	.5821	.5832	.5843	.5855	.5866	.5877	.5888	.5899
3.9	.5911	.5922	.5933	.5944	.5955	.5966	.5977	.5988	.5999	.6010
4.0	.6021	.6031	.6042	.6053	.6064	.6075	.6085	.6096	.6107	.6117
4.1	.6128	.6138	.6149	.6160	.6170	.6180	.6191	.6201	.6212	.6222
4.2	.6232	.6243	.6253	.6263	.6274	.6284	.6294	.6304	.6314	.6325
4.3	.6335	.6345	.6355	.6365	.6375	.6385	.6395	.6405	.6415	.6425
4.4	.6435	.6444	.6454	.6464	.6474	.6484	.6493	.6503	.6513	.6522
4.5	.6532	.6542	.6551	.6561	.6571	.6580	.6590	.6599	.6609	.6618
4.6	.6628	.6637	.6646	.6656	.6665	.6675	.6684	.6693	.6702	.6712
4.7	.6721	.6730	.6739	.6749	.6758	.6767	.6776	.6785	.6794	.6803
4.8	.6812	.6821	.6830	.6839	.6848	.6857	.6866	.6875	.6884	.6893
4.9	.6902	.6911	.6920	.6928	.6937	.6946	.6955	.6964	.6972	.6981
5.0	.6990	.6998	.7007	.7016	.7024	.7033	.7042	.7050	.7059	.7067
5.1	.7076	.7084	.7093	.7101	.7110	.7118	.7126	.7135	.7143	.7152
5.2	.7160	.7168	.7177	.7185	.7193	.7202	.7210	.7218	.7226	.7235
5.3	.7243	.7251	.7259	.7267	.7275	.7284	.7292	.7300	.7308	.7316
5.4	.7324	.7332	.7340	.7348	.7356	.7364	.7372	.7380	.7388	.7396
x	0	1	2	3	4	5	6	7	8	9

TABLE 1 (*continued*)

x	0	1	2	3	4	5	6	7	8	9
5.5	.7404	.7412	.7419	.7427	.7435	.7443	.7451	.7459	.7466	.7474
5.6	.7482	.7490	.7497	.7505	.7513	.7520	.7528	.7536	.7543	.7551
5.7	.7559	.7566	.7574	.7582	.7589	.7597	.7604	.7612	.7619	.7627
5.8	.7634	.7642	.7649	.7657	.7664	.7672	.7679	.7686	.7694	.7701
5.9	.7709	.7716	.7723	.7731	.7738	.7745	.7752	.7760	.7767	.7774
6.0	.7782	.7789	.7796	.7803	.7810	.7818	.7825	.7832	.7839	.7846
6.1	.7853	.7860	.7868	.7875	.7882	.7889	.7896	.7903	.7910	.7917
6.2	.7924	.7931	.7938	.7945	.7952	.7959	.7966	.7973	.7980	.7987
6.3	.7993	.8000	.8007	.8014	.8021	.8028	.8035	.8041	.8048	.8055
6.4	.8062	.8069	.8075	.8082	.8089	.8096	.8102	.8109	.8116	.8122
6.5	.8129	.8136	.8142	.8149	.8156	.8162	.8169	.8176	.8182	.8189
6.6	.8195	.8202	.8209	.8215	.8222	.8228	.8235	.8241	.8248	.8254
6.7	.8261	.8267	.8274	.8280	.8287	.8293	.8299	.8306	.8312	.8319
6.8	.8325	.8331	.8338	.8344	.8351	.8357	.8363	.8370	.8376	.8382
6.9	.8388	.8395	.8401	.8407	.8414	.8420	.8426	.8432	.8439	.8445
7.0	.8451	.8457	.8463	.8470	.8476	.8482	.8488	.8494	.8500	.8506
7.1	.8513	.8519	.8525	.8531	.8537	.8543	.8549	.8555	.8561	.8567
7.2	.8573	.8579	.8585	.8591	.8597	.8603	.8609	.8615	.8621	.8627
7.3	.8633	.8639	.8645	.8651	.8657	.8663	.8669	.8675	.8681	.8686
7.4	.8692	.8698	.8704	.8710	.8716	.8722	.8727	.8733	.8739	.8745
7.5	.8751	.8756	.8762	.8768	.8774	.8779	.8785	.8791	.8797	.8802
7.6	.8808	.8814	.8820	.8825	.8831	.8837	.8842	.8848	.8854	.8859
7.7	.8865	.8871	.8876	.8882	.8887	.8893	.8899	.8904	.8910	.8915
7.8	.8921	.8927	.8932	.8938	.8943	.8949	.8954	.8960	.8965	.8971
7.9	.8976	.8982	.8987	.8993	.8998	.9004	.9009	.9015	.9020	.9025
8.0	.9031	.9036	.9042	.9047	.9053	.9058	.9063	.9069	.9074	.9079
8.1	.9085	.9090	.9096	.9101	.9106	.9112	.9117	.9122	.9128	.9133
8.2	.9138	.9143	.9149	.9154	.9159	.9165	.9170	.9175	.9180	.9186
8.3	.9191	.9196	.9201	.9206	.9212	.9217	.9222	.9227	.9232	.9238
8.4	.9243	.9248	.9253	.9258	.9263	.9269	.9274	.9279	.9284	.9289
8.5	.9294	.9299	.9304	.9309	.9315	.9320	.9325	.9330	.9335	.9340
8.6	.9345	.9350	.9355	.9360	.9365	.9370	.9375	.9380	.9385	.9390
8.7	.9395	.9400	.9405	.9410	.9415	.9420	.9425	.9430	.9435	.9440
8.8	.9445	.9450	.9455	.9460	.9465	.9469	.9474	.9479	.9489	.9489
8.9	.9494	.9499	.9504	.9509	.9513	.9518	.9523	.9528	.9533	.9538
9.0	.9542	.9547	.9552	.9557	.9562	.9566	.9571	.9576	.9581	.9586
9.1	.9590	.9595	.9600	.9605	.9609	.9614	.9619	.9624	.9628	.9633
9.2	.9638	.9643	.9647	.9652	.9657	.9661	.9666	.9671	.9675	.9680
9.3	.9685	.9689	.9694	.9699	.9703	.9708	.9713	.9717	.9722	.9727
9.4	.9731	.9736	.9741	.9745	.9750	.9754	.9759	.9763	.9768	.9773
9.5	.9777	.9782	.9786	.9791	.9795	.9800	.9805	.9809	.9814	.9818
9.6	.9823	.9827	.9832	.9836	.9841	.9845	.9850	.9854	.9859	.9863
9.7	.9868	.9872	.9877	.9881	.9886	.9890	.9894	.9899	.9903	.9908
9.8	.9912	.9917	.9921	.9926	.9930	.9934	.9939	.9943	.9948	.9952
9.9	.9956	.9961	.9965	.9969	.9974	.9978	.9983	.9987	.9991	.9996
x	0	1	2	3	4	5	6	7	8	9

TABLE 2

TRIGONOMETRIC RATIOS

TABLE 2 Trigonometric Ratios

ANGLE	SIN	COS	TAN	ANGLE	SIN	COS	TAN
0°	.0000	1.000	.0000	45°	.7071	.7071	1.0000
1°	.0175	.9998	.0175	46°	.7193	.6947	1.0355
2°	.0349	.9994	.0349	47°	.7314	.6820	1.0724
3°	.0523	.9986	.0524	48°	.7431	.6691	1.1106
4°	.0698	.9976	.0699	49°	.7547	.6561	1.1504
5°	.0872	.9962	.0875	50°	.7660	.6428	1.1918
6°	.1045	.9945	.1051	51°	.7771	.6293	1.2349
7°	.1219	.9925	.1228	52°	.7880	.6157	1.2799
8°	.1392	.9903	.1405	53°	.7986	.6108	1.3270
9°	.1564	.9877	.1584	54°	.8090	.5878	1.3764
10°	.1736	.9848	.1763	55°	.8192	.5736	1.4281
11°	.1908	.9816	.1944	56°	.8290	.5592	1.4826
12°	.2079	.9781	.2126	57°	.8387	.5446	1.5399
13°	.2250	.9744	.2309	58°	.8480	.5299	1.6003
14°	.2419	.9703	.2493	59°	.8572	.5150	1.6643
15°	.2588	.9659	.2679	60°	.8660	.5000	1.7321
16°	.2756	.9613	.2867	61°	.8746	.4848	1.8040
17°	.2924	.9563	.3057	62°	.8829	.4695	1.8807
18°	.3090	.9511	.3249	63°	.8910	.4540	1.9626
19°	.3256	.9455	.3443	64°	.8988	.4384	2.0503
20°	.3420	.9397	.3640	65°	.9063	.4226	2.1445
21°	.3584	.9336	.3839	66°	.9135	.4067	2.2460
22°	.3746	.9272	.4040	67°	.9205	.3907	2.3559
23°	.3907	.9205	.4245	68°	.9272	.3746	2.4751
24°	.4067	.9135	.4452	69°	.9336	.3584	2.6051
25°	.4226	.9063	.4663	70°	.9397	.3420	2.7475
26°	.4384	.8988	.4877	71°	.9455	.3256	2.9042
27°	.4540	.8910	.5095	72°	.9511	.3090	3.0777
28°	.4695	.8829	.5317	73°	.9563	.2924	3.2709
29°	.4848	.8746	.5543	74°	.9613	.2756	3.4874
30°	.5000	.8660	.5774	75°	.9659	.2588	3.7321
31°	.5150	.8572	.6009	76°	.9703	.2419	4.0108
32°	.5299	.8480	.6249	77°	.9744	.2250	4.3315
33°	.5446	.8387	.6494	78°	.9781	.2079	4.7046
34°	.5592	.8290	.6745	79°	.9816	.1908	5.1446
35°	.5736	.8192	.7002	80°	.9848	.1736	5.6713
36°	.5878	.8090	.7265	81°	.9877	.1564	6.3138
37°	.6018	.7986	.7536	82°	.9903	.1392	7.1154
38°	.6157	.7880	.7813	83°	.9925	.1219	8.1443
39°	.6293	.7771	.8098	84°	.9945	.1045	9.5144
40°	.6428	.7660	.8391	85°	.9962	.0872	11.4330
41°	.6561	.7547	.8693	86°	.9976	.0698	14.3010
42°	.6691	.7431	.9004	87°	.9986	.0523	19.0810
43°	.6820	.7314	.9325	88°	.9994	.0349	28.6360
44°	.6947	.7193	.9657	89°	.9998	.0175	57.2900
45°	.7071	.7071	1.0000				

ANSWERS TO ODD-NUMBERED PROBLEMS

CHAPTER 1

Exercise 1-1

1. 15
3. 200
5. 2,098
7. 176
9. 194
11. 233
13. 143
15. 1,602
17. 1,966
19. 18,106
21. 11,427
23. 13,923
25. 101,570
27. 27,271,294
29. 828 ft 0 in.
31. 31 gal 2 qt
33. 19,547 lb
35. 50 hr 35 min

Exercise 1-2

1. 52
3. 817
5. 2,482
7. 61,305
9. 56,566
11. 468,379
13. 86,151
15. 1,345 kwhr
17. 20,250
19. $4,395
21. 787
23. 5 years 6 months

Exercise 1-3

1. 1 is the multiplicative identity element.
3. Commutative property of multiplication.
5. Commutative property of addition.
7. Associative property of multiplication.
9. Commutative property of addition.
11. 980
13. 2,688
15. 9,000
17. 56
19. $168 = 168$
21. 2,688
23. 61,884
25. 20,297,894
27. 37 ft 6 in.
29. 60 yd 2 ft
31. 440 hr
33. 591 yd 4 in.
35. 5,184,000
37. Plot A. 6,069
39. 3,743

Exercise 1-4

1. 280
3. 404
5. 174
7. $913\frac{16}{572}$
9. 318
11. $4{,}509\frac{3}{222}$
13. $5{,}133{,}234\frac{2{,}978}{3{,}704}$
15. 411,117
17. 8 ft $10\frac{3}{4}$ in.
19. $16\frac{2}{177}$
21. 53
23. 12

Exercise 1-5

1. 63
3. 206
5. 3
7. 6
9. 13
11. 5
13. 8
15. 4
17. 7
19. 41
21. 26
23. 10
25. 5

CHAPTER 2

Exercise 2-1

1. 15
3. 32
5. 3
7. 52
9. 47
11. $6\frac{1}{3}$
13. $11\frac{1}{4}$
15. $1\frac{5}{12}$
17. $3\frac{7}{12}$
19. $3\frac{1}{4}$
21. $\frac{15}{8}$
23. $\frac{50}{11}$
25. $\frac{25}{9}$
27. $\frac{31}{4}$
29. $\frac{48}{5}$
31. $\frac{3}{5}$
33. $\frac{5}{6}$
35. $\frac{2}{9}$
37. $\frac{5}{9}$
39. $\frac{37}{53}$

Exercise 2-2

1. $2 \cdot 2 \cdot 3 \cdot 3$
3. $2 \cdot 2 \cdot 2 \cdot 3 \cdot 5$
5. $2 \cdot 3 \cdot 3 \cdot 5 \cdot 5$
7. $3 \cdot 3 \cdot 7 \cdot 11$
9. $2 \cdot 2 \cdot 2 \cdot 5 \cdot 5 \cdot 13$
11. 60
13. 240
15. 1,200
17. 210
19. 1,848
21. $\frac{2}{3}$
23. $\frac{7}{12}$
25. $\frac{3}{5}$
27. $1\frac{2}{3}$
29. $1\frac{11}{24}$
31. $\frac{11}{20}$
33. $\frac{137}{270}$
35. $4\frac{5}{24}$
37. $4\frac{323}{720}$
39. $\frac{421}{7,350}$
41. $40\frac{5}{8}$
43. $15\frac{5}{24}$
45. $14\frac{233}{270}$
47. $8\frac{1}{12}$
49. $10\frac{191}{270}$
51. $49\frac{1}{2}$ yd
53. $33\frac{3}{4}$ lb
55. $1\frac{19}{64}$ in.
57. $\frac{9}{16}$ in.

Exercise 2-3

1. $\frac{1}{2}$
3. $\frac{35}{96}$
5. $\frac{7}{12}$
7. 8
9. $\frac{16}{27}$
11. $\frac{3}{10}$
13. 1
15. $\frac{368}{4,059}$
17. 22
19. 3
21. $\frac{132}{7}$
23. $\frac{264}{5}$
25. 3 ft $7\frac{1}{2}$ in.
27. 55 ft 10 in.
29. $4\frac{4}{5}$ hr

Exercise 2-4

1. $\frac{2}{7}$
3. $\frac{32}{3}$
5. $\frac{8}{7}$
7. $\frac{25}{27}$
9. $\frac{15}{28}$
11. $\frac{9}{10}$
13. $\frac{10}{7}$
15. $\frac{1,116}{161}$
17. 12
19. $\frac{11}{24}$
21. $\frac{21}{32}$
23. $\frac{3}{8}$
25. $\frac{105}{2}$
27. $\frac{178}{87}$
29. 5
31. $\frac{35}{2}$
33. $\frac{25}{27}$
35. $\frac{85}{98}$
37. 1 ft $6\frac{3}{5}$ in.
39. 2 lb $9\frac{1}{3}$ oz
41. $\frac{68,000}{63}$
43. 108
45. 150

Chapter 3

Exercise 3-1

1. (a) 0.642 (b) 14.108 (c) 18.004 (d) 0.056 (e) 87.0605 (f) 2,002.200 (g) 8.3 (h) 47.013 (i) 0.73 (j) 0.9451 (k) 4.8039
3. 1.803, 1.083, 1.08, 0.1083, 0.10380, 0.01803, 0.01083, 0.0108

5. 0.625
7. 0.024
9. 1.625
11. $0.\bar{5}$
13. $1.4\overline{818}$
15. $0.\overline{05}$
17. $\frac{7}{50}$
19. $\frac{1}{16}$
21. $\frac{23}{32}$
23. $\frac{67}{8}$
25. $\frac{127}{25}$
27. $\frac{57}{16}$

Exercise 3-2

1. 110.435
3. 7.623
5. 30.710
7. 457.855
9. 9.345
11. 6.2866
13. 59.211
15. 20.2583
17. 126.934
19. 3.524
21. 6.082
23. 2.66
25. 4.9996
27. 10.45
29. 3.649
31. 823.693
33. $0.0\bar{3}$
35. $0.010\bar{1}$
37. 1235.4
39. 10,050.4
41. 12.73484
43. 1.683879
45. 0.918271809
47. 0.01028376
49. 2300.361 lb

Exercise 3-3

1. (a) 32 is approximate.
   (b) 6 is exact; 3 is approximate.
   (c) Dozen is exact; 3 is approximate.
   (d) 240,000 is approximate.
   (e) 14.7 is approximate.
   (f) 1,728 is exact.
   (g) 1,535 is approximate.
   (h) 12 is approximate.
   (i) 100 is approximate; 9.4 is approximate.
   (j) $\sqrt{5}$ is exact; 2.236 is approximate.
   (k) Three is exact; 2.49 is exact.
   (l) 25 is approximate; 1,963 is approximate; 75,000 is exact.
3. (a) 87.235 and 87.245
   (b) 0.3085 and 0.3095
   (c) 12.35 and 12.45
   (d) 84,500 and 85,500
   (e) 0.00010365 and 0.00010375

5. (a) 87,200
   (b) 4,640 ft
   (c) 20,000 bottles
   (d) 3,020 trees
   (e) 99,300 gal
   (f) 183,000,000 people
7. 0.075
9. 0.50
11. 89.012
13. 2,000
15. 75.8125 in.^{2}
17. $1.65

Exercise 3-4

1. 14.1
3. 63.90
5. 15
7. 1.7
9. 315
11. 846
13. 8.5
15. 1.63×10^6
17. 0.220
19. 3.91
21. 14.35
23. 2.876×10^5
25. 4.37×10^2
27. 8.03×10^{-1}
29. 1.07×10^{-2}
31. 83,000,000
33. 701,000
35. 0.00000427
37. 0.0000005130

CHAPTER 4

Exercise 4-1

1. 0.11
3. 0.084
5. 1.375
7. 0.0025
9. 0.004
11. 0.00007
13. 3.745
15. 0.000625
17. 0.9%
19. 25%
21. 0.04%
23. 760%
25. 1,200%
27. 8.0045%
29. 28,300%
31. $58\frac{1}{3}$%
33. 31.25%
35. 14.0625%
37. 65%
39. $791\frac{2}{3}$%
41. $231\frac{1}{4}$%
43. 178%
45. $2{,}312\frac{1}{2}$%
47. $\frac{5}{4}$
49. $5\frac{12}{25}$
51. 15
53. $\frac{7}{8}$
55. $\frac{33}{80}$
57. $\frac{191}{400}$
59. $\frac{1}{150}$

Exercise 4-2

1. 32
3. 9.6
5. $4,500
7. 113.6
9. 800
11. $3\frac{1}{3}$%
13. $195\frac{35}{37}$
15. 1,057.548
17. 200%
19. 1,176
21. 3,387.6
23. 432
25. 3.20
27. $504
29. $122\frac{2}{3}$
31. 0.00907872
33. $14\frac{6}{7}$
35. 1,200%
37. 675
39. $21\frac{7}{833}$%

Exercise 4-3

1. 23%
3. $7,128
5. 37.5%
7. $6,250
9. $3\frac{1}{2}$%
11. $3,182.50, $477.38
13. $367.96
15. $80.00
17. $1,080
19. $13.33
21. 400 lb
23. 30,000 ft

CHAPTER 5

Exercise 5-1

1. -4; 2
3. -5
5. 2
7. $1\frac{1}{2}$
9. t
11. r
13. 0
15. -25
17. -5
19. -3
21. $<$
23. $>$
25. $<$
27. $>$
29. $=$
31. $>$
33. $=$
35. $=$
37. $>$

Exercise 5-2

1. 48
3. -27
5. -13
7. 0
9. 0
11. -11
13. -5
15. 10
17. 30
19. -45
21. -35
23. -33
25. 7
27. -0.9
29. 0
31. $+9$
33. -8
35. -18
37. $+20$
39. $+54$

Exercise 5-3

1. 13
3. -6
5. 18
7. -334
9. -59
11. -24
13. 9
15. -12.11
17. -13
19. 22
21. 39
23. -19
25. -87
27. -9
29. 23
31. -20
33. -6
35. -59
37. 12
39. -1.20

Exercise 5-4

1. -30
3. 63
5. 4
7. $\frac{1}{4}$
9. $\frac{4}{9}$
11. 20
13. -28
15. $-\frac{3}{4}$
17. 120
19. -5
21. -72
23. -15
25. $-4\frac{3}{4}$
27. -6
29. -20
31. -2
33. $-\frac{128}{175}$
35. 4
37. -6
39. -30
41. -6
43. -4
45. -6

Exercise 5-5

1. T
3. T
5. T
7. F
9. T
11. F

13. T
15. T
17. F
19. F
21. F
23. T
25. F
27. F
29. T
31. T
33. T
35. T

CHAPTER 6

Exercise 6-1

1. 3.02
3. 845
5. 0.0721
7. 857
9. 4,020
11. 0.272
13. 130.9
15. 4.36
17. 0.00001554
19. 0.0563

Exercise 6-2

1. 7.68
3. 5.55
5. 2,770
7. 164,400
9. 433,000
11. 53,900
13. 10.24
15. 0.01266
17. 2.12
19. 5.04

Exercise 6-3

1. 1,186
3. 0.259
5. 0.1145
7. 0.1948
9. 0.306
11. 0.746
13. 0.01824
15. 1.326
17. 57.5
19. 0.000363

Exercise 6-4

1. 22.0
3. 181.6
5. 0.0252
7. 2.39
9. 0.1617
11. 21.3
13. 2,230
15. 0.524
17. 119.1
19. 0.0839
21. 0.696
23. 0.1262
25. 1.533

Exercise 6-5

1. 43.5
3. 0.0476
5. 14.90
7. 52.6
9. 22.10
11. 69,900
13. 63.3
15. 500
19. 0.0286

Exercise 6-6

1. 29
3. 148
5. 429
7. 719
9. 97.4
11. 0.878
13. 0.0615
15. 0.0293
17. 387
19. 0.571

Exercise 6-7

1. 5.29
3. 75.7
5. 1,815
7. 77,300
9. 0.00728
11. 7.48
13. 16.06
15. 28.8
17. 214
19. 0.340
21. 55.2
23. 2.44
25. 3.63
27. 0.783
29. 8,090

Exercise 6-8

1. 22.0
3. 0.238
5. 0.0566
7. 576,000
9. 0.000691
11. 1.965
13. 8.31
15. 0.879
17. 0.330
19. 44.3
21. 8.82
23. 1.704
25. 0.443
27. 7.72
29. 0.588

CHAPTER 7

Exercise 7-1

1. 0.283 m
3. 0.0593 mm
5. 8.732 g
7. 6.115 l
9. 3,150 in.
11. 15,150 ft
13. 1.440 short tons
15. 7.325 bushels
17. 115.5 in.^3
19. 0.0593 dm^2
21. 0.0000835 m^3
23. 54 in.^2
25. $831\frac{7}{8}\ \text{yd}^3$
27. $0.0144\ \frac{\text{m}}{\text{hr}}$
29. $875{,}000\ \frac{\text{kg}}{\text{m}^3}$
31. $\frac{35}{54}\ \frac{\text{oz}}{\text{in.}^3}$
33. 31.75 m
35. 5.45 lb
37. $110{,}000\ \frac{\text{cm}}{\text{min}}$
39. $0.0742\ \frac{\text{lb}}{\text{in.}^2}$
41. $2.17\ \frac{\text{gm}}{\text{cm}^3}$
43. $18{,}070\ \frac{\text{mg}}{\text{mm}^2}$
45. $14.1\ \frac{\text{km}}{\text{min}}$
47. 0.000132 in.

CHAPTER 8

Exercise 8-1

1. (a) $9ab$
   (b) 6
   (c) $5yz$
   (d) $6x^2y$ and $3x^2y$
3. $y - x$
5. $d + 8$
7. $c - 6$
9. $y - z$
11. $\frac{r + s}{rs}$
13. $\frac{uw}{u + w}$
15. $(r + s) - (h + k)$
17. $(2x)(3y)$
19. -5
21. 4
23. 10
25. $\frac{11}{7}$
27. -7
29. 9
31. 20
33. 0
35. 0
37. 11

39. -1
41. -13
43. (a) $15-n$
(b) $4(15-n)$
45. (a) $n-12$
(b) $5(n-12)$
47. $10d$ cents or
$0.10d$ dollars
49. (a) $0.40x$
(b) $0.60x$

Exercise 8-2

1. $7a$
3. $7xy$
5. $-8b^2c$
7. $-xy^2$
9. $-7m^2n$
11. $4x^2y^3$
13. $-2uv^3$
15. $8x^2-4xy-2y^2$
17. $2x^2-xy-10y^2$
19. $6a^3-3a^2b+ab^2+b^3$
21. $4x$
23. $-12w$
25. $11xy$
27. $45a^2b$
29. $13x$
31. $5x^3+6x$
33. $d^4+2d^3+2d^2$
35. $2a^3+12a^2b+4ab^2-b^3$
37. x^3-x-3
39. $4a^3+a^2b+6ab^2-5b^3$

Exercise 8-3

1. 8^5
3. a^3b^2
5. $2x^2y^2$
7. $x^3s^4yz^3$
9. 5^5
11. $-15y^2$
13. $6abcd$
15. $-x^2y^3z^2$
17. a^2+b^2
19. k^3+k^2
21. r^3s^3
23. $4a^2b^2$
25. $-27c^6d^3$
27. $-2xy^2$
29. $-8rt^3$
31. $-8ce^2$
33. x^5
35. x^2
37. $-\frac{D^3}{c^4}$
39. x^4

Exercise 8-4

1. $2a^3b-2ab^2c+2abc^2$
3. $3x^4y-6x^3y^2+9x^2y^3$
5. $30x^4-12x^3+18x^2$
7. $10x^2y-10xy^2+30xy$
9. $x^2+7x+12$
11. $y^2-13y+36$
13. $2x^2-5x-12$
15. $6-9y+3y^2$
17. $6x^4+7x^2y-3y^2$
19. $a^3+2a^2b-7ab^2+4b^3$
21. $6k^4-2k^3m-17k^2m^2+3km^3+12m^4$
23. $x^4-x^3-x^2-11x-12$
25. $x^4+x^2y^2+y^4$
27. $10x^4-27x^3y+36x^2y^2-21xy^3-4y^4$
29. $6x^4+13x^3-9x^2-31x-14$

31. $2x^3 - x^2 - 13x - 6$
33. $4x^3 + 8x^2 - x - 2$
35. $2x^3 - 3x^2y - 3xy^2 + 2y^3$

Exercise 8-5

1. $3x^4 - 4x^3 - 2x + 1$
3. $6a^3 - 4a - 3$
5. $2a^2b^2 - 3ab - 4$
7. $a - 5$
9. $5x - 4y$
11. $2x - 5y$
13. $-3a - 8$
15. $2a^2 + 3a - 4$
17. $2x^2 + xy - 2y^2$
19. $3a^2 - 2ab + b^2$
21. $5k^2 - 3k - 4$
23. $16a^2 + 36ab^2 + 81b^4$
25. $a - 3b + 2c$
27. $7a^2 + 5a + 2 + \dfrac{-3}{2a^2 + 5a + 7}$
29. $3x^2 + 4xy - 2y^2$

CHAPTER 9

Exercise 9-1

1. 8
3. -3
5. 6
7. $\frac{1}{2}$
9. -4
11. 4
13. -2
15. $\frac{10}{3}$
17. 2
19. $\frac{1}{2}$

Exercise 9-2

1. -2
3. 4
5. 1
7. -4
9. -7
11. 2
13. -10
15. All real numbers
17. -4
19. All real numbers

Exercise 9-3

1. 54
3. 12
5. 10
7. 9
9. -11
11. 7
13. -1
15. 4
17. 3
19. $-\frac{18}{7}$

Exercise 9-4

1. $P = 2x + 2y - a - b$; $A = xy - ab$
3. $P = 4x + 2z - 2a$; $A = 2ax + az - 2a^2$
5. $P = 2x + \frac{1}{2}\pi x$; $A = \frac{1}{2}x^2$
7. $A = 21\pi a^2$
9. $-\dfrac{5a}{2}$
11. $\frac{3}{10}d$
13. $5a$
15. $\dfrac{ab + ac}{5}$
17. $\dfrac{V}{lh}$
19. $\dfrac{S - 2\pi r^2}{2\pi r}$

21. $\frac{3V}{\pi r^2}$

23. $\frac{v^2}{2g}$

25. $\frac{6M}{h^2}$

27. $\frac{33{,}000H}{LAN}$

29. $\frac{l-a}{n-1}$

31. $2A - b$

33. $\frac{Wv^2}{Fg}$

35. $v_0 + at$

37. $\frac{RR_1}{R_1 - R}$

39. 32 in.^4

41. $\frac{5}{6}$ in.

43. 3,200 ft

Exercise 9-5

1. Rod, \$15; reel, \$20
3. Length, 31 yd; width, 28 yd
5. $\frac{6}{10}$
7. $54\frac{3}{4}$ in., $51\frac{3}{4}$ in., $48\frac{3}{4}$ in., $45\frac{3}{4}$ in.
9. Father, 40 years; son, 20 years
11. \$4,920
13. $41\frac{2}{3}$ gal; $58\frac{1}{3}$ gal
15. $33\frac{1}{3}$ gal
17. 48 mph
19. $6\frac{3}{7}$ hr
21. (a) 16 hr
    (b) 240 mi
23. 28 days

Exercise 9-6

1. (a) $>$, (b) $>$, (c) $<$, (d) $>$, (e) $<$, (f) $>$, (g) $>$, (h) $>$, (i) $<$
3. $x < 5$
5. $x > -4$
7. $x < -5$
9. $x > -4$

CHAPTER 10

Exercise 10-1

1. $6a^2 - 8ab$
3. $-8x^3y^2 + 20x^2y^3 + 24xy^4$
5. $a^2 - 4$
7. $16 - k^2$
9. $49x^2 - 36y^2$
11. $9y^2 - \frac{1}{4}$
13. $a^4b^2 - d^2$
15. $\frac{k^2}{25} - 25m^2$
17. $\frac{4x^2y^2}{9} - \frac{1}{16}z^2$
19. $5(x + y)$
21. $3(4h - 5k)$
23. $\frac{1}{4}xy(3x - y)$

25. $(x + 4)(x - 4)$
27. $(6r + 11)(6r - 11)$
29. $(5x^2 + 4y)(5x^2 - 4y)$
31. $(1.1x + 0.8z)(1.1x - 0.8z)$
33. $(a + \frac{1}{4})(a - \frac{1}{4})$
35. $(2x - 3y + 2z)(2x - 3y - 2z)$

Exercise 10-2

1. $x^2 + 6x + 9$
3. $x^2 - 14xy + 49y^2$
5. $x^2 - 4b^2$
7. $16c^2 - 4c + \frac{1}{4}$
9. $36x^2 + 60xy + 25y^2$
11. $9 + 12xy^2 + 4x^2y^4$
13. $r_1^2 - 6r_1r_2 + 9r_2^2$
15. $(x + 6)^2$
17. $(p - 5q)^2$
19. $(4a - 5b)^2$
21. $(8 - b^2)^2$
23. $(1 - 15rs^2)^2$
25. $(ab - \frac{3}{2}c)^2$
27. $(0.1ab^2 + 0.8c^3)^2$
29. $6a$
31. y^2
33. 36
35. $22ab$
37. $0.09b^2$
39. $\dfrac{cd}{3}$

Exercise 10-3

1. $x^2 + 11x + 30$
3. $6x^2 + 19x + 10$
5. $10a^2 + 31a - 63$
7. $4k^2 + 5k - 6$
9. $25x^2 - 60x + 36$
11. $6 - x - 12x^2$
13. $4 + 27x - 81x^2$
15. $x^2 - ax - 2a^2$
17. $3y^2 - 7by + 2b^2$
19. $0.06x^2 + 0.10x - 35$
21. $(x + 2)(x + 3)$
23. $(a - 5)(a + 2)$
25. $(r - 4)^2$
27. $(5x + 1)(2x + 3)$
29. $(2y + 3)(y + 2)$
31. $2(k - 3)(3k + 1)$
33. $(x - 2)(4x - 3)$
35. $4(2x + 3)^2$
37. $(9a - 4b)(a - b)$
39. $(4x - 3z)(3x - 20z)$

Exercise 10-4

1. $(x + 3)(x^2 - 3x + 9)$
3. $(2a + b)(4a^2 - 2ab + b^2)$
5. $8(a + 2y)(a^2 - 2ay + 4y^2)$
7. $2x(y + 2)(y^2 - 2y + 4)$
9. $5(r + 5)(r - 5)$
11. $(x^2 + y^2)(x + y)(x - y)$
13. $(2 + a)(y + x)$
15. $2(x - 2y)(3a - 5b)$
17. $(a + b)(a - b)(a^2 - ab + b^2)(a^2 + ab + b^2)$
19. $(x^2 + 1)(x - 1)$

CHAPTER 11

Exercise 11-1

1. $\frac{4xy}{5}$

3. $\frac{4x^2y^2}{7}$

5. $\frac{t^2}{3r^2}$

7. $\frac{5x^4y^2}{4y-7x}$

9. $\frac{2(x-2)}{x-6}$

11. $\frac{3}{a+b}$

13. $-(x-4y)$ or $4y-x$

15. $\frac{y(x-y)}{x+y}$

17. $-\frac{x+2}{x^2+2x+4}$

19. $\frac{5b-6a}{4b+3a}$

21. $\frac{b+x}{c-y}$

23. 1

Exercise 11-2

1. $\frac{3ay}{2b}$

3. $-\frac{7v^2}{5u^2w}$

5. $\frac{a^3b^3c^3}{x^2y^3z^4}$

7. $\frac{cx^4y^5z^6}{a^3b}$

9. 1

11. $\frac{(a+3)(a+1)}{a+2}$

13. $\frac{d(a+c)}{ac}$

15. $\frac{5a-2}{a-2}$

17. $\frac{1}{3(a-b)^4}$

19. $\frac{x+2}{x-2}$

21. $\frac{2}{d}$

23. $\frac{4(m^2-mn+n^2)}{3mn}$

25. $-\frac{(5x-4y)(2x+3y)}{(4r-s)(2r+3s)}$

Exercise 11-3

1. $\frac{5x}{4}$

3. $\frac{33x-32}{20}$

5. $\frac{bc+2ac+3ab}{abc}$

7. $\frac{1}{4a}$

9. 0

11. $\frac{14x-1}{(2x+1)(2x-1)}$

13. $\dfrac{-2a^2 + 45a - 23}{(a+5)(a-4)}$

15. $\dfrac{x^2 - 4xy}{(x+y)(x-y)}$

17. $\dfrac{a^2 + 2a^2b + 2ab^2 - 2ab + b^2}{(a+b)^2(a-b)}$

19. $\dfrac{2}{(x-y)(x-z)}$

21. $\dfrac{-4x^2 + 20x - 22}{(x-1)(x-2)(x-3)(x-4)}$

CHAPTER 12

Exercise 12-1

1. K, L, and M
3. $\angle L$
5. An infinite number
7. $\overleftrightarrow{AB}$, $\overleftrightarrow{AO}$, $\overleftrightarrow{BO}$, $\overleftrightarrow{BA}$, $\overleftrightarrow{OA}$, $\overleftrightarrow{OB}$
9. OA
11. CD
13. $\angle OAB$ or $\angle CAB$
15. ABC
17. $\angle COB$; $\angle BOA$
21. Lines r and s are parallel.
25. Draw P, Q, R, S as collinear points (in any order).
31. Draw $l \parallel m$ and n intersecting both l and m.
33. 30°
35. 130°
37. 109°
39. 164°

Exercise 12-2

1. Scalene, obtuse
3. Isosceles, acute
5. Isosceles, acute
7. Equilateral, equiangular
9. Scalene, right
11. Scalene, right
13. $\overleftrightarrow{AB}$ and $\overleftrightarrow{CD}$; $\overleftrightarrow{AC}$ and $\overleftrightarrow{BC}$
15. $\angle H$ and $\angle K$
25. (a) 150°
    (b) 115°
    (c) 90°
    (d) 45°
    (e) $180 - \alpha°$

Exercise 12-3

1. Perpendicular
3. Obtuse
5. Scalene
7. Isosceles
9. 90°
11. 360°
13. Acute
15. Midpoint (or bisector)

17. F
19. F
21. F
23. F
25. T
27. F
29. T
31. 3
33. $1\frac{2}{3}$
35. $2\frac{1}{2}$
37. 1
39. 50°
41. 20°
43. 55°
45. 110°
47. $m\angle\alpha = 95°$; $m\angle\phi = 95°$
49. 55°

CHAPTER 13

Exercise 13-1

1. Yes
3. No
5. Yes
7. Yes
9. Yes
11. Yes
13. No
15. No

Exercise 13-2

1. ASA
3. No
5. Hypotenuse-acute angle
7. Hypotenuse-acute angle
9. Hypotenuse-leg
11. ASA
13. SAS
15. No

Exercise 13-3

1. Exterior
3. Corresponding
5. Corresponding
7. An isosceles triangle
9. Hypotenuse
11. T
13. F
15. F
17. F
19. T
21. F
23. F
25. T
27. (a) $\angle BAP \cong \angle CAP$ because it is given that $\overrightarrow{AP}$ bisects $\angle BAC$.
    (b) $\angle ABP$ and $\angle ACP$ are right ⦞ because $\overline{PB} \perp \overline{AB}$, $\overline{PC} \perp \overline{AC}$ and perpendicular lines form right angles.
    (c) $\angle ABP \cong \angle ACP$ because all right ⦞ are $\cong$.
    (d) $\triangle ABP$ and $\triangle ACP$ are right ⨺, by the definition of a right $\triangle$.
    (e) $\overline{AP} \cong \overline{AP}$ because any quantity equals itself.
    (f) Then $\triangle ABP \cong \triangle ACP$ by hyp-acute theorem.
    (g) $PB = PC$ because corresponding sides of $\cong$ ⨺ are $\cong$ to each other.
29. (a) $ST = SR$ is given.

(b) $\overline{ST} \cong \overline{SR}$, because if $ST = SR$, then $\overline{ST} \cong \overline{SR}$.
(c) $\angle STP$ and $\angle SRP$ are right ∠s because $\overline{PT} \perp \overline{ST}$, $\overline{PR} \perp \overline{SR}$ and $\perp$ lines form right ∠s.
(d) $\triangle STP$ and $\triangle SRP$ are right △s because each has one right ∠.
(e) $SP = SP$, because any thing equals itself.
(f) $\triangle STP \cong \triangle SRP$ by hypotenuse-leg theorem.
(g) $\angle RSP \cong \angle TSP$ because corresponding parts of $\cong$ △s are $\cong$ to each other.
(h) $\overline{SP}$ bisects $\angle RST$, because statement (g) satisfies the definition of angle bisector.

CHAPTER 14

Exercise 14-1

1. Yes
3. Yes
5. Yes
7. No
9. No
11. 45°
13. 140°
15. $l_2, l_3; m_1, m_2, m_2$
17. (a) $\overline{RP} \| \overline{TQ}$
(b) Then $\angle R \cong \angle T$ and $\angle P \cong \angle Q$ (by Theorem 14-3).
(c) $RP = TQ$ (given).
(d) $\triangle RPS \cong \triangle TQS$ (by ASA).
(e) $\therefore RS = TS$ and $PS = QS$ (because corresponding parts of $\cong$ △s are $\cong$ to each other).
19. (a) $\overline{PT} \| \overline{RS}$ (given).
(b) $\angle y \cong \angle x$ (by Theorem 14-3).
(c) $PT = RS$ (given).
(d) $RT = RT$ (Axiom 1: Equality is a reflexive relation).
(e) $\triangle RPT \cong \triangle TSR$ (by SAS).
(f) $PR = ST$ (because corresponding parts of $\cong$ △s are $\cong$ to each other).

Exercise 14-2

1. 120°
3. 70°
5. 140°
7. 63°
9. 82°
11. 28°
13. 40°
15. 80°
17. $m\angle KAB = 71\frac{1}{2}°$; $m\angle KMB = 108\frac{1}{2}°$

Exercise 14-3

1. T
3. T
5. F
7. T
9. F
11. F
13. T
15. T
17. F

19. T
21. F
23. F
25. F
27. Trapezoid
29. Rhombus
31. Congruent
33. Perpendicular
35. 15

CHAPTER 15

Exercise 15-1

1. $60°$
3. $20°$
5. $m = 106°$; $m\angle\alpha = 37°$
7. $45°$
9. $m\angle\alpha = 69°$; $m\angle\beta = 69°$
11. $44°$
13. $50°$
15. $110°$
17. $m = 50°$
19. $15°$
21. $58°$
23. $m = 100°$; $m\angle\alpha = 25°$; $m\angle\beta = 25°$

Exercise 15-2

1. $61°$
3. $15°$
5. $108°$
7. $32°$
9. $92°$
11. $m\angle\alpha = 90°$; $m\angle\beta = 25°$
13. $m = 120°$; $m\angle\alpha = 25°$
15. $m = 80°$; $m\angle\alpha = 20°$; $m\angle\beta = 55°$
17. Radii
19. Right angle
21. Congruent
23. $5°$
25. Perpendicular
27. F
29. F
31. F
33. F
35. F
37. F
39. F
41. F

CHAPTER 16

Exercise 16-1

1. (a) $\frac{2}{3}$ (b) $\frac{16}{9}$ (c) $\frac{8}{9}$ (d) $\frac{2}{3}$ (e) $\frac{1}{6}$ (f) $\frac{20}{1}$ (g) $\frac{24}{5}$ (h) 1.5:1 or 381:250

3. (a) $\frac{a}{b} = \frac{c}{d}$ (given).

   (b) $ad = bc$ (by Theorem 16-1).

   (c) $\frac{ad}{cd} = \frac{bc}{cd}$ (by Axiom 7—see Sec. 9-2).

   (d) $\frac{a}{c} = \frac{b}{d}$

5. (a) 18
   (b) 25
   (c) 10
   (d) $\frac{15}{8}$
   (e) 14.4 in.
   (f) $\frac{7}{18}$ ft
7. (a) 4:3
   (b) 3:4
   (c) 5:2
   (d) $\frac{1}{4}$:1 or 1:4
   (e) $b:a$
   (f) $x:y$
9. (a) 6
   (b) 8
   (c) 10
   (d) 8
   (e) $\frac{1}{3}$
   (f) $\sqrt{20}$
11. 14,400
13. 256 rpm

Exercise 16-2

1. 12
3. 6
5. 19.2
7. 14
9. 48
11. 9
13. 8
15. 24
17. $16\frac{2}{3}$
19. 25
21. $15\frac{3}{7}$

Exercise 16-3

1. 20
3. 8
5. 12
7. 30
9. 20
11. 10
13. 12
15. 12
17. 10
19. 19.2
21. 12
23. 15.6
25. 8 ft

CHAPTER 17

Exercise 17-1

1. (a) 12 yd
   (b) 15.9 in.
   (c) 6.007 mi
3. (a) 1.4 ft
   (b) 0.800 in.
   (c) 2.000 yd
5. 18 ft^2
7. $199\frac{3}{4}$ ft^2
9. $323.49
11. 54 $in.^2$
13. 512 $in.^2$
15. $7\frac{7}{8}$ ft^2
17. $139\frac{1}{4}$ $in.^2$
19. $361\frac{7}{8}$ ft^2
21. 560 yd^2

Exercise 17-2

1. 432 $in.^2$
3. 15.7 in.
5. 432 $in.^2$
7. 80 ft^2
9. 24 ft
11. 240 ft^2
13. 46.5 $in.^2$
15. 299 ft^2
17. 448 yd^2
19. $31\frac{1}{9}$ ft^2
21. 152 $in.^2$

Exercise 17-3

1. (a) 9.42 in.
   (b) 28.3 in.
   (c) 11.31 in.
   (d) 22.6 in.
3. (a) 13.04 $in.^2$
   (b) 41.1 $in.^2$
   (c) 88.5 $in.^2$
5. 164.4 $in.^2$
7. 30.9 $in.^2$
9. 83.4 $in.^2$
11. 506 $in.^2$
13. 490 $in.^2$

15. 411 in.^{2}
17. 39.3 in.^{2}
19. 374 in.^{2}

CHAPTER 18

Exercise 18-1

1. 160 yd^3
3. 3,375 in.^{3} or $\frac{125}{64}$ ft^3
5. 2,295 in.^{3}
7. $\sqrt[3]{3{,}456} \approx 15.12$ in.
9. 640 in.^{2}
11. $20\sqrt{455} \approx 427$ in.^{2}
13. 610 lb
15. 1,132 in.^{2}
17. 112,200 gal

Exercise 18-2

1. 2,356 ft^3
3. 603 in.^{2}
5. 22.14 in.
7. 5,300 gal
9. 1,380 lb
11. 785 ft^2

Exercise 18-3

1. 144 in.^{3}
3. 1,080 ft^3
5. 1,440 ft^3
7. 10.4 ft^3
9. 1,732 in.^{3}

Exercise 18-4

1. 3,050 in.^{3}
3. 2,720 ft^2
5. 27,100 ft^3
7. 3,850 ft^2
9. 427 ft^2
11. 6.73 ft^3 or 11,620 in.^{3}
13. 4,520 in.^{2}

Exercise 18-5

1. 524 ft^3
3. 300 in.^{2}
5. 8 in.
7. 3,054 in.^{3}
9. 2,490 ft^3
11. 168 in.^{2}

CHAPTER 19

Exercise 19-1

1. 5
3. -2
5. 16
7. $\frac{2}{5}$
9. $\frac{1}{12}$
11. $\frac{1}{8}$
13. $\frac{6}{5}$
15. 3
17. 8
19. $-\frac{1}{2}$
21. 1
23. 1
25. a^2
27. x^2
29. $y^{5/6}$
31. $x^{1/2}$
33. $6^{5/12}$
35. x
37. y^2
39. $\dfrac{1}{a^4}$
41. $\dfrac{b^2c^6}{a^4}$
43. $\dfrac{1}{4x}$
45. $\dfrac{4c^6b^4}{a^2}$
47. x^n
49. $\dfrac{b+a}{ab}$

51. $\dfrac{1}{a+b}$

53. 0

55. $\dfrac{x^2+y^2}{xy}$

57. $\dfrac{x^2y^2}{y^2-x^2}$

59. $\dfrac{-1}{x^2y^2}$

61. $a+2a^{1/2}b^{1/2}+b$

63. $a-\dfrac{1}{a^2}$

65. b^{-2m+8}

67. (a) 2.03×10^{-1}
(b) 1.07×10^{-4}
(c) 8.312×10^{-8}
(d) -5.37×10^{-2}

Exercise 19-2

1. $\sqrt[3]{k}$
3. $\dfrac{1}{\sqrt{a}}$
5. $4\sqrt[3]{b^2}$
7. $\sqrt{\dfrac{b}{a}}$
9. $2\sqrt{x}-5\sqrt{y}$
11. $y^{3/4}$
13. $6^{1/3}a^{5/3}b^{2/3}$
15. $(x^2+y^2)^{2/3}$
17. $x^{3/2}y^{2/3}$
19. $5\sqrt{2}$
21. $6\sqrt{2}$
23. $2\sqrt[3]{3}$
25. $2a\sqrt{2}$
27. $2a\sqrt[4]{2b}$
29. $\dfrac{2xy}{3b}\sqrt{\dfrac{y}{ab}}$
31. $x\sqrt{1+x}$
33. $2x\sqrt{2y-4y^2}$
35. $x\sqrt[n-1]{x}$
37. $\sqrt{12}$
39. $\sqrt{16x}$
41. $\sqrt{5a^3}$
43. $\sqrt[3]{b^4}$
45. $\sqrt{a^2-b^2}$
47. $\frac{1}{7}\sqrt{7}$
49. $\dfrac{1}{b}\sqrt{ab}$
51. $\frac{2}{5}\sqrt{5}$
53. $\frac{1}{3}\sqrt[3]{18}$
55. $\dfrac{1}{x}\sqrt[3]{x}$
57. $\dfrac{1}{3a}\sqrt[3]{18a^2}$
59. $\frac{4}{9}\sqrt{3}$
61. $\dfrac{4a}{x^2b}\sqrt[3]{2ab}$
63. $\dfrac{x^5}{y}\sqrt[n]{y}$
65. $\sqrt[12]{64}, \sqrt[12]{625}, \sqrt[12]{27}$
67. $\sqrt[12]{\dfrac{x^8}{16}}, \sqrt[12]{\dfrac{y^9}{27}}$
69. $\sqrt[6]{\frac{9}{16}}, \sqrt[6]{\frac{4}{5}}$

Exercise 19-3

1. $5\sqrt{2}$
3. $-\sqrt{3}$
5. $17\sqrt{2}$
7. $20\sqrt{5}+14\sqrt{7}$
9. $5a\sqrt{y}$
11. $2\sqrt{3}$
13. $8\sqrt[3]{6}-12\frac{5}{6}$
15. $\sqrt[3]{4}$
17. $-8\sqrt{2x}$
19. $\dfrac{1}{b}\sqrt[3]{a^2b^2}+\dfrac{1}{b}\sqrt[3]{ab}+\sqrt[3]{a^2b^2}$

Exercise 19-4

1. $2\sqrt{3}$
3. $6\sqrt{35}$
5. $a\sqrt{b}$
7. $\sqrt{35}$
9. $\sqrt[3]{36}$
11. $60\sqrt{30}$
13. $7x\sqrt{21x}$
15. $8\sqrt[3]{20}$
17. $x+4\sqrt{x}+4$
19. $k-2\sqrt{k}+1$
21. $4a-12\sqrt{a}+9$
23. -1
25. $-1-\sqrt{15}$

Exercise 19-5

1. $\sqrt{2}$
3. $\frac{6}{5}$
5. $\dfrac{4\sqrt{5}}{15}$
7. 2
9. $\frac{3}{2}$
11. $\dfrac{1}{b}\sqrt{ab}$
13. $\frac{8}{25}\sqrt{55}$
15. $\dfrac{3\sqrt{3}}{2}$
17. $\dfrac{4\sqrt[6]{108}}{3}$
19. $-\frac{1}{2}\sqrt[6]{32}$
21. $3ab$
23. $4(\sqrt{3}-\sqrt{2})$
25. $\sqrt{5}+\sqrt{2}$
27. $\dfrac{13+7\sqrt{3}}{22}$
29. $4+\sqrt{15}$
31. $-4+3\sqrt{6}$
33. $\dfrac{-(24+7\sqrt{42})}{38}$

Exercise 19-6

1. 4
3. −8
5. 7
7. $\frac{75}{4}$
9. −3
11. No root
13. 3
15. 20 ft
17. 9.77 m/sec^2

Exercise 19-7

1. $-j$
3. 1
5. $4j$
7. $2j\sqrt{2}$
9. $4j$
11. $5j\sqrt{3}$
13. $\frac{1}{2}j\sqrt{10}$
15. $\frac{2}{3}j\sqrt{15}$
17. $-\frac{3}{5}j\sqrt{15}$
19. $-\dfrac{1}{b}j\sqrt{ab}$
21. $-j\sqrt{2}$
23. $56j\sqrt{2}$
25. $-15j\sqrt{5}$
27. $-\frac{25}{8}j\sqrt{6}$
29. $\frac{59}{5}j\sqrt{6}$
31. −8
33. −20
35. $30\sqrt{10}$
37. 2
39. $-34-2j$

Exercise 19-8

1. 11
3. −49
5. $41+2j\sqrt{6}$
7. $-12+16j\sqrt{6}$
9. 98
11. $x+y$
13. $18-17j\sqrt{6}$
15. $\dfrac{6+2\sqrt{6}}{10}+\dfrac{4-3\sqrt{6}}{10}j$
17. $\frac{8}{13}-\frac{25}{13}j$
19. $\frac{23}{27}+\frac{10}{27}j\sqrt{2}$
21. $-\frac{1}{2}j\sqrt{6}$
23. $\frac{14}{9}+\frac{7}{9}j\sqrt{5}$
25. $\dfrac{15\sqrt{5}}{8}+\dfrac{15}{8}j\sqrt{3}$

CHAPTER 20

Exercise 20-1

1. $5x^2-4x-6=0$
3. $2x^2+x+3$
5. $2x^2+(-2k-5)x+(-17k+9)=0$

7. $3x^2 - 9x + 36 = 0$
9. $73x^2 - 349x + 208 = 0$
11. ± 9
13. $\pm\frac{5}{3}j$
15. $\pm\frac{1}{3}a\sqrt{3ab + 3ac}$
17. $\pm\frac{2}{5}\sqrt{30}$
19. ± 3
21. $\pm\frac{3}{4}a\sqrt{2}\,j$
23. $\pm j\sqrt{6}$
25. $\pm\frac{10}{3}$

Exercise 20-2

1. $\pm\frac{3}{2}$
3. $\pm\frac{3}{4}$
5. 0, 4
7. 3, 3
9. $-2, -3$
11. $-3, -\frac{7}{4}$
13. $\frac{2}{5}, \frac{7}{9}$
15. $-\frac{7}{8}, \frac{4}{3}$
17. $-\frac{4}{7}, -\frac{2}{3}$
19. $\frac{2}{15}a, \frac{3}{2}a$
21. $3 \pm 3a$
23. $2, -5$
25. $\frac{1}{5}, \frac{1}{2}$
27. $\frac{1}{4}, 3$

Exercise 20-3

1. $\dfrac{7 \pm \sqrt{37}}{2}$
3. $4, -\frac{5}{4}$
5. $-\dfrac{3}{2} \pm j\dfrac{\sqrt{7}}{2}$
7. $2 \pm \sqrt{2}$
9. $\dfrac{-1 \pm \sqrt{2}}{3}$
11. $1 \pm \frac{3}{2}j$
13. $\dfrac{-5 \pm \sqrt{129}}{4}$
15. $4, \frac{7}{3}$
17. $\dfrac{1 \pm \sqrt{33}}{4}$
19. $\dfrac{-k \pm \sqrt{k^2 + 16m}}{8}$
21. $\dfrac{b}{a}, \dfrac{c}{b}$
23. $2 \pm \sqrt{-2a^2 + 6a}$
25. $\dfrac{-\pi h \pm \sqrt{\pi^2 h^2 + 2\pi T}}{2\pi}$
27. $\dfrac{-CR \pm \sqrt{C^2R^2 - 4CL}}{2CL}$

Exercise 20-4

1. 10
3. 4
5. $\frac{1}{2}$
7. $2, -1$
9. 16
11. 0
13. 8, 4
15. 5
17. $-1, 3$
19. 4
21. No solution

Exercise 20-5

1. 49
3. $\pm\sqrt{5}, \pm j\sqrt{3}$
5. $\frac{1}{4}, -\frac{1}{5}$
7. $\pm\dfrac{1}{2}, \pm\dfrac{\sqrt{2}}{4}$
9. 8, 64
11. 81
13. $\pm\frac{1}{2}, \frac{1}{4}(1 \pm j\sqrt{3}), \frac{1}{4}(-1 \pm j\sqrt{3})$
15. $0, \pm 1, \pm 2, \pm\sqrt{3}$
17. $1, 1, -1, 3$
19. $\pm 2, \pm 3$
21. $\pm 2, -\frac{3}{2}, \frac{5}{2}$
23. $-3, \frac{1}{3}$
25. $\dfrac{1 \pm \sqrt{33}}{4}$

Exercise 20-6

1. 16 ft × 28 ft
3. $2\frac{3}{4}$ in.
5. 6 ft
7. 12 ft
9. Base = 12 in.; alt = 18 in.
11. 12 in.
13. 5 in.
15. $\sqrt{10}$ in. × $3\sqrt{10}$ in.
17. 15 ft
19. $9\frac{3}{8}$ in.
21. 4 in.

CHAPTER 21

Exercise 21-2

3. $A(2,0)$, $B(-4,0)$, $C(5,0)$, $D(-5,0)$, $E(0,3)$, $P_1(5,3)$, $P_2(-4,2)$, $P_3(-5,-3)$, $P_4(3,-2)$.
5. $A(6,0)$, $B(-12,0)$, $C(15,0)$, $D(-15,0)$, $E(0,9)$, $P_1(15,9)$, $P_2(-12,6)$, $P_3(-15,-9)$, $P_4(9,-6)$
13. 0
15. (a)

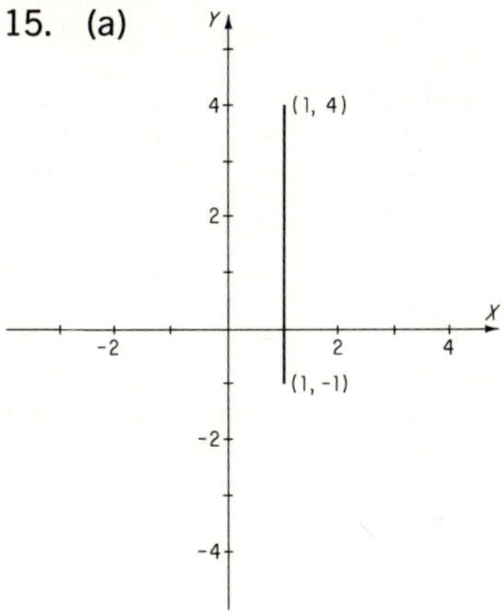

(c)

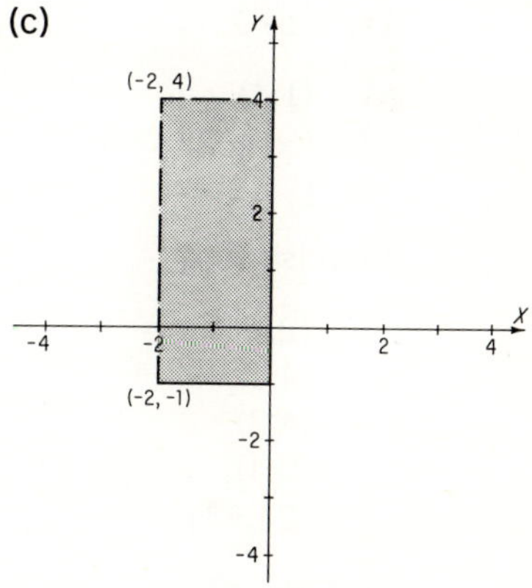

(b)

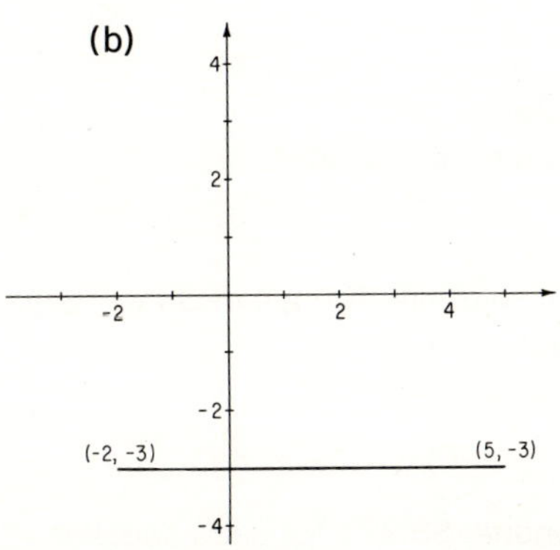

(d)

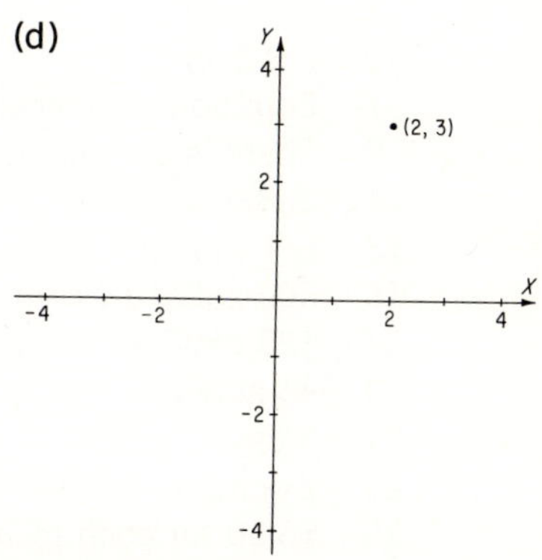

17. 6
19. $y_1 - y_2$ or $y_2 - y_1$
21. $(\frac{9}{2}, 2)$
23. $\left(\frac{x_1 + y_2}{2}, y_1\right)$

Exercise 21-3

1. (a) $\sqrt{13}$ (b) $2\sqrt{5}$ (c) $\sqrt{29}$ (d) $\sqrt{5}$ (e) $\sqrt{89}$ (f) $5\sqrt{2}$ (g) $3\sqrt{2}$ (h) $2\sqrt{13}$ (i) $\sqrt{26}$
3. $10 + \sqrt{2}$
5. $AB = 6$, $BC = 8$, $AC = 10$. Then show that $(AC)^2 = (AB)^2 + (BC)^2$.
7. $DE = \sqrt{89}$, $DF = \sqrt{356}$, $EF = \sqrt{445}$. Show $(EF)^2 = (DE)^2 + (DF)^2$.
9. (1,–2)
11. Show that $0A = 0B = 0C = \sqrt{50}$.
13. $(7,-\frac{7}{2})$, $d_1 = d_2 = d_3 = \frac{1}{2}\sqrt{445}$

CHAPTER 22

Exercise 22-1

1. (1,4)
3. (–3,–2)
5. (–1,1)
7. Null set
9. (4,–2)
11. (–2,0)
13. (–2,–2)
15. $(1\frac{1}{2},3\frac{1}{2})$
17. (5,1)

Exercise 22-2

1. (2,–1)
3. (4,1)
5. (4,3)
7. (5,4)
9. $(2\frac{1}{2},-1)$
11. $(\frac{1}{3},4)$
13. $(\frac{1}{3},-\frac{1}{2})$
15. $(\frac{5}{4},-\frac{3}{2})$
17. $(-\frac{5}{4},\frac{2}{3})$
19. $(12\frac{1}{5},18)$
21. (15,12)
23. (2,3)
25. (–5,7)
27. (4,–3)

Exercise 22-3

1. (2,1,4)
3. (2,0,–3)
5. (–1,1,2)
7. Equations are inconsistent.
9. There is an infinite number of solutions.
11. (5,8,–6)
13. $(-\frac{3}{2},\frac{2}{3},1)$
15. 600 gal of 34 cents/gal solution; 400 gal of 39 cents/gal solution
17. 123 dimes; 117 nickels
19. 35 acres
21. 3 days
23. $2\frac{4}{13}$ mph
25. \$6.50 for each microphone; $8.37\frac{1}{2}$ for each speaker
27. A: 48 hr; B: $\frac{48}{7}$ hr; C: $\frac{48}{5}$ hr

CHAPTER 23

Exercise 23-1

1. $\log_2 16 = 4$
3. $\log_{10} 1{,}000 = 3$
5. $\log_{10} 0.01 = -2$
7. $\log_8 4 = \frac{2}{3}$
9. $\log_8 \frac{1}{32} = -\frac{5}{3}$
11. $\log_b 1 = 0$
13. $2^5 = 32$
15. $(64)^{1/2} = 8$
17. $(27)^{4/3} = 81$
19. $9^{-3/2} = \frac{1}{27}$
21. $(\frac{1}{16})^{-5/4} = 32$
23. $10^{0.77815} = 6$
25. 3
27. $\frac{3}{2}$
29. -3
31. -5
33. 2
35. -2
37. 36
39. 32
41. -3
43. 11
45. $\frac{\sqrt[3]{9}}{9}$
47. 5

Exercise 23-2

1. -12
3. $\frac{2}{3}$
5. 0
7. 6
9. -3
11. 1
13. $\frac{3}{2}$
15. 7
17. $x + y$
19. e^2
21. $\frac{1}{2}$
23. $\frac{3}{2}$
25. $\log_b \pi + 2 \log_b r$
27. $\log_a 2 + \log_a x + 2 \log_a y - \log_a 3$
29. $\log_e 1 - \log_e 2 - \log_e \pi - \log_e f - \log_e C$
31. $\log_c K + \log_c (t_2 - t_1) + \log_c \alpha + \log_c T - \log_c d$
33. $\log_b \frac{xz}{y}$
35. $\log_c y^5 \sqrt[3]{z}$
37. $\log_a \sqrt{\frac{gl}{2\pi}}$
39. $\log_a \frac{Mc}{c - v}$
41. F
43. T
45. T
47. F
49. T
51. T
53. F
55. F

Exercise 23-3

1. 1
3. 3
5. -3
7. -2
9. 6
11. 1
13. 1.5378
15. 4.7760
17. $7.8751 - 10$
19. 8.3820
21. $5.7782 - 10$
23. $5.5705 - 10$
25. 75.1
27. 30,400,000
29. 0.00957
31. 900,000
33. 3,990
35. 9.20×10^{-6}

Exercise 23-4

1. 0.7954
3. 3.4448
5. 5.9033
7. $9.5462 - 10$
9. $8.5962 - 10$
11. $7.5831 - 10$
13. 4.536
15. 8,026
17. 9.074×10^9
19. 0.4082
21. 6.556×10^8
23. 1.442×10^{-5}
25. 3.061
27. 2.075×10^3
29. 1.053×10^4
31. 8.533×10^{-3}
33. 0.0164

Exercise 23-5

1. 2,060
3. 0.0470
5. 45.9
7. 534.0
9. 0.4873
11. 1,340
13. 230.5
15. 0.0760
17. 16.8
19. 147
21. 18.14
23. 22.18
25. 0.4286
27. 0.3037
29. 7.023

Exercise 23-6

1. 220
3. 0.364
5. 20.46
7. 0.8286
9. 2.013×10^{-9}
11. 0.3663
13. 4.887
15. 0.1402
17. 5.119
19. 3.557
21. 85.84
23. 46.20

CHAPTER 24

Exercise 24-1

1. 50 ft
3. 266 ft
5. 12°
7. 1,000 ft
9. 1,000 ft

Exercise 24-2

1. 0.4802
3. 0.7701
5. 1.6775
7. 0.0436
9. 0.6959
11. 0.9875
13. 71.6°
15. 78.8°
17. 56.5°
19. 26.0°
21. 34.1°
23. 69.1°
25. 21°
27. 65°
29. 67°
31. 59°

Exercise 24-3

1. 20.6
3. 42.6
5. 92.07
7. 60°
9. 63°
11. 63.4°
13. 22
15. 33
17. 44°
19. 479 ft
21. 1,840 ft^2
23. 699 ft
25. 3,990 ft
27. 125 ft
29. 164 ft

Exercise 24-4

1. 0.470
3. 0.891
5. 0.445
7. 0.0523
9. 0.0349
11. 0.198
13. 0.900
15. 0.678
17. 0.894
19. 0.0491
21. 0.384
23. 30°

25. 14°
27. 39.2°
29. 62.8°
31. 23°
33. 69.3°
35. 1.4°
37. 57°
39. 67.2°
41. 35
43. 1.3
45. 2.3
47. 4.26
49. 6.75
51. 1.73

INDEX